Process Risk Management Systems

Process Risk Management Systems

William F. Kenney

VCH

William F. Kenney
33 Village Road
Florham Park, NJ 07932

This book is printed on acid-free paper. ∞

Library of Congress Cataloging-in-Publication Data
Kenney, William F.
Process risk management systems / William F. Kenney.
p. cm.
Includes bibliographical references and index.
ISBN 1-56081-045-9 (acid-free)
1. Chemical plants—Risk assessment. I. Title
TP155.5.K4 1993
660'.2804—dc20 93-7879
CIP

Printed in the United States of America

ISBN 1-56081-045-9 VCH Publishers, Inc.
ISBN 3-527-28116-9 VCH Verlagsgesellschaft

Printing History:
10 9 8 7 6 5 4 3 2 1

Published jointly by

VCH Publishers, Inc.
220 East 23rd Street
New York, New York 10010

VCH Verlagsgesellschaft mbH
P.O. Box 10 11 61
69451 Weinheim, Germany

VCH Publishers (UK) Ltd.
8 Wellington Court
Cambridge CB1 1HZ
United Kingdom

Preface

Since the tragedies at Bhopal and Seveso, thousands of trees have been cut down to make the paper for Federal, state, and local regulations controlling industries that use "hazardous" chemicals. Thousands more have been sacrificed so that technical papers, company policies, and risk management procedures could be published. Tens of thousands more have been pulped so that citizens, politicians, and columnists could express their outrage at each industrial incident and the companies that perpetrated them. Only the complete absence of incidents seems to be tolerable to society, yet perfection seems an unrealistic and certainly an elusive goal to industry. Without a practical solution to this dilemma the viability of the process industries is in question.

In response to societal expectations several states in the United States have passed laws and the Occupational Safety and Health Agency (OSHA) has established regulations requiring industrial and municipal producers and users of hazardous chemicals to have in place systems that manage the risks imposed on workers and the community by processing these hazardous materials. The chemicals covered by law in New Jersey, California, and Delaware and by regulation by OSHA are not the same. Sometimes it seems that everything is classed as "hazardous."

In an attempt both to aid companies in preparing their risk management programs and to forestall more onerous regulation, various professional and industry groups have developed guidelines for risk management programs. The American Institute of Chemical Engineers (AICHE) has been a leader in providing technical input for process companies. In addition, the Chemical Manufacturers Association (CMA) has adopted a program called "Responsible

Care" that provides a series of guidelines in key areas. In addition to all of this, major companies that handle hazardous substances have internal standards and guidelines that they use to deal with their own materials. In many cases, these programs are added to each time a safety or environmental engineer gets a new insight or there is another incident somewhere.

This leaves the vast majority of process companies, packagers, and distributors plus most sewerage and water-treating plants; and those who handle only what are perceived as mildly hazardous materials awash in a sea of paper without firm ideas about what might be appropriate for their problems and their geographical locations. It may also leave environmental groups and state agencies convinced that industry is in need of more help to operate as a good corporate citizen, but without any real practical approaches to suggest. Of course, consultants abound who hope to help both the companies and the agencies to be responsible and to deal with current and proposed regulations.

The purpose of this book is to describe the principles essential to a successful risk management system and to illustrate practical methods of implementing those principles. It draws together the individual components of a system in a way that is intended to allow any reader to identify those components that are pertinent to the risks at his or her plant. Many examples are given, but it is not intended to be a comprehensive handbook from which specific applications can be drawn for every plant situation. These examples are given both to demonstrate the practical application of the principles and to encourage risk managers to see that many of the historical practices at their plants are good, and can be integrated into their own comprehensive risk management system.

The book can be used in several ways. Taken as a whole, it attempts to provide the background and principles critical to a sound risk management program and the specifics needed to define and implement practical systems. The middle chapters describe in some depth applications of these principles to specific areas of risk management—that is, hazard identification, inherent safety, ensuring containment, and so on. Thus, it is possible to explore only one or a few specific topics if other aspects of a reader's program are well established. Emphasis in these chapters is on practical applications and action plans. At the end, an attempt is made to draw these individual activities together into a comprehensive system and compare it to the various regulations now in force in the United States. Thus, it is also possible to read the systems chapter first and turn to the central chapters for more data only in those areas where one's current practices may be found in need of enhancement to meet the system objectives chosen.

I have attempted to use a consistent set of words to describe principles and actions. All of these descriptions involve some judgment on my part. When I say "must" I mean that the action described is essential either legally or to have success. "Needs to" and "should" are used to describe the scope of program elements and to define things that are important to consider. Things that "may be worth considering" normally have to do with optimizing a program to make it easier to implement and more cost-effective. Let the reader note that these judgments come with no guarantees. Each risk manager must evaluate his or her own situation specifically. What is merely nice at one location may be essential at another, even within the same company. This is because a large part of the evaluation of potential impact of any accident (its risk) is subjective. It depends on the perceptions of the workers, the community, and the government agencies. *In many cases there is no right and wrong;* there are only informed choices. These are likely to be based partly on scientific data and partly on management's judgment of how best to solve the total risk management problem within the constraints imposed upon it by itself and others. Recognizing that if a hazard exists some risk will always remain is management's first step. Dealing with that risk in a responsible, economically viable way is the challenge management faces, knowing that the company' detractors will always have the wisdom of hindsight if an incident occurs.

A broad definition of "incident" is needed in planning. One dead duck may not seem important to an industrial manager, but it may to a concerned citizen. Catastrophes are in the eye of the beholder. It is no longer safe to assume that what seems minor to you will be treated that way by the community. Obviously some dialogue is needed.

This book does not discuss in detail already-well-documented procedures (e.g., fault tree analysis) that might be used to implement any principle. Some description of how these techniques might be used is given along with examples and comment.

I attempt to develop a practical risk management system that can be used for both simple and complex situations. In addition, I believe that a firm grasp of the fundamentals will allow flexibility in dealing with variations in local regulations, in geography and population, and in the severity of individual hazards.

In closing this preface, let me repeat the caveat mentioned earlier: This is not a cookbook for success. Blind adherence to any suggestion does not guarantee a satisfactory result. There is no substitute for careful analysis of each specific situation.

William F. Kenney
March 1993

Acknowledgments

Special thanks are owed to many. I will start by thanking Exxon Chemical Company for permission to publish. Day-to-day discussions with Bill Warvel and Liz Garcia of the Bayway Chemical plant were extremely valuable in helping to generalize from practice to principle. The input, support, and patience of Attlio Bisio were critical. Special thanks go to Debbie Trippiedi, who typed most of the manuscript at least twice, in addition to her other duties, and always smiled.

Contents

Glossary of Acronyms

ACS	American Chemical Society (Washington, D.C.)
ACGIH	American Congress of Governmental Industrial Hygienists (The group which sets TLVs)
AIChE	American Institute of Chemical Engineers (New York City)
AIHA	American Industrial Hygienists Association
API	American Petroleum Institute
CIC	Chemical Industry Council (an industry lobbying group, usually for a state; e.g., CIC of New Jersey)
CMA	Chemical Manufacturers Association (Washington, D.C.)
CCPS	Center For Chemical Process Safety (Sponsored by AIChE)
CSSBA	Critical Safety System Bypass Authorization
DEP	Department of Environmental Protection (a state agency)
DEQ	Department of Environmental Quality (usually a state or city agency comparable to DEP)
EHS	Extraordinarily Hazardous Substance (under New Jersey TCPA)
EPA	Environmental Protection Agency (a federal agency)
ETA	Event Tree Analysis (quantitative probability assessment method)
FCA	Field Change Authorization
FMEA	Failure Mode and Effects Analysis (a method of identifying process risks)
FTA	Fault Tree Analysis (a method of quantifying the probability of a given event happening)

HAZOP	Hazard and Operability Study (a method of identifying process risks)
HFL	Higher Flammability Limit (above which the mixture is too rich to burn)—a.k.a. HEL
LEPC	Local Emergency Planning Commission (an agency set up by SARA charged with developing effective local emergency plans for all emergencies)
LFL	Lower Flammability Limit (no combustion below this concentration of flammable)—a.k.a. LEL
NJDEP	New Jersey DEP (now NJDEPE: Dept. of Environmental Protection and Energy)
NJ TCPA	New Jersey Toxic Catastrophe Prevention Act (a New Jersey law requiring risk management plans for companies that make or use certain quantities of 104 substances)
OSHA	Occupational Safety and Health Agency (a U.S. federal agency)
P&ID	Process and Instrument Diagram
SARA	U. S. Superfund Amendments and Reauthorization Act of 1988
SOC	Safe Operations Committee
STEL	Short-Term Exposure Limit
TACB	Texas Air Control Board (State agency)
TLV	Threshold Limit Values (maximum allowable workplace concentrations for chemicals)
QRA	Quantitative Risk Assessment (procedures for quantifying the severity and probability of a given event and presenting the risk numerically)

CHAPTER

1

Fundamentals and Principles

Introduction

As recently as 1980, a number of major process companies still viewed emergency preparedness as a financial problem. I once happened on a letter from an operating-company vice president to his managers that moved the responsibility for auditing and reviewing emergency response plans from the law department security group to the insurance department. The development and maintenance of these plans was to be the responsibility of the operating department, but the insurance arm would advise on their adequacy.

The law department would still be responsible for some aspects of the plan. For example, it would worry about how adequately the plan protected company assets from damage due to "unlawful entry, civil commotion, theft and looting," and so on.

The insurance department would worry about whether the plans would "insure immediate and proper response to any foreseeable incident to minimize loss of life, injury, property damage, business interruption, liability and adverse publicity." Those risks that could not be controlled below an acceptable level were to be transferred to an insurance company.

What struck me about the list of problems and proposed actions was its internal focus. There is nothing wrong with protecting your own people and property, but from today's viewpoint, guidance was sadly lacking concerning the scope of possible hazards, any impact on the environment, the potential reaction of governments and communities, and immediate attention to community needs resulting from any incident. While it may be naive to hope that an insurance department could offer much in the way of technical guidance on possible hazards, is it naive to ask why they were given the audit responsibility if technical capability was lacking?

My answer to this question is "inertia." Historically, most industries have viewed accidents as financial issues. Fringe-benefit costs went up as a result of injury or accident, as did workers' compensation costs, insurance rates, and so on. Damage to property and people outside the plan fence were "liability issues." Business might even be interrupted—also a financial factor. These factors might be important to the bottom line, but they were not viewed as essential to the viability of the enterprise.

In the last ten years, things have changed. Risk management has become a much more comprehensive and vital challenge. As seen by this author, there are four major parts to the challenge:

- The comprehensive scope of the technical issues
- The role played by perceptions and assumptions, both inside and outside the plant
- The much higher level of performance now expected
- The increased activity of government in regulating industries perceived as risky

Some perspective on these four factors is essential before attempting to approach a system for dealing with the overall challenge.

Technical Scope

Safety programs in the petroleum, chemical, and other process industries have evolved through several stages over the 100 years or so during which the process industries have made a significant impact on our society. Initially, safety programs were very much involved in protecting workers from accidents. The New England paper mills situated on rapidly flowing streams derived all of their power from the water. Machinery was often run by an impressive collection of belts, pulleys, and flywheels. It was not unusual for workers to suffer injuries from all of these mechanical marvels as well as from the saws, chippers, and other mechanical equipment used to process the raw material for the plant—namely, the local trees.

These largely mechanical hazards gradually were supplanted by more chemical and environmental problems.

The potential for fires and explosions increased rapidly as the refinery and petrochemical industries grew in size. The nature of the chemicals that were discharged into the passing stream became important and received appropriate scrutiny because of the impact on fish in the water and on the flora along the sides of the stream. I remember one of my chemical engineering professors, thirty-five years ago, describing an idea—install a fish tank in a plant effluent stream to warn operators of dangerous concentrations before they reach the river. From concern about this potential, it was only a step to consider what happened to the people who ate the fish that were caught from the river. The community health aspects of environmental hazards became a major focus. This hazard phase also has an historical aspect. Wastes were disposed from these plants over a period of years into what once were thought to be safe and appropriate places. No one anticipated that slow, natural processes would contaminate ground and surface water, rendering even some legal dumps unsafe. Thus, Superfund-site public-health issues arose many years after the acts that were the sources of the toxins.

These days, more subtle matters may also be classified as hazards by the process industries. Noise and odors impact on the quality of life in surrounding communities and are often recognized in state and/or local regulations as in need of control. In many cases, odor thresholds for chemicals, for example, mercaptans, are 100 times lower than the concentrations of these chemicals known to be bad for the health of human beings. One part per million of hydrogen sulfide in the air will not cause any deleterious health affects, but the smell is outrageous. As a result, communities have risen up to protest such assaults on their ability to enjoy life.

These protests have been upheld in various legal forums even though the protester established his or her home or business long after the process plant was in operation and with full knowledge of the potential for odors and/or noises. According to a special report in the *The New York Times* (11/21/90), in some areas process companies are attempting to buy out their close neighbors to establish a "green belt" around their plants in hopes of providing an extra margin of community safety in the event of an incident. The final phase of this evolution has manifested itself over the last ten years or so in the form of huge cleanup costs for past and current releases of materials considered to be unacceptable for one reason or another.

The accident at Bhopal was the catalyst, and rightfully so, for the enactment of major legislation in many states and countries and the establishment of major programs at most process-industry installations. No community in recent memory has been so damaged by an industrial accident as Bhopal. Other incidents, which had no immediate impact on the health of human beings, have also resulted in similar financial penalties for the erring corporations. The *Valdez* cleanup by Exxon will cost in the range of $2 billion to complete before any civil penalties are assessed or criminal charges are litigated. Clearly, incidents have moved into the "major cost" category, regardless of whether human health has been endangered.

As a result of all this, managing the hazards and risks of a process operation (including personnel safety, environmental issues, community health, and quality of life) is now a major part of the business, along with marketing, distribution, production, and maintenance. In the United States, laws have been passed in New Jersey and other states, and federal regulations are now in place, which will require risk management programs. The cost for government audit of these risk management programs will be borne by the industrial concern that creates the hazard.

In an effort to forestall even more onerous legislation in other parts of the country, the Chemical Manufacturers Association has begun developing *responsible care principles,* adherence to these principles will become a requirement for membership in the association. These are discussed in more detail in Chapter 12.

These principles go significantly beyond the requirements of NJTCPA, in that they require a positive outreach into the community as well as responsibility for the products and wastes of the operation that extends through the distribution of the product and includes the use by the customer. SOCMA, the Synthetic Organic Chemical Manufacturers Association, a group that includes many smaller companies, is considering adopting part of the Responsible Care Program as a condition of membership in that organization. Thus, the nature of what is now required from a company that uses hazardous processes and substances has evolved from the protection of workers to a full assessment of its potential impact on the surrounding community *and* the need to explain this effectively.

Perceptions/Assumptions

Public perception (the truth as the community sees it) is a major factor in current risk management activities. If an industrial site uses chlorine, the community perceives it as a risk. If the community sewage treatment plant uses chlorine, it may never be given a second thought. The scientific facts have no validity in this perception. Thus, a subjective, emotional variable is involved in what industrial concerns would like to think is a manageable, measurable aspect of their business. In fact, nothing could be further from the truth. Chauncey Starr, writing in *Risk Analysis* (1985), provides an interesting example. He considers a tiger in the town zoo. There is indeed a finite, perhaps calculable, probability that the tiger will escape from the zoo and injure local citizens. Regardless of that probability, citizens come to see the tiger and bring their children. The public acceptance of the tiger in the zoo is obviously not influenced by risk assessment but rather by confidence that the management of the zoo is keeping the risks associated with having a tiger within reasonable bounds. This is a matter of public trust as opposed to quantitative analysis. Mr. Starr goes on to opine that public acceptance of nuclear power could be increased markedly if the public were somehow assured of a reasonable risk management program by the operators of these plants. He goes further to say that the Nuclear Regulatory Commission programs aimed at establishing safety goals based on quantitative risk assessments are counterproductive to achieving public acceptance and trust because of their complexity.

The CMA carried out a public opinion survey for the chemical industry. Only twenty-eight percent of the people polled felt favorable toward the industry, which placed it ahead of only the tobacco industry in the public's eye. The summary of the report says that in the minds of the general public, the chemical industry lacks credibility, cannot be trusted, is foot-dragging in its efforts to improve itself, and is neither honest nor ethical in its dealing with others. Interestingly enough, those living near chemical plants gave six percent more favorable ratings to the industry, and employees and their families felt they knew a lot about chemical risks and were comfortable. The public at large is calling for tougher regulations of the industry and does feel that the Responsible Care principles are good. Again, an interesting observation was that half the people polled felt that the industry was already following some of the Responsible Care principles. To me, this means that Responsible Care will not be enough to deal with the whole issue.

The study also showed that "opinion leaders" were "down on" the industry and knew little about Responsible Care or other industry efforts. Thus, it is not only the community near each plant that must be dealt with but also the "opinion leaders" in the area and perhaps the entire region.

An interesting example of the power of public perception was reported in *The New York Times* on October 19, 1990. The story was entitled, "Mobil Settles California Suit Over Toxic Acid at Refinery." The city of Torrance had filed a public nuisance suit against the company, seeking assurance that the refinery, which is located in a densely populated suburb, would operate "safely." The article reported a settlement of the suit that focused on the use of hydrofluoric acid (HF—a catalyst) even though there *had been no incident involving this material at the site.* Apparently, the combination of the overall safety and environmental record at the site and the HF release at the Marathon Refinery in Texas City, Texas, was enough to convince the residents of Torrance, California, that there was potential for a real disaster in their community.

Under the agreement Mobil agreed to rework its process to ensure that the potential for a major HF release would be eliminated by 1994. If this proves to be impossible, they will switch to the sulfuric acid catalyst process by 1997. An outside consulting firm will be hired for the seven years to work under the court's jurisdiction and certify the results of the study.

The refinery manager was quoted as follows: "We think this is a fair settlement. While hydrofluoric acid can be used safety, it is also necessary to maintain good community relations." In short, Mobil felt that this was the best arrangement it could get in dealing with public perception if Mobil wanted to stay in business in Torrance. Would this (as yet) hypothetical disaster have become a legal issue if the plant's record on other, smaller issues had been better? Who knows, but clearly, nonperformance on small issues creates a climate in which public response can escalate.

The "opinion leaders" who instigate public responses are not necessarily knowledgeable. Physicians and environmental groups were listed in the CMA survey as the most trusted sources, even though respondents felt doctors were among the least knowledgeable about environmental health issues. The next-most-trusted groups were news reporters and local emergency planning groups, followed by government officials, with industry management bringing up the rear. This makes it fairly obvious that each facility must not only have a reliable risk management program but also convince its community, and perhaps a broader cross section of citizens and government agencies, of this. Using a plant's own employees as a means to communicate effectively has often been productive in this regard.

In the United States, certain trends in society also affect the challenge faced by industry. John Thielke, writing in the May 1990 issue of the *Journal of the Air and Waste Management Association,* lamented the lack of mutual problem solving exhibited in today's society compared to his experience as a child. He noted how few "How can I help?" responses there are now when an emergency occurs.

He used the *Valdez* oil spill as an example and wrote,

> Unfortunately for us all, true concerns about the environment, about our total community, were not very evident during and after the disaster. Instead of putting the community and the environment first, it appears to me that many people and groups put their special interests well in front of everything else. . . . But the whole affair is tainted by those who appear to be trying to get as much money as they can by filing lawsuits in order to tap into this disaster for special and limited interests. All told, the *Exxon Valdez* oil spill has been used by many as a way to seek publicity and money, with the environment simply serving as the vehicle for this self-aggrandizement.

Thielke's solution to this situation clearly involves a culture change. He writes, "We must work together to build a true sense of community, one where we do the right thing rather than wasting precious resources pursuing narrow and special interests." This tendency manifests itself in a number of areas other than safety and the environment. Consider the issue of burning the American flag. In a TV program over the weekened of June 15, 1990, I watched two modern lawyers earnestly debate what the founding fathers of the country meant when they wrote the word "speech" in the First Amendment to the Constitution. Winning the debate on TV was the only objective. Never was any thought given to what was right for society today or what the authors of the Constitution would have thought about burning the flag that symbolized their hard-fought victory.

Consider also the attitude of regulatory agencies in writing and interpreting regulations. The usual approach is to take as aggressive a stance as possible, assess chances for "victory," and then compromise on a solution. Industry follows the same path to save litigation costs, rather than stand on principle. One example might be instructive. In debating the content of a work plan for a state-mandated environmental study at a refinery, the agency lawyer interpreted the word "toxic" as including the possibility of fire/explosion from an LPG (liquified petroleum gas) leak. This argument was driven by concerns about the Pemex disaster, where multiple LPG storage vessels exploded in a residential suburb of Mexico City. Accepting such a definition would have meant a huge expansion of the scope of the work for the refinery in question since many of the process units handled LPG components, but only a few contained "toxic" chemicals by the conventional definition. After some blustering by both sides, a compromise was reached. The refinery agreed to study its LPG storage vessels only, and left breathing a sigh of relief. The agency patted itself on the back for bringing some additional facilities under scrutiny regardless of the rationale or the extent of the gain.

Does this approach rest on any fundamental logic or scientific principle? The answer is obvious.

Is society any better off as a result of the compromise? This answer is not at all clear. Some additional facilities were studied, but was the result worth the cost? The cost included further bastardizing of the language, reinforcing animosity between the agency and the regulated community, encouraging setting policy and enforcement practice by bluff and charisma, and the general undermining of trust in that policy.

This author's opinion is that society was the loser. Philosophy aside, what this methodology means is that "truth" and "right" are not absolute. It means that those who make honest attempts to comply with laws and regulations must anticipate basically amoral debates based on unpredictable interpretations and semantics.

Lest the reader rush to change jobs or fall on his or her sword, let me introduce a ray of hope, courtesy of *Business Week*. In its June 25, 1990, issue, it told of what was going on in Ohio, Illinois (population 600). In Ohio, mothers sew the school band uniforms, local businessmen arrange to shovel snow from everyone's driveway, no police department is needed, and incentives are offered to move into town. There is a waiting list to move into this town, no one drops out of school, and a cooperative effort called the Ohio Growth Foundation helps with mortgages, taxes, and promoting the town's crime-free lifestyle. All this a mere two hours from Chicago.

It is possible to forge a cooperative community spirit given a sincere effort to work the real problems the people perceive. Perhaps, if solving general problems took precedence over attending to individual interests, a similar sense of mutual trust and cooperation could develop among an industrial concern, its neighbors, and its regulators.

Increasingly Stringent Standards

Not too long ago, an industrial concern that took care to meet the conditions of its environmental permits was regarded as a good corporate citizen. There was little concern about whether these conditions were adequate to protect the health of workers and the community or about whether all of the necessary conditions and/or controls had been imposed.

As we begin the 1990s, we see far more stringent standards being established. Part of this trend is the result of increased scientific capability. Concentrations of most chemicals can routinely be measured down to one part per billion (equivalent to one second in thirty-two years) and some to one part in a trillion. We are finding that "chemicals" are everywhere, that concentrations are higher in our homes than outside, that trees give off "chemicals," and that our bodies manufacture "chemicals." Thus, consciousness and concern are raised at the same time that the scientific education of the general public is widely acknowledged to be increasingly ineffective. "Science" comes from the TV and newspaper headlines.

Another part of the drive for stricter standards comes from the effort to "continuously improve"; one set of targets is digested and another is set. As industry or automobiles or athletes reach one performance plateau, the expection that higher levels can and should be reached is enhanced. If the worldwide competitive business situation requires continuous improvement to survive, why should not citizens expect the same in safety and environmental issues?

A third part of the trend toward stricter standards is a consequence of the increase in emission data now available to the public. The reports required under SARA Title III have engendered many public questions. While many of these questions have been handled very well and cooperation with community agencies has improved, a number of advocacy groups have worked to convince legislators that more laws are needed.

In response, most companies have tightened their own standards on allowable amounts of known hazards and begun classifying other things as hazards. For example, certain glass and ceramic fibers are beginning to be recognized as lung hazards. Asbestos fibers have long been known to damage lungs and possibly cause lung cancer. Small asbestos fibers are very inert and remain in the lung tissue, irritating it until a cancer can form. It turns out that similarly sized glass and ceramic fibers and have a similar, but much less potent, effect. They are less potent because they are much less inert than asbestos. As a result of this finding, internal company standards have been developed for the few types of nonasbestos fiber insulations that might produce similar hazards. While the risk from these is probably several orders of magnitude smaller than that from asbestos fibers, it is being addressed to protect workers now and to avoid future problems.

Internal standards are also sensitive to community issues. Many of these are implicit rather than explicit. In March of 1990, Exxon shut down its dock operations in New Jersey after two loading mishaps, even through this would mean the imminent shutdown of a refinery from lack of crude. The company's judgment was that citizens would not tolerate another incident in the area, no matter how small. A worldwide task force was assembled to review operations and recommend improvements. Tests, witnessed by the interested state and federal agencies, were carried out before final conclusions were reached. The results of this study were announced in June 1990 and, with the concurrence of various agencies, the basis for resuming operations was set forth and made public, all without any formal requirement to do any of the above. Inherent in this story are two new standards accepted by the company: (1) "State-of-the-art" facilities and procedures were to be employed, and (2) "zero incidents" was to be the only acceptable level of performance from that point forward. But these sorts of voluntary standards are not the end of the story. Both industry and government agencies will be expected to do yet more. In May of 1990, Ashland Oil began to defend itself against a class action suit by the residents of Huntington, West Virginia, seeking $800 million in damages. The suit claims that emissions from Ashland's Catlesburg, Kentucky, refinery have damaged property and the "quality of life" of people living nearby. Some of the complaints center around fears that even very low concentrations of certain chemicals in the air pose health problems long term, and some simply claim that odors deny people the right to enjoy life.

Regardless of the outcome of this particular case, odor in the community will now become a risk to be managed. In addition, the safe-threshold concentrations of many chemicals in the air—about which too little is known—is now subject to legal debate, despite what government agencies may or may not have said on the subject. These will become two more standards by which corporate citizenship will be judged.

I believe it is sound to conclude that as time goes on industry will be faced with continuingly changing standards. To survive, it will need to prepare for these and involve itself in their formulation

Regulatory Action

There is a clear perception by legislators that they *must* respond to any significant (as perceived by the public) incident with a regulatory initiative if they are to continue their political careers. By incident I mean any unanticipated event that catches the public eye, whether it be small or large, substantive or ephemeral. This perception applies at both the state and federal level. Aside from private comments by the legislators themselves, there is ample testimony to this point in the large number of laws proposed after Bhopal and the 1989/90 oil spills.

Thus, another facet of the risk management problem will continue to grow in importance. There will be more rules to comply with and especially more documentation to be provided. In addition, there will be greater incentive to influence proposed legislation and even to preclude some by voluntary industry action.

Consider proposed legislation in New Jersey entitled Hazard Elimination Through Local Participation (HELP). This bill would establish, among other things, local committees that would have the right to *inspect* industrial facilities AND *cause them to be shut down* if the committee judged them to pose a threat, regardless of what OSHA, state agencies, or the Local Emergency Planning Commission (LEPC) said about the site. In spite of much bureaucratic duplication that would be caused by the proposed law, major questions about the qualifications of the committee, the potential for erroneous and/or misunderstood data to be used, and many other logical arguments advanced by industry lobbyists, the bill is receiving serious consideration by the legislature.

This testifies to the level of anxiety over and the urge to control the unknown hazards perceived by the public. Indeed, even the attitude of the most intrepid industry lobbyist is affected by the public outcry. In spite of branding the HELP bill as "irrational, bizarre, etc.," there is an acceptance that *some bill* will be passed on the subject, and the effort is to get an "acceptable" bill. There seems to be little hope of stopping it.

The public (and hence politicians) is also perceived as demanding more enforcement of all regulations. There is a clear sentiment that pollutors should be sent to jail. New Jersey now has an environmental prosecutor and some laws that *mandate* criminal penalties. The federal EPA is being bashed for not setting more air pollution criteria. Self-appointed watchdogs, such as the Public Interest Research Group (PIRG), are taking full advantage of the right of citizens to sue companies that violate their own permits. The net result is that the climate is becoming more risky for business management. Effective risk management may be the only way to keep plant managers out of jail in the future.

The situation is part of the problem a comprehensive risk management system must deal with. A plant must not only do its best to forestall onerous legislation; it also must attack the root cause of such proposals by effectively communicating the truth to the public and building trust therein.

To summarize, an effective risk management program must deal with a very broad range of issues. Its solution must be comprehensive, dealing with both the hard facts and the subjective aspects—that is, not just science, but also perception. Fortunately, many of the actions and most of the analysis that form the response are relevant to all aspects of the situation. Let us call these elements of the program the core activities. I believe that a reasonable program for managing process hazards must include all of these core activities. Once this is established, the myriad, seemingly specific, action programs required to address all the facets of reliable operation will be seen to stem from these core values. Figure 1.1 attempts to save 1,000 words with a schematic representation of this point.

The core activities and information form the basis for the many reports and actions needed to manage the entire risk scenario effectively. By concentrating on these fundamental issues, a company may (hopefully) avoid a great deal of ad hoc response effort and ensure cohesiveness in its program. This book deals with principles and gives practical examples of how these principles have been applied over a wide range of hazards and risks.

It will not, however, be a cookbook with a recipe for every situation. Some creative effort will be needed to apply these principles to all the types of risks discussed earlier.

Some Definitions

A wide range of terms have been used to describe the process that is discussed in this book. As a result, confusion sometimes exists as to what is meant by the terms *hazard* and *risk,* and *probabilistic* and *quantitative risk assessment,* as to

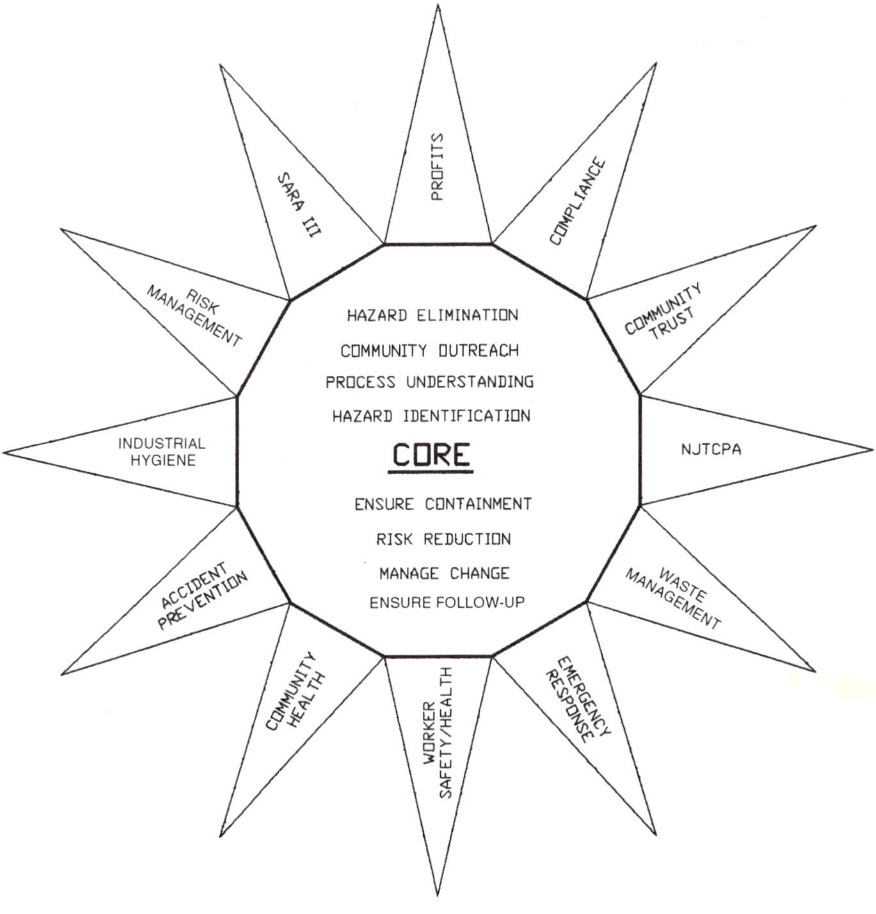

Figure 1.1. Interaction Schematic. Similar core action addresses many needs.

what constitutes hazard analysis, and so on. As a start, we will use the following definitions:

Hazard:	An intrinsic property, characteristic, or condition of material or the system that has the potential to cause an accident. Thus, having a toxic chemical in your facility constitutes a hazard.
Risk:	The combination of the probability of an abnormal event or failure and the consequences of that event for workers and community. There can be no risk without a hazard.
Critical Safety Equipment:	Equipment that controls risks.
Critical Safety Procedures:	Procedures and actions that are relied upon to control risks.
Risk Management:	The system that minimizes risks, communicates those remaining to all involved, plans emergency response, and mitigates the impact of any incident.

The characteristics of hazards and risks are different. Hazards are facts. There can be no dispute about the presence of toxic or flammable materials at a site.

Risks are more subjective. When the Three Mile Island nuclear plant released a certain amount of radioactivity into the air after its failure, some people fled the area with all the possessions they could carry. Others went about their day-to-day business. Clearly, there was some risk of long-term health effects present from exposure to radioactivity. The perception of risk ranged from panic to nonchalance. Even the scientists involved made assessments of the risk to citizens on the basis of different assumptions depending on their point of view. As a result, the management of risk involves education of the workers and communities that might be exposed to a hazard in an accident, as well as dealing with the scientific and technical aspects of handling the incident.

Fundamentals and Principles

The principal element involved in managing risks can be boiled down to a single sentence: Good process risk management results in perfect containment and safe handling of the hazard. This is clearly a performance standard. It does not spell out every procedure or every definition of what constitutes success. The word "perfect" will upset many. Yet that seems to be the standard to which any plant in the process industry will be held accountable, at least as far as accidental breaches of containment go. A functioning risk management program can be established relatively simply once the following requirements are met:

- The hazards of a process must be capable of being defined at any time.
- The risks resulting from these hazards must be controllable by equipment, by procedures, or by some combination thereof.
- Management must uncompromisingly maintain controls over the equipment and procedures that are identified to control the risks (safety critical).

How these simple tasks are addressed will be the subject of the rest of this book. Eight underlying principles are identified in this chapter and explored in more detail individually in subsequent chapters.

1. *Commit Management*

Management's role in the overall process might be glibly categorized as follows:

VISIBILITY, ACCOUNTABILITY, CREDIBILITY, AND LEADERSHIP

The commitment starts with a policy that addresses safety and the environment as a priority in the operations in the company. Air Products Corporation, one of the current leaders in risk analysis on the industrial scene, began its improvement in personnel safety from the top down. The chairman of the company decided that an industry-average safety record was not good business for Air Products (*I. Chem. E Symposium Series No. 110,* 1988). He instituted a change in philosophy and increased the rigor of the management review of safety incidents. It took awhile for the individual plant managers to come to believe that safety was a priority. After several of them had been called before the board of directors to explain how injuries occurred in their plants, it became clear that this was indeed a new priority, of equal standing to production. Once this message began to spread, systems were put into place.

Such embarrassing moments before the board of directors quickly became few and far between. The corporate safety record improved until Air Products was among the leaders in the industry, and the definition of safety began to broaden and meld with risk management.

This was not the only action taken by the company, however. Hundreds of people were trained in hazard identification and quantitative risk assessment. Even laboratory experiments were subject to formal safety review, and multistage reviews were carried out on projects that progressed. The entire fabric of the company was saturated with the results of the safety policy.

This example clearly demonstrates visibility and accountability. In addition, while credibility took a while to develop, it is currently clearly present.

The essentials of credibility involve follow-through. Obviously, the resources must be allocated, but also activities must be evaluated and stewarded. When management is responding to an incident or trying to convince the community that efforts to improve a situation are sincere, there is nothing more damning than a long list of unacknowledged recommendations. Credibility is also demonstrated by the fullness of compliance to the various regulations that apply to any operation. Regulations being what they are, it is often necessary to interpret how they apply to a specific operation. Management earns its credibility in this area by establishing a standard of judgment in the gray areas that consistently manifests its stated policy and avoids embarrassment. Any decisions must pass what might be called a "red face test" under the scrutiny of an investigative reporter. *To be legally correct is not adequate.*

The hardest part of management's role is providing the necessary leadership in managing risks. This does not mean the stewardship of various safety or environmental indices only. It means taking the trouble to analyze experience, searching for potential problems and eliminating those that are believed to be potentially serious. Management cannot put its head in the sand when confronted with such issues. Indeed, in these days of militant enforcement in high-visibility cases, there is a very large financial incentive associated with prudent foresight.

2. *Identify Hazards*

As defined earlier in this chapter, hazards are inherent properties of materials or systems that have the potential to do harm. Basically, there are three types of hazards:

- Energy related
- Toxicity related
- Reactivity related

Energy-related hazards represent the potential for fire and explosion. The general public lives with many such hazards in their day-to-day life and, therefore, is fairly sanguine about them. However, when a major event occurs, such as the Pemex disaster in Mexico, the power of fire and explosion to wreak havoc on the lives of ordinary citizens is emphasized. One need only cite the reaction of New York City to social-club safety violations after the fire that caused the deaths of people in March of 1990. The result, which came too late for the victims, was increased prosecution of fire code violations. Most users of highly flammable materials have learned a great deal about how to manage these hazards in the time since their original facilities and procedures were installed. In places where safety is a priority, improvements to facilities and procedures are made whether or not there is an incident that prompts such measures.

Toxicity-related hazards incite the most public outrage these days. The fears include not only acute injury to health, but also the fear of injury to offspring and the ultimate development of cancer or other disease. Toxicity is very broadly defined. We are not just talking about the release of agents that are highly toxic to human beings, such as at Bhopal.

We must also include agents that are toxic to wildlife, marine biologicals, and plant life as well. Indeed, oil constitutes a toxic release, not only by legislative fiat, but also because of its impact on a number of sensitive species. The news-media pictures of oiled sea gulls unable to fly and perhaps subject to death are merely a reflection of the public attitude toward any chemicals that are environmentally hostile.

Reactivity hazards are often not as well appreciated as the two foregoing categories. The release of materials that of themselves are not a major hazard, but that can react with a commonly available material, such as water, can pose some serious questions for those who manage hazards. When aluminum chloride is spilled on the ground, sooner or later it will encounter some water. This can produce a very large hazard by generating hydrogen chloride (HCl). Similarly, the inadvertent mixing of reactive chemicals inside a normally secure containment facility can produce monstrous results. This is what happened at Bhopal. Identifying this type of hazard requires more imagination but is still very important.

Hazards are generally identified through a chemical screening study. A great deal of data must be assembled about the amounts and properties of every chemical on the site as well as the health effect of the materials themselves and their reaction products with many common materials.

Once these data are assembled, it is necessary to evaluate the magnitude of the hazards they represent. This invariably involves a judgment about the significance of that particular hazard at the particular site in question. *In many cases the answer is not clear-cut.*

Various indices have been suggested (for example, the Dow index) to help assess the significance of the hazards. However, a subjective judgment is required in almost every case. A great deal will be said in the chapter on identifying hazards about techniques that can be used to answer two questions: First, what are the hazards, and second, how bad are they?

3. *Eliminate Nonessential Hazards*

Having identified the hazards at the site, the next step is to eliminate those that can be removed and minimize those that must remain. The first step is to assess whether the business really needs to make or use a given chemical. Such a decision can be complex. The business implications of the chemical need to be balanced against the magnitude of the hazard. In addition, a large research-oriented component is involved in making these judgments. The possibilities for alternate chemistry and improvements in raw material and product purity need to be considered. In a plant environment, it is not unusual to assume that this type of item cannot be changed because the research department made the decision and it cannot be overruled by the plant staff. This is a mistake.

There are a number of plant-related aspects involved in eliminating or minimizing hazards. These include the location of the facility, the design basis and conditions of the facility, and a conscious effort to maximize the inherent safety of the system. This does not mean adding gadgets to prevent an accident but rather reducing the inherent magnitude of the hazard. Some people refer to this as *intrinsic safety* as opposed to the *extrinsic* variety. One might pithily summarize this notion by saying that *"What you don't have can't explode."*

4. *Maximize Containment and Safe Handling*

Many elements are involved in this fundamental principle. They are treated in three separate chapters.

The first step in this process, having already identified significant hazards in the plant, is to identify the risks that these hazards represent in the facility in question. This means that techniques must be applied to identify the nature of the risk and the extent to which it is controlled by existing systems. This usually implies a HAZOP (hazard and operability) or FMEA (failure modes and effects analysis) study (see Chapter 5), the analysis of previous incidents, and the use of other techniques to identify the realities of the risks inherent in operating the facilities in question. As indicated earlier, the facility has an obligation to reduce these risks and the potential impact on the community.

Risks can be reduced by a number of approaches. One might be categorized as the mechanical approach. Mechanical approaches include design standards, construction practices, inspection procedures, and maintenance programs. Risk reduction can also be accomplished through procedures. These are described in operating manuals, learned and validated by training and certificaiton, and retained through constant practice in dealing with incipient incidents. The discipline of performing preventive maintenance on critical safety equipment involves very important procedures, similar to those that control the bypass of critical safety equipment.

Safe handling is a topic that requires considerable discussion. The objective here is to prevent a significant incident even though the primary containment system has failed. The flare stacks in refineries and chemical plants are the most common example of release prevention structures but other forms of secondary containment are also important such as scrubbers on vents and safety-valve discharges. In recent years, much more sophisticated information has become available through the Center For Chemical Process Safety (CCPS) and other sources on how to design effective, safe, handling systems. Many people need to update the analysis of their current systems based on these technology advances.

5. *Control Changes*

In many cases, new facilities are well documented and operate in excellent conformance with the original operating procedures and conditions. As time goes on, this situation changes as the process and the operating procedures evolve in line with the experience gained by the staff. Minor bottlenecks develop, efficiency upgrades are installed, and before long, the original designer of the plant might not recognize it or how it is being run.

All changes to facilities, procedures, materials, and operating conditions must be evaluated and documented to maintain the same level of control over the risks inherent in the plant as was originally approved by management and the various government oversite agencies. This is a complex system to install and maintain and considerable time will be spent in discussing specific approaches.

6. *Assess Risks*

The proper approach to risk assessment requires considerable thought. There are those who maintain that the quantitative risk assessment and consequence analysis approach undertaken by the nuclear industry is the only scientific way to represent risk management in the chemical industry. With the wisdom of hindsight, it would appear that this approach has contributed to the slow death of the nuclear industry because no one can decide how safe is safe enough. Some risk always remains. There are those who conclude that this is a deserving end for such a hazardous business, but I believe that a more intelligent use of risk assessment information is essential to the continued viability of the process industries.

Various techniques for risk assessment are discussed ranging from qualitative to the highly quantitative. I believe that risk assessment can only be viable as a relative risk evaluation tool. At the moment, the absolute value calculated in any quantitative risk assessment procedure is so dependent on the validity of the data base and the judgment of the evaluator as to be invalid on an absolute scale.

On a relative basis, conclusions can be drawn, however, and management can be given guidance in the business of making what would previously have been only intuitive judgments about the safety of any given operation.

7. *Upgrade/Prevent*

A number of states are now calling for *state-of-the-art* controls on industrial risks, and mandating the upgrade of facilities and procedures to prevent incidents. The definition of state of the art sometimes includes an economic cost/benefit ratio, but the phrase is sometimes defined as the best available technology regardless of cost. The fundamental question is, How do you define *cost-effectiveness* in human health situations? Government agencies are tempted to define absolute standards but are at a loss as to how to cover the myriad possibilities with which they are faced. The key element in an effective upgrading program would seem to be that the individual facility must take complete responsibility for both managing its risks in an effective way and communicating how this is accomplished. The possible impact on the community needs to be communicated to the local emergency planning commissions and other responsible community administrators. This requires a depth of analysis and understanding that many sites have not yet developed, and an openness and honesty in communications with the community that most plants have not attempted.

What little history there is so far indicates that communities are tolerant of reasonable risks if they trust the people that are managing them. The fundamental need then is to build that trust.

8. *Ensure Follow-up*

We are all familiar with the proverb about what happens to the best-laid plans of mice and men. What is essential in maintaining the effectiveness of these plans is to ensure that that the facilities, systems, and procedures in place to manage the risks remain fully operational. This has many implications, but an effective follow-up and auditing program is an essential. Risk management must be a line management responsibility and followed up as stringently as management does production, costs, and product quality. Cooperative relationships with state oversite agencies can also contribute to effective follow-up. Since one way or another we pay for the activities of government agencies, every effort should be made to obtain value from this investment rather than simply to keep the agencies at a safe distance.

To accomplish all of this, it is necessary to create a functioning system that deals with the principles of a risk management program. This text attempts to define the outlines of such a system by identifying subsystems and sub-subsystems needed for support. Obviously, the more complex the facility, the more complex the system needed to manage the increased number of risks. However, in terms of ensuring the operability of the system, simplicity has a great merit. As in the fundamental thermodynamic analysis of processes, there is a fundamental risk management analysis of the facility that maximizes the inherent safety and adds only those extrinsic factors that are essential.

Summary

Managing risk in the process industry is a complex challenge. The risks are broader in scope, the expected performance higher and the need to communicate and negotiate outside the plant fence more critical than ever. In the ideal system, risk management begins with the first idea for a commercial venture and flows through the development, construction, and operating phases right through the safe dismantling/disposal phase. The principles that must be applied to accomplish the task are known but are so general that how they are applied becomes the critical step. Core values that apply to many facets of the problem must be found and used to knit together a comprehensive approach that avoids waste while achieving both the "no incidents" target and also explaining what is going on to the public and government agencies. As in the management of many other aspects of business, continuous improvement must be committed-to as regulations and performance standards will surely evolve to more stringent configurations with time.

CHAPTER 2

Management Role

Philosophy

In Chapter 1, I described management's role as consisting of visibility, accountability, credibility, and leadership. One of the advantages of writing on the subject of risk management is that it lends itself to characterization by apparently simple principles. The challenge comes in reducing these principles to practice in an effective, cost-efficient way. Fundamentally, management has to recognize that every incident results from a failure of, or the absence of, a management system. This is the essence of visibility and accountability. Technical information is not enough. Harold Kemp, a recent past president of the American Institute of Chemical Engineers (AIChE), was asked whether the existence of the guidelines developed by the Center For Chemical Process Safety (CCPS) would have prevented the accidents that occurred in Flixborough, Seveso, Mexico City, or Bhopal. His answer was a resounding no. Guidelines are only useful if they are transformed into local communication and control systems which are monitored by management. In each of these incidents, the well-known technical requirements were either overlooked, ignored, or bypassed, and essential information was not communicated to the workers at the plant.

In these and in other plants, most major process incidents result from the breakdowns in management control. These breakdowns may be associated with:

- Incomplete, or bypassed, emergency procedures or instrumentation.
- Unauthorized or inadequately designed equipment modifications.

- Inadequate inspection and maintenance programs.
- Inadequate supervision.
- Failure to communicate essential process safety information. (A clear example of this would be failing to provide an adequate understanding of the hazards associated with the process to all of those people who work at it. Simply handing out a copy of the Materials Safety Data Sheets (MSDS) for the chemicals in use is not adequate communication.)

Management's highest priority in conducting its day-to-day business must be the protection of its employees, its customers, and its neighbors. To do otherwise threatens the viability of the enterprise, regardless of current profits. This involves training and motivating everyone in the entire organization to make safety a top priority, to find out what might possibly go wrong, and to be prepared to deal with it. Failure to communicate this philosophy amounts to management failure.

Nor can management discharge this responsibility by espousing idealistic standards or indulging in wishful thinking. It must visibly back up its words with action and with constant follow-through. This is the long-held management philosophy of the DuPont Company, the acknowledged leader in process-plant safety. In fact, DuPont has created a significant business out of its longstanding reputation for excellence in the safety area. Many companies in the process industry have attended seminars on the subject put on by DuPont and/or have hired DuPont engineers to turn around their own process safety situations. The turnaround story presented by Air Products (*I. Cheme. Symposium Series No. 110,* 1989, and elsewhere) in terms of their safety record was based primarily on the DuPont approach and top management initiative. Top management visibly demanded that all lower levels recognize, as a first priority, the need to improve the process safety performance of the company. Plant managers were called to headquarters to explain the circumstances of every disabling injury. Gradually, they began to realize that they were accountable for the behavior of the people who worked in their plants. In addition, they realized that they had to use the entire system to achieve a safe operation. By this, I mean that the equipment must be well designed and properly maintained, and the operations must be controlled within safe parameters. In addition, the human components of the enterprise need to be understood, educated, and motivated. Only when management realized that the effort and expenditures required to achieve a safe operation paid off in economic benefits did the full realization of what it meant to have a "safe" plant become clear to the local management. Costs were lower for workers' compensation, repairs to damaged equipment, lost production, and legal fees. Worker morale, productivity, and product quality all improved. It was not only the cost of non-compliance that generated economic benefits for the safe plant, but also the positive impact of working in a safe environment that added to the bottom line.

In many cases, smaller companies will argue that it is well and good for the managers of large corporations to make this investment in process safety, but that medium- and small-size companies cannot afford it. This is the proverbial Catch-22: we cannot afford to be a better, more profitable company, because we cannot make the investment to achieve it. The ultimate projection of this impasse is inevitably an ever-downward spiral until the enterprise is no longer viable. Indeed, if one reviews the list of sponsors for CCPS, he finds a number of smaller companies represented. Behind each of these sponsors stand many more, even smaller companies that also use the information to create improvements in their own companies and in the industry as a whole.

One of the challenges for management is to recognize risks explicitly before the fact rather than waiting to deal with them after an incident. The costs of dealing with an incident—for example, an oil spill—are incredible. These are not limited to the cleanup of the oil on waters or beaches, albeit these are substantial, and in some cases, excessive. There are also very significant expenses involved in the public relations aspects of the situation and in lobbying against, and subsequently dealing with, the myriad new regulations conceived by elected representatives in an effort to protect the public from further mishaps. In addition, there is a horrendous dislocation of management and technical manpower required to deal with the incident per se.

Management is often hard pressed to justify investment and manpower aimed at preventing something that has never happened in the past, but might. Thus, there is a constant balancing of perceived cost/benefit ratios based on management's perception of the magnitude of the risk. In many cases, management does not have an objective definition of these risks. It goes instead on the basis of an historical, subjective, intuitive understanding of the probability of a certain incident occurring. In many cases, management does not even have a qualitative understanding of the sequence of the events that might lead to the undesired result. This makes it very difficult to make judgments on the benefits of upgrading systems. In addition, outside pressures can bias managers' decisions. I include in this pressures from headquarters to reduce manpower, capital expenditures, and maintenance budgets in an effort to maximize profits. Outside pressures come from legislation and/or the activities of civic groups as well. These days, the latter source of outside pressures will usually merit a higher place in the decision-making process if the proper balance is reached.

In many cases, management's main challenge is to earn the right to operate in the eyes of the community. No longer is the argument that risks must be taken if jobs are to be provided a viable response. This situation not only introduces new demands on management, it also provides considerable opportunity for leadership.

In companies managed by technically trained people, leadership in these subjective areas requires new skills and new information. A knowledge of the community's attitudes toward the operation is necessary, as is a better understanding of the risks associated with that operation. The new techniques involve building trust in the community by sharing information with it and gaining its concurrence in the company's risk management decisions.

Visibility

Management can be visible in many ways. Ideally, it is visible on the preventive front by such actions as endorsing CMA's Responsible Care principles, rather than being visible in front of television cameras explaining why an incident occurred.

The first step in becoming visible is to develop a meaningful safety and environmental policy that is at the same time both idealistic and practical. Endorsing motherhood is not an acceptable policy. A general statement such as, "Safety and the environment are our first priority" falls far short of the mark. Somehow this simple principle must be related to plant operations and to real activities in a way which is recognizable to the workers. Figure 2.1 shows an example of a policy that addresses the operative issues in one organization. This approach may or may not work in any one specific organization. However the policy is addressed, in whatever language is found most effective, no confusing body language can be permitted. Management must live its policy in

---

The health and safety of all employees, contractors, and the surrounding community are of primary importance in the conduct of the plant's business. We believe that safety is good business, an integral part of a successful and respected enterprise. Consequently, production is never so urgent as to require the use of unsafe approaches. We also believe that safety is a line responsibility and that all injuries can be prevented.

It is the policy of the plant to provide a safe work environment for those on-site with no potential hazards to the community. Each employee is responsible for helping to implement this policy by working safely, by being alert to help co-workers, and by helping to develop and following the appropriate safety practices for each task. In addition, management makes the following commitments to ensure continued improvement in safety standards:

1. With the help of all the work force, to continue to identify, evaluate, and control safety hazards in the workplace
2. To assure the safe planning, design, construction, operation, and maintenance of all facilities
3. To educate and train all members of the workforce to ensure the ability to do (1) and (2) above, and to be prepared to respond effectively to emergency situations
4. To recognize contributions to the safety of the site in evaluating an individual's job performance

Only by accepting our shared responsibilities and working together can we have a safe and healthy work environment for ourselves, our fellow workers, and the community. By giving safety primary emphasis both on and off the job, we develop an attitude toward safety awareness that enhances our commitment to excellence in all aspects of our operations.

Site Safety Committee

Figure 2.1. Example safety policy.

Table 2.1. "Hawk" Project Philosophy Statement —Section II: Environmental

- The HAWK project philosophy will be consistent with the stated safety, health, and environmental Standards. Specifically:
  - We are committed to operations, products, and practices that protect safety, health, and the environment
  - As a minimum, we will comply with all applicable laws, regulations, and other standards as necessary to provide adequate protection
  - We will work to minimize risks associated with manufacturing, transportation, use, and disposal of our products
  - We are committed to conserve natural resources and further increase margins by reducing waste generation and contamination of air, water, and land. We will ensure that any remaining waste is disposed of in a safe and environmentally sound way
- The HAWK project environmental plan will be developed to be consistent with the base manufacturing "Five Year Plan" to reduce emissions by 50%
- A plan to approach "zero emissions" will be developed during the design basis phase of the project, though high-cost/minimal-impact portions may be deleted or deferred until later years
- Special consideration will be given to one-time installation opportunities (example—sealless pumps, $\frac{1}{4}$-turn ball valves, etc.) that are not conducive to retrofit/deferred installation

everything it does, not just when it is convenient. The policy must be enforced and become part of the fabric of the enterprise or it will be worthless. Consider the example for the "HAWK" project philosophy shown in Table 2.1. Specific objectives are set and realistic limitations are listed. A cost/benefit relationship was also developed (see Fig. 2.2) that plotted reductions in plant emissions for specific design improvements vs. their cost. In such ways management's input is made sufficiently explicit so as to ensure that the correct information is

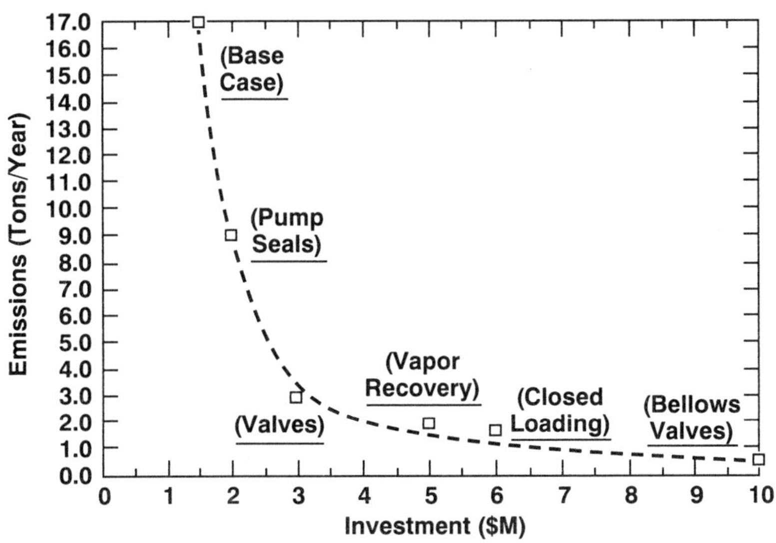

Figure 2.2. Project Hawk environmental basis. Summary—air emissions

available to those who make design decisions and that the appropriate questions are brought to management's attention.

Management can be visible in other ways as well. In June of 1989, President Richard J. Stegemeier, chairman of Unocal, wrote to each employee about the importance of the company's safety policy. In this letter, he reiterates that safety is not just a question of "having good technology and sound operating procedures." Rather, safety is "a state of mind." He goes further to say, "in particular, I want to stress that when it comes to safety, there can be no compromises. As a company, we insist on safe and lawful operations for the protection of our employees, the public and the natural environment." He goes on to point out some of the facts and figures associated with the Unocal safety improvement program and accident reporting system. He cites also the safety audit program, the fire-protection training sessions, the back-injury prevention program, and the net result in terms of numerical performance improvements. Clearly, measurable and visible action had been taken and measurable results had been achieved. Through the letter, both the action and the results were being made more visible to each employee. At the same time, management's responsibility for continuing improvement was acknowledged. In addition, the role of each employee in helping to achieve those goals while helping to protect him/herself and the environment was clearly elucidated.

Many other companies have taken steps to communicate with their employees on the question of policy and expected behavior.

Davis et al. ("Accidental Releases of Air Toxics," Noyes Data Corp., Park Ledge, NJ, 1989) point out the need to overcome "safety burdens." By this, they mean routine (and boring) transmission of safety data, canned exhortations to better performance, and so on. Rather, it is necessary for management to participate directly with all workers in improving safety and to respond actively to initiatives from those in the field.

One of the major thrusts of modern management practice embodies one form or another of the principle called "managing by wandering around." Thus, many managers visit many separate work sites for many reasons. Some of these involve very formal visits; others are much more informal. Invariably, managers ask about the safety record of the operations they visit. Oftentimes, almost as a religious observation, managers will ask about safety as the inital topic on the agenda. But, how many times does visiting management probe safety activities to the same depth to which it probes operating performance? How many times do managers merely dwell on the statistics or ask for a full explanation of the most recent incident and then turn to subjects that more obviously affect financial performance? How many times do they probe what management systems are in place to prevent accidents and to provide improvement in the safety culture rather than merely look at the numerical results? How many times do they walk out into the plant to do their own personal audit of safety practices, the permit procedure, the state of training of the operators and mechanics, and the physical integrity of the equipment? Admittedly, this depth of inquiry will cause a great deal of attention. The investment in this will certainly add a great deal to the credibility of management's safety policy.

A fundamental principle that lies behind all these efforts is that communications need not be boring. They need to be sincere and effective attempts to relate the desired policy to the expected actions of the employees.

Endorsement of and support for a plant's activities in the local community are also a major part of being visible in these days of greater public scrutiny of what goes on behind the plant fence. SARA III provides both a challenge and an opportunity for process plants to deal more effectively with the public. Not only is it necessary to confess one's emissions to the community: One can and must also explain them and assist in the formulation of the community's emergency plans. Many companies had already been participating in Community Awareness Emergency Response (CAER) programs sponsored by CMA that evolved quite naturally into local emergency planning committees (LEPC). Others had to form a new initiative to join the planning activities. Some companies were highly visible; others contributed quietly. The temptation to reach for headlines by contributing a computer or other element of modern technology to the LEPC is often hard to resist. Quiet contributions of professional services, secretarial help, or even duplicaiton services are also of value. In all of these cases, management has had, and will continue to have, the opportunity not only to make visible its commitment to process safety but also to work harmoniously with its community host to build trust and earn the right to operate.

One of the greatest benefits of joint activities with the community is the enhanced opportunity in the area of simulated emergency exercises—or, if you will, drills. Supporting and/or hosting such emergency exercises can be a frightening experience for those who are new to it. To many, it will seem that they are baring their souls to the public about what might go wrong inside their plant. Such plants (and their legal staffs) live in fear that such exercises will result in pickets at their gate and onerous local regulation of their operations. In general, this has not been the case. Local emergency responders are generally overjoyed at the prospect of being better prepared to deal with each emergency they might encounter and of having information suitable for preparing the community for such events. In addition, part of their budgets (for police, fire, and other emergency services) may come from participating in such drills, and they welcome the opportunity to do so on a realistic basis.

Drills require manpower to help plan and execute. Sometimes, it is even necessary to expend an appreciable amount of money to videotape the exercise for later training use. In most cases, the benefits of these investments have far outweighed the costs. Aside from the necessary exercise of both the plant and the community's emergency procedures, bonds are formed with the first responders and communications are established of a sort that have stood most plants in good stead when real emergencies have occurred.

Accountability

There is a common element in all of this: It is possible to measure whether the organization did what it said it would do. In addition, the responsibility is clear.

As a result, the management is *accountable; the report card is defined.*

The fundamental philosophy is that process safety, in all its aspects, is a line responsibility. Staff groups help, provide technology and innovative ideas, do analysis, and so on *but line management must implement.* This includes allocating the necessary resources, following up on the recommendations, providing time for the training, and stewarding progress just as is done with any other business objective. This implies that progress will be audited as appropriate and reported to upper management, to plant employees, and to the community.

Dan Peterson, a noted safety consultant, summarizes the accountability principle in these words: "What gets measured and rewarded, gets done." What can be measured are safety, environmental, and risk management activities. These need to become part of each employee's performance assessment, and he or she must be trained to do these activities well.

One key point in this message is that safety performance (the *statistics*) introduces the element of luck. It is not fair to hold an employee accountable for luck. To do so merely creates resentment in the unlucky and carelessness in the lucky. But safety *activities* can be stewarded and have no element of luck. The principle is that if the appropriate *activities* are well done, better performance *statistics* will follow.

The desired activities will occur when the following four elements have been put in place:

1. Tasks have been clearly defined.
2. People are taught how to do each task.
3. Performance of the tasks is validly measured.
4. Workers are rewarded for high-quality completion of the tasks.

If these elements are in place at each level of supervision/management, then each person is clearly accountable for his or her share of the whole effort. Rewards can be positive or negative, but they will have a firm basis.

While reporting on safety performance will probably never go away, it is not just a numbers game. It is not appropriate to be creative about the OSHA injury logs (as major fines assessed by OSHA will attest). Nor is it accepted to taxi injured employees to work to avoid a disabling-injury charge (albeit transparently). An effective, sincere management no longer accepts such subterfuges for reasons discussed in the next section.

What management must do is work the "incident triangle" from the bottom up. Consider Figure 2.3. The peak of the triangle involves the statistics usually reported. They represent only a small percentage of the unsafe acts, "near misses," and minor mishaps that constitute the root causes of these peak events. The difference between the number of events at the bottom of the triangle and that at the top is mostly *luck.* In the long run, to reduce the number of peak events, one must reduce the number of unsafe acts. The luck (or the extraordinary skill of a responder) that prevents an unsafe act from becoming a peak event is hardly a management system. Even if the company does not hold management accountable, the public will.

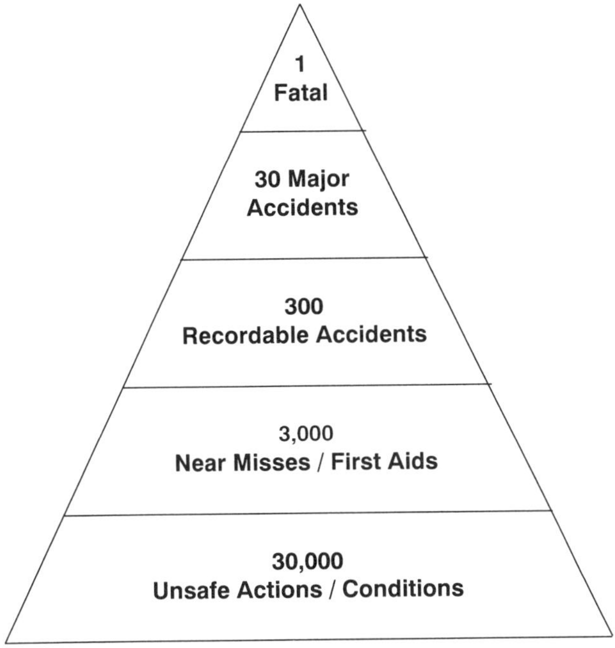

Figure 2.3. Safety triangle.

Credibility

Management will be perceived as credible or not by the majority of observers based on the perceptions and prejudices of the observers, as well as the scientific facts. To be credible is essential these days, both with employees and with outsiders. Without the trust and commitment of employees, management's efforts to operate safely can be only partially successful at best. Without credibility from the community and the regulators, management can do little right in the public arena.

Recognizing this, the Chemical Manufacturers Association (CMA), a 180-member trade association of larger chemical companies, launched its Responsible Care program. Acceptance of the program is a condition of membership in the CMA. The program is based on performance—not public relations. The foundation of credibility was pinpointed explicitly: Tell people what you will do, then do it (and tell them about that, too).

Davis et al. list some specific responsibilities of management that speak to the credibility issue. This list is by no means complete but it is a good starting point. Slightly paraphrased and annotated:

- Ensure worker competency—that is, invest the time and talent to train and test plant workers.
- Develop and enforce standard operating procedures.

- Document policy and procedures—that is, communicate expectations to the work force—and keep them up to date.
- Identify, assess, and control hazards. (This cannot be done without participation by a wide cross section of the work force.)
- Communicate and promote feedback regarding safety issues. (Face the unsafe acts, report results, and seek input on how to improve.)
- Perform regular audits and independent checks of procedures and compliance (a risk to the ego perhaps, but a good way to ensure against complacency, provide for technology transfer, and show responsibility to everyone).

The specifics of how these responsibilities apply to any individual plant will vary according to local factors such as union contracts, the nature of hazards present, and the geographical relation to the nearest community. Some useful methods that can be used to meet such responsibilities follow:

- Establish a viable communication system so workers in the field can express their concerns and receive an answer. This might include regular safety meetings, joint labor/management safety audits, "tickets" (like those for illegal parking of your automobile), for unsafe equipment, encouragement of intervention in unsafe acts, an so on.
- Provide adequate resources (people, time, and money) to the task of accomplishing safety objectives.
- Put high-quality, effective, and respected people into safety-related jobs.
- Set clear priorities—even when some profit is lost by taking the safe course. The following words of a profit manager to a plant manager whose facilities suffered an accident make the point clearly: "Even if we get the go-ahead from the authorities, we should *not* start up if we do not feel 100% sure that we can start up and operate the plant safely."
- Dedicate effort to comply with government regulations and publicize results of inspections.

  Compliance does not mean that there will be no debates or even no lawsuits between the plant and the regulatory agencies. Nor does it mean that there will never be a notice of violation. It does mean that there will be a consistent effort to comply with clear regulations and a good-faith effort to interpret gray areas in the spirit of the law. Big "task force" efforts after a citation or an incident are a sign to many that the ongoing effort is now viewed as inadequate. Investing the manpower to work with the agencies while they develop the regulations is far more productive in terms of both reasonable regulations and a good understanding of how to comply.

  Compliance also applies to in-house standards, national codes, CMA Responsible Care practices, and other voluntary standards that characterize "good citizenship."

  Some sort of Red Face Test around compliance needs to be operative at the plant. It boils down to: would the plant manager be embarrassed to admit in a public forum (or to his or her boss) the basis for any decision in case an incident results?

- Include safety-related activities in performance appraisal discussions.
- Provide the necessary training to all employees and contractors who are exposed to normal hazards in the plant. Dow (*Chemical Engineering Progress,* March 1990) reports that every new employee receives forty to eighty hours of training (more than thirty directly related to safety) before being allowed to work inside the plant. As another example, a refinery/petrochemical complex in New Jersey, a half-day safety training program for contract personnel was developed into a series of lively videotapes. Computerized records are used to ensure each individual's training was current (even if he or she switched contractors) and decals were issued to those trained to display on hard hats.
- Perform effective audits at reasonable intervals. No management enjoys being audited, but there are positive aspects. Not the least of these is that audits represent a sincere effort to obtain objective input and measure local standards against those of others. Given intelligent follow-up of audit recommendations (this does not mean blind implementation, as will be discussed later), audits can go a long way toward establishing a plant as a responsible citizen in the public and regulatory arenas.
- Test programs in the public arena. Discuss them in technical papers at society meetings. Share them with the community at public meetings. Explain them to the LEPC and the individual emergency response organizations. Write them down in a newsletter for the community at large. Establish a "hotline" through which the community can ask questions *and get prompt answers.* Above all, recognize that if you do this, you must be prepared to evaluate and respond to any feedback received.

Credibility involves taking some risk. One cannot be credible by claiming credit for meeting goals known only to the achiever. Objectives need to be shared beforehand with those to whom they are important.

Credibility involves listening and responding to input from all levels of the organization and from the community.

Credibility involves trust. If the vast majority of the time you do what you commit to, *most people* will believe you, provided you take the time to discuss it with them personally.

Leadership

Providing adequate leadership is the most difficult management responsibility. A great deal has been written in an effort to differentiate management from leadership. Very briefly, management plans, organizes, staffs, follows-up, rewards, and stewards. It is concerned with the control of the activities of the enterprise. Management is a very important function in any activity be it private or public.

Leadership is made of different stuff. It stems from a vision of the direction and values that the organization should pursue. Leadership deals with understanding what the right course of action is and catalyzing the organization to follow it. Leadership involves risk.

While there are courses that deal with various aspects of leading, such as understanding one's own style, and recognizing the culture in which one seeks to lead, it is my perception that learning to lead is an implicit process. It evolves from experience in athletics, in social and civic organizations, on playgrounds, and from the development of self-respect in individuals. Thus, in adulthood, there are those who are capable of leading and those who are only capable of management. Aside from the limitations imposed by an individual's capabilities, there are those who could lead, but choose not to for reasons of personal gain. An editorial in *Business Week* (10/29/90) castigates both governmental and industrial "leadership" for their self-indulgence. There are many opportunies but few takers.

Many of our industrial enterprises are headed by those with extensive technical training of one kind or another. In many cases, such managers lack either the vision or the tolerance for risk that is essential for leadership. In some cases they may believe that their technical expertise gives them a special insight that makes their judgment superior to that of people outside the organization. In such cases, the concerns of the public can be ignored rather easily, until an incident occurs. It is important to recognize limitations in this area and find a way to structure an effective mechanism with which to verify that the organization is on the right path.

Table 2.2 lists a vision and a set of values for one department in one plant of a multiplant corporation. The department has a fairly narrow mission, which roughly coincides with the objective of this book. You will note that there are no specific systems or procedures written down. Rather, the list consists of

Table 2.2. SERM* Vision: Safety, Legality and Profitability Through Teamwork and Innovation

SERM Values
• Providing leadership in promoting process and personnel safety and protection of the environment
• Ensuring compliance with internal standards and external codes, laws and regulations, intelligently interpreted. (i.e., in good faith, with insight and common sense)
• Being a helpful, creative consultant to the organization
• Being a champion for better ways and better service to our clients and to the community
• Providing clear communication of options, reasons, and risks
• Being effective and persistent on the details; for example, foresight, cost consciousness, technical competence, and follow-through
• Obtaining first-hand knowledge of what is really going on out in the plant as a basis for realistic input
• Anticipating rather than reacting. Getting and staying ahead of possible problems and legal requirements

*Safety Environment and Risk Management

performance standards against which the department will judge itself in pursuit of its mission. They are flexible, idealistic, and subject to interpretation by the individuals involved in any specific situation. Such is the nature of providing guidance to leaders.

Increasingly, leadership means involving all of the people in the enterprise in one way or another. In the trite phrasing of current management texts, the people are "empowered," but in a very real sense, those people who do the work are the experts in any particular job and should contribute to the strategy and tactics for accomplishing it. By operating in this way, the leader has the benefit of their commitment and of their motivation to accomplish the task, as well as their wisdom on how to go about it. The essence of leadership is to have the buy-in of those who are led.

This is not to say that leadership of itself can accomplish the whole of any task. Management of the implementation of the strategy and tactics developed by the leadership group is still a very essential function. Without effective management follow-up, the implementation phase will soon flounder. Some key leadership functions are discussed in more detail in the following sections of this chapter.

Setting Standards

A major role of leaders is to define what constitutes excellence in their specific enterprise. In addition, continuous improvement of this defintion must occur. These definitions of excellence should follow directly from the vision and values of the enterprise. It is difficult to comprehend the meaning of the statement above without some specific example of how it might be applied. Table 2.2, which presents the vision and values of a single department, provides a definition of excellence and examples of certain behaviors aimed at achieving it. Department members can evaluate each day whether they are behaving according to the values they espouse. However, no list of examples can be comprehensive enough to capture all of what it means to be excellent. Thus, the creative effort of each individual comes into play.

In the risk management area, one significant opportunity for excellence involves beginning to identify and control risks at the earliest phase of process development. Very often, laboratory chemists pursue the production of the desired product by any way at hand. In many cases, this involves the use of highly toxic reagents that are convenient and effective to use in the laboratory. Often, the type of reactors studied are those typical of laboratory investigations; namely, relatively large-volume stirred vessels. Often reaction conditions are those convenient to create with lab equipment. In short, no thought is given to the commercial risks associated with the chemical pathways being explored to produce the product.

The opportunity for leadership comes in educating the bench chemist to the fact that excellence in process development includes early consideration of the commercial hazards that are inherent in his or her selection of reagents and in the volume of reactants that must be dealt with in a commercial facility. Various management systems have been discussed in the literature that can be used to accomplish this vision. Union Carbide, for example, initiates what it calls a Waste Minimization Opportunities Review (WMOR) as part of the early definition of the technology for any new product or process. This management system focuses on the minimization of hazardous waste, which has an implicit affect on the selection of reagents and on the efficiency and completeness of the various reactions employed to produce the product.

Another approach to achieving this objective involves carrying out what is called *an inherent safety review* very early in the design of a new process or product facility. This is based on the concepts espoused by Trevor Kletz (see Chapter 4), wherein he points out that what you do not have cannot either blow you up or poison you.

The sophistication and completeness of the technical work used as a basis for facilities design or procedures development are another area in which excellence often needs to be defined. In many plants, technical staffs are pressed for time and wish to do problem analysis in the simplest and least time-consuming way. "Back of the envelope" calculations are the norm for many decisions about operating procedures and facilities changes. Detailed analysis and, worse yet, the use of partial differential equations tend to be avoided because of lack of familiarity and the significantly larger time requirements needed to do that type of analysis. In addition, the existence of similar precedents plays a large part in technical decisions.

The leadership opportunity comes in knowing when more detailed analysis is required. For example, transient conditions in reactors are often dealt with by designing the facilities for the worst case and living with it. The challenge comes when a new process is used in existing equipment, and a far more detailed understanding of the transient behavior of the process, the peak reaction conditions, and the interaction of various components in the reactor system is needed before a judgment can be made about whether the process is safe. Since the equipment exists, the business people are often impatient to get on with the production of the new material, or the implementation of the new process conditions that will increase the capacity of the unit. A leader will intercept this momentum and direct it toward a thorough technical understanding of the risks before proceeding. In many plants, this might mean there is a need for outside expertise and an unbudgeted expense. It might also mean the use of specialized tests such as Accelerating Rate Calorimitry (ARC) or Vent Sizing Package (VSP).

A mechanism is needed to ensure that this happens. In many plants, a group is charged with approving any new operation or facilities and empowered to say no where appropriate. If this group has the necessary technical understanding, it can also define the standards that need to be met before approval is given. At the very least, it must be able to raise the questions that have to be answered before it will grant approval. Obviously, this group must have top management's complete trust and support. Just as obvious, the group must be seen as a partner in the enterprise and not merely as a roadblock, or ways will be found around it. At many Exxon plants this group is called the Safe Operations Committee (SOC) and more will be said about its role later. Other plants have other names, but the function is the same.

Sometimes, excellence in safety activities needs definition as well. These activities can become uninspired or uni-directional and therefore less effective. Leadership plays an important role in the continued viability of these activities. For example, safety meetings are one of the primary means of providing two-way communication on safety issues. Excellence in safety meetings includes:

1. *Good preparation:* Pertinent material, good aids/props, guest speakers, etc.
2. *Worker involvement:* Allowing team members time to prepare, soliciting their input, etc.
3. *Providing new information*
4. *Follow-up and feedback:* Timely response to the ideas submitted and concerns raised at these meetings is critical.

Effective leadership will see to it that excellence is pursued in all critical areas.

Outside Audits

While "audit" is generally a dirty word to most plant managers, the need to expose his or her organization to outside scrutiny is important. Leaders will take the risk in order to test in-house judgments and to verify that the organization is truly doing what it says it is in every area, everyday.

Audits are often imposed by corporate headquarters and/or government agencies. They represent, if you will, the final examination for the plant in a specific area. Failure implies embarrassment and perhaps worse.

Yet one can view audits from a different perspective. Like an examination, an audit can show areas in which the staff's judgment or expertise needs to be improved to live up to the vision and values that the organization espouses. In the preparation for an audit, the staff will learn many things. Often, this preparation would not be undertaken in such a comprehensive way if only internal stewardship requirements were to be met. The audit itself will provide further insights because the judgment of the outside auditors will not always coincide with that of the plant staff. The assumptions of the plant staff will be tested and those found wanting will be replaced. The natural tendency to excuse small failures within the organization will also be evaluated. If local management approaches an audit as a learning opportunity, as opposed to an obstacle to be surpassed, the experience can be valuable. It takes courage to approach it in this way, but the audit is going to happen whether the plant likes it or not, so one might as well get some benefit from it.

Much will be said about community outreach later, but there is also a public aspect to consider with audits and their results. Sharing the results of audits with the community can build trust with plant neighbors. At the very least, the impression that the plant cares about doing things right can be developed. Of course, progress reports on any remedial action plan will be needed to build on the initial impact. In the case of audits by government agencies, which are likely to be a matter of public record anyway, there may be some significant benefits to making the results public first, along with your interpretation and action plan. Subsequent interpretations by special-interest groups are often blunted by this strategy if effectively implemented.

Compliance With Regulations

Most companies make it policy that they will comply with all laws and regulations that apply to their operations. Some companies even include this kind of a statement in their value system with the further comment that where there is ambiguity, the company will follow the strictest regulatory interpretation.

The difficulty with such policy statements is that they are easy to apply when the regulations are clear but fail to provide specific guidance when ambiguity is involved. In most such cases, the agency takes an aggressive position about the scope and requirements of the regulation and companies disagree. The inevitable debates, negotiations, and even litigation are sometimes interpreted as an attitude of noncompliance on the part of the company.

I do not believe that this is a reasonable interpretation. Experience with many regulations indicates that there are often gray areas over which regulated groups and regulators of good conscience may debate. There is a credibility hurdle to be crossed. Over time, a company with a good understanding of the intent of the regulations and an objective view of how they apply to its business will win more of the debates with the agencies than it loses, and such a company can often develop a competitive advantage as a result of this understanding. In general, considering the conservative bent of most corporate lawyers, they will probably win the vast majority of cases that they choose to litigate. I assert that this represents a good record of compliance and should be shared with both the employees and the community. A recent article by John Wood, a senior consultant for C-E Environmental, Inc., in C-E's newsletter supports the view that reasonable debate about regulations fosters compliance.

A good example of this type of litigation involves the J. T. Baker Company, which operates a plant on the Delaware River in New Jersey. After rather long debates with the NJDEP about studies conducted on the contamination of the site, J. T. Baker was charged with many violations and assessed a fine of several million dollars. J. T. Baker contested all of the charges and was vindicated by the courts on the vast majority of them. The company ended up paying a fine that was about ten percent of the initial assessment by the DEP.

Another result of this litigation was clarification by the court of some incomplete legislation and regulations. A number of other companies are now aggressively responding to that decision by upgrading their own reports to the

DEP of contamination on their sites. In general, these had previously been interpreted differently by corporate lawyers, but the judgment of the court changed that.

In litigating this particular case, J. T. Baker not only justified its own interpretation of many of the regulations that applied to its situation but also clarified other important regulations for other companies. The net result will be improved compliance actions by other companies in response to the clarified decision on the interpretation of the regulations. All of this, I believe, had a much more positive impact on the environmental situation in New Jersey than if Baker had quietly paid the fine.

The key element in such a situation is that the company support, by its actions, compliance with laws and regulations. In addition, it is essential to explain the company's position in negotiations and litigations to employees and to the public so that the position in a specific action is not interpreted as noncompliance or antiregulation.

To provide a view external to the plant in question many companies have instituted the management practice of compliance audits by knowledgeable people from other sites in the corporation. At single plant sites, outside consultants are often used, not only to assess current compliance, but to suggest ideas on how to deal with future regulations.

The quid pro quo for testing a company's position on a regulation in public (whether it be via litigation or outside audit) must be a willingness to change the company's approach if it is found to be inadequate. Again, the J. T. Baker case is a good example of a company willing to do just that. One finding of the judge was that the interpretation by many companies that historic soil contamination in the absence of a traceable current spill did not need to be reported to the state as a "discharge" was wrong. The judge's logic was that the state has the right to determine whether something spilled on a site represents a danger to the public. The company cannot exercise this judgment on its own, nor can it deprive the state of the right to exercise the judgment by not reporting the evidence of contamination. While the evidence around the J. T. Baker case was very specific and involved hazardous substances (DDT), the decision would appear to apply to all sites and all potentially hazardous substances at this point. As a result, many companies are initiating site contamination studies and dialogue with the DEP to react to this new interpretation of the law.

In my view, the state should interpret this reaction as a positive compliance step, and companies should be open with employees and the community about these new initiatives. No doubt other lawsuits will result from these changes, and only time will tell whether the current ruling indeed applies as generally as the DEP would have it.

The Cost of Noncompliance

As already inferred, the cost of noncompliance as measured by the response costs to an incident can be very high. There almost seems to be a requirement that some government agencies find some violation if an industrial incident occurs. For example, in a recent explosion in an oil refinery in Louisiana where two people were killed, the resultant OSHA investigation, while complimenting the company for its emergency response procedures and other fundamental programs, also fined the company $6,000 for what might be termed nuisance violations—for example, failure to immediately institute urinalysis for firefighters and to immediately replace eyewash stations damaged by the explosion, and the absence of a signature in the 1988 injury report. In addition, at a polyethylene-plant explosion in Houston, a number of the response organizations were fined for violations of OSHA training requirements for hazardous-incident response teams. While it certainly is important for emergency responders to be properly trained, the hazards at the Phillips explosion and fire were not exotic in nature. It struck me that OSHA was working a separate agenda unrelated to the major issue in both cases. This does not make the monetary values of the fines nor the record of the violations any less significant. If there is an incident, there will be a presupposition of violations. It is a new facet of Murphy's Law.

The monetary cost of even small incidents can also be very high. In 1985, gases containing odiferous mercaptans were released from the Exxon site in New Jersey. Because of the atmospheric conditions, the odors were noticeable up to 100 miles away. There were justified fines for air pollution, there were claims for expenses for response to gas leaks from a number of local utilities, and there were direct payments to business and home owners who could substantiate losses because of the foul-smelling environment. In addition, the unit in which the incident occurred was shut down pending a significant investigation in which the DEP was very active.

Those were only the direct costs, however. As a result of the situation and the history of other incidents at the site, a major environmental-accident risk-assessment study was mandated by the state for the entire site. The study involved the use of contractors and a number of senior technical resources at the site and cost in the area of a million dollars to complete. In addition, several million dollars of improvements were recommended by the study over the course of the next several years to upgrade facilities. The net cost was probably in the range of $20M, in addition to all of the technical and managerial time that was required to negotiate the settlement with the state and for follow-up on the results of the study. All of this for an incident that involved no loss of life and no injuries to human or animal but did certainly degrade the quality of life in the area. The cause of this incident was primarily ignorance of the potential reactivity of a system that had operated for two decades. Could management have known more about what was necessary to maintain the the stability of the product they produced? The answer could only be yes and, of course, it now knows infinitely more about that stability for the whole family of products after this painful experience.

The lesson here is that management must exert leadership to develop information *before* an incident occurs, not after. Clearly, it would have been infinitely more economic to spend the money to do the research to understand in minute detail the stability of the product in question rather than to pay for the cost of the incident. It is probably true that developing the information for a host of products that might be capable of causing an incident would still be highly economic relative to a single incident with any one product.

Sentences for environmental crimes have also become more severe. The Sentencing Reform Act of 1984 set up a commissioner to draw up sentencing guidelines for federal judges. Suspended sentences and probation are on the way out and jail is much more likely. The state of New Jersey listed specific jail terms in a recent water pollution control bill. In many cases and places, it is possible to go to jail for an accidental discharge of a pollutant that causes no lasting environmental damage. Protection against overly zealous prosecution lies not in the language of the law, but only in the judgment of the prosecutor.

Clearly, personal as well as corporate disaster is a real possibility. A corporation might be barred from bidding on government contracts, but an individual manager may face both financial disaster and personal disgrace. This is indeed a very large incentive to ensure compliance with all safety and environmental regulations.

Involving the Experts

In modernizing manufacturing procedures to reflect the the approach of the Japanese, many companies have found that the true experts in many jobs are those who actually perform the task in the plant, not the technical staff or the management. With this in mind, many manufacturing facilities have mounted a significant effort to obtain input from the plant operators and mechanics about the real nature of a problem and their ideas about possible solutions. In recognition of this, CMA's Responsible Care Program makes a major point of getting feedback from employees about plant hazards and ways to reduce them. The primary thrust of this effort is to understand what worries the "experts" and obtain their views on the solution that would do most to reduce their risk. A fundamental principle is that if the workers in the plant are safe, then those outside the fence are even safer by virtue of their greater distance from the hazard.

There are many possible ways to get input from the "experts." These include regular safety meetings at which they are invited to express their concerns; frequent interactive safety surveys in which conversations with the workers not only evaluate what procedures they actually use, but also their view of the hazards; and various written forms of communication such as near-miss reports, safety action tickets, and so on. *The key element in the success of any of these techniques is a prompt and open response to each comment/suggestion.* Management must listen carefully to the message being conveyed and take on the obligation to evaluate it objectively and respond promptly. The timeliness of the response is perhaps the most important element of the program. A long technical study is generally not an acceptable answer to this type of input.

To involve the "experts" effectively, management, in addition to finding a way around apparent roadblocks, has to be willing to view conditions through the eyes of those who work with the facilities and procedures every day in a hands-on fashion. This does not mean that management will agree with every comment or implement every suggested solution. However, management does take on the obligation of an objective reply. As the process goes on, trust can be built up so that a more effective dialogue can proceed.

The principle here is that management cannot afford to ignore potential trouble spots. If it seeks out these before they become critical, they will be much easier to deal with in advance of an actual incident.

Procrastination—the Biggest Danger

One of the easiest traps in which to become ensnared in the incident prevention business is procrastination. It is easy because management usually perceives no compelling urgency to do something *now*. Often, the risk identified has existed for a long time without incident. It is also easy to procrastinate because of budget pressures. How does one justify spending scarce dollars on preventing a potential incident that has never seemed imminent? It is often just as difficult to obtain funding for technical efforts to find out how serious the risk is. The "head in the sand" approach is often pursued in the face of budget pressures.

Legal advice often supports procrastination as well. In some cases, it is necessary to report any data gathered to a government agency. Faced with the data, the agency has little choice but to order remedial action (and therefore, an unbudgeted cost) of one kind or another. This often applies in the environmental area. In New Jersey, the DEP's position is that hydrocarbon in the soil is evidence of a "discharge." All discharges must be reported so that the state can determine whether there is a danger to the health of the people. The discharger is not permitted to do the evaluation on its own. As a result, the short-term bottom line often will look better without the expense of soil-study costs imposed by the state agency, particularly when the agency oversight costs must also be paid. Is it any wonder that corporate lawyers advise management not to look for potential trouble—that is, to keep their heads in the sand?

What is left out of the equation, which leads to the "procrastination solution," is a realistic look at the cost impact of an incident. This applies to both industry and state agencies and must include the costs of a muddied reputation as well as the direct financial impact.

Consider the waste fire that caused the shutdown of Route 78, a major east/west link in New Jersey in 1989. A "recycling company" had illegally piled tons of construction trash under a bridge on this six-lane highway over the course of many months. A state agency had ordered the company to remove the garbage but got no response. The agency subsequently went to court to seek an order forcing the company to move the refuse. There was confusion over state and local government permits. The debate went on for months. One day, the trash caught fire. There was great difficulty getting a water supply to fight the fire. The net result was extensive damage to the road, which reduced it to two lanes for months. The cost to the public at large was immense in terms of delays for people and goods, increased congestion on local streets as drivers sought alternates, increased fuel consumption, and general abuse of the quality of life.

The aftermath of the fire was also instructive. The owners of the company were suddenly found to have criminal backgrounds and were immediately arrested. The agencies scurried to explain why the site had not been cleaned up before the fire and cited legal issues. The transportation department saw an opportunity and created short cuts through all sorts of standard procedures to hire a repair contractor in record time and get the extensive damage to the road fixed on a very accelerated schedule. This was a triumph for the transportation department, in severe contrast to the regulatory process that proved so ineffectual in preventing the disaster.

It seems obvious to me that the company could have been forced to clean up the site and the disaster could have been avoided if one-quarter of the energy that was expended after the fire had been directed toward preventing it. Clearly, the state proved it could move rapidly when motivated, but business-as-usual on the preventive end was totally inadequate.

The same failed perception and diminished energy may often be noticed in cases of industrial incidents. Bhopal is the most famous example. Refrigeration systems had been turned off, alarms disabled, and a scrubber down for maintenance for weeks prior to the incident. The logic was that there was no obvious reason to expect the chemical to decompose—that is, there was no imminent danger perceived. No one stepped back and really asked whether, with several lines of defense down, there was *any way* a decomposition could occur.

We know the answer to that one. Deliberately or accidentally, water was introduced into a chemical intermediate storage tank, a decomposition occurred with no backup protection available, and 2,500 neighbors died. This could not happen at our companies—right?

The rest of the process industry is still paying for the Bhopal incident in spite of belated efforts to face up to prevention issues. This is not meant to disparage these much-needed efforts, nor those of the lawmakers who responded to this horrible incident with stringent measures. The industry deserves all it gets in this area for not having faced up to the potential for such a disaster long before.

Is that a fair indictment of process industry leadership or merely the lament of hindsight? Some data may help. The Davis et al. text previously cited tabulates twenty-eight known serious air-borne toxicity incidents between 1950 and 1980. In 1950, twenty-two people died and 320 more were injured when hydrogen sulfide (H_2S) leaked in Mexico. More than 430 people were injured when a chlorine leak occurred in Philadelphia in 1983. Ammonia leaks in 1968,

1969, 1973, 1976, and 1977 in the United States, South America, and France killed sixty-seven people and injured more.

Governments did not respond until the infamous Seveso incident in Italy in 1976. The "Seveso Directives," a series of regulations requiring risk analysis when inventories of certain materials exceeded certain amounts in a plant, resulted in increased regulation for the European industry. However, this still did not catalyze a general effort by U.S. management to look comprehensively at potential incidents of its own.

What industry did do was review specific operations. I remember the big NH_3 tank reappraisal. Company-wide, each NH_3 tank was studied for possible catastrophic failure. All design parameters, instrumentation, and operating procedures were revisited and some weak points were shored up. Natural disasters were considered, but sabotage and airplanes falling from the sky were considered too remote to deal with. Time will tell whether this was a sound judgment or not. Time will also tell whether, after this big reappraisal effort, continuing maintenance of the facilities and the competency of the operators is effective, whether changes to the facilities and procedures are controlled, whether management continues to place a high priority on the integrity of the overall containment and mitigation system, and whether plans jointly with the community to deal effectively with the event that they clearly believe will never happen will prevent another disaster. This remains a challenge—perhaps the most difficult type of challenge—for local leadership.

Procrastination can also take the form of waiting for government and/or "higher management" approvals either before or after an incident. The most visible cases occur after an incident.

Contrast the *Valdez* actions with those after the Norco, Louisiana, light hydrocarbon explosion in about the same time frame. In the *Valdez* case, Exxon, appalled by the magnitude of the oil spill and anxious not to do anything else "wrong," waited for approval of its cleanup plan. The net result was a severe public indictment for delay, for missing a window in the weather that might have reduced the size of the area impacted, for being slow in the work even after starting, and for disinterest/arrogance on the part of "top management." All this, in spite of spending $2 billion on the cleanup!

In the Norco incident, the refinery manager was unable to get guidance from "top management" (or was too wise to seek it) and was in the streets personally very shortly after the explosion, checking on the damage, authorizing hardware stores to dispense supplies to those needing them, and was generally seen to be concerned, helpful, caring, and responsible.

Several people died in the Norco explosion and not a one did in Alaska, yet the Shell manager was seen as a hero and Exxon's management as arch villians.

I believe there is a lesson in this: Managers must have the courage to lead, even if there is some risk involved, especially when they believe they know what is right. Some condemnation is inevitable. At least the leaders will be seen as caring and proactive.

Let us consider a hypothetical example of this kind of procrastination on the preventive side. Many states require an air permit to be issued before operation (and in some states construction) of a new air-pollution control device can begin. Suppose you discovered that it was possible for one of your process or storage vessels to emit a toxic gas into the atmosphere in an accident—for example, a fire. A scrubber would solve the problem but an air permit would take six months. It is perfectly legal to operate as is but Bhopal is on your mind. What do you do?

Three courses of action come to mind at first:

1. File for the permit and wait to begin construction. Continue operations.
2. Shut down and empty the vessel until permit issues.
3. Install the scrubber and negotiate a special consent order with the permit authority in a rush.

The procrastination solution is legal but not ethical. The shutdown is both legal and ethical, but not profitable. The negotiation solution involves some embarrassment perhaps and probably a fee or fine but can preserve the opportunity for profits as well as legal and ethical values. There may be other more creative approaches depending on local laws but the primary guidance for choice must come from local management. It is interesting to speculate on how many times decisions of this type get made at relatively low levels in the operations or environmental organization without even being exposed to the plant manager. Such oversights occur not to protect the manager from hard questions, but because those making the decision do not realize the possible consequences of their decision and that the manager wants to be involved.

Other procrastination solutions abound in the day-to-day decisions in a plant. Consider the bypass of a critical safety device for routine maintenance—for example, a safety relief valve protecting a vessel. Many plants allow the removal of a safety valve for a shift for maintenance and provide a backup method to verify that pressures remain within safe limits. Often this involves increased or even full-time operator surveillance, this being judged more economic than maintaining a spare safety valve. What happens if the valve cannot be returned to service before the end of the day for some unanticipated reason? Is overtime authorized for either the repair or the safety watch? Or is it necessary to get approval from "higher management" to authorize overtime? (And of course they have already left for the day.) Is the unit shut down? Or is the policy simply set aside at the first line-supervisor level with the hope that nothing will go wrong overnight?

Another example involves technical information. Suppose there was an explosion at a competitor's plant and you learned that a safety valve appeared to be too small to handle an unanticipated runaway reaction. You now have doubts about the adequacy of your own safety valves but do not have the equipment or the expertise at the plant to reevaluate their sizes. You call the research department and they do not have any more data either. Furthermore, they have other priorities right now and are not interested in your potential problems. You find that outside contractors could make the evaluation for you but there is no money in either your budget or that of the research department for this testing.

Many plant people would throw up their hands at this point or merely agitate to get the tests into next year's budget. My belief is that the public would expect the bearer of the above knowledge to motivate his management to find the money for the tests or to change the research group's priorities. From an internal standpoint, if an incident did occur, how would you explain to the public and state agencies why your company did nothing even though it knew of the competitor's problem?

Leadership is most valuable when it prevents incidents. Each new piece of data needs to be evaluated and, if necessary, new priorities need to be established to accommodate a response. To do this, it is necessary to step back and take a look at each situation through the public's eyes to ensure that nothing significant is missed. What may seem trivial before an incident can become a major issue after one.

Community Outreach

Based on the reaction of the public to industrial plant accidents in recent years, we can conclude that it is no longer adequate to explain what happens after the fact. Armed with the wisdom of hindsight, the public can always find errors of judgment, violations of regulations, lack of required certifications, and other failures from which to build a cross on which to crucify plant management. Several examples of OSHA situations that appeared trivial relative to the severity of accidents were cited earlier as evidence of this tendency. The objective of management must be to communicate its message to the public in an ongoing way before it needs to explain the causes of any accident. Since many choices are neither black nor white, management faces a real task in getting its position understood.

This is a long-term continuing activity. It requires nothing less than a plan to create trust in the community and in the regulatory agencies with which each plant has to deal. There is significant risk involved because by exposing the plant hazards and the potential impact on the community to the public, the plant manager may be inviting more assistance than he wishes. In return, however, he or she does get to share the responsibility for the situation with the community as opposed to bearing it all on his/her own shoulders.

This is not a traditional public relations program. The communications with the public must deal with the substance of its questions and perceptions. Knowledgeable, credible spokespersons, who are willing to listen and respond openly and honestly, must be involved. In many cases, it is necessary to face up to opposing views from environmental groups, angry neighbors, and the media in an effective way to begin the process.

As indicated in Chapter 1, communities that are familiar with the risks associated with industrial facilities seem to be more sanguine about these risks than is the case when what goes on behind the plant fence remains a mystery.

The presentation of facts by a familiar face goes a long way toward reducing outrage and building trust over a period of time. The key is that credible spokespeople be available to answer questions and deal with minor annoyances over time so that if a more significant accident occurs, the community will have both a mechanism through which to ask questions and a responder who is familiar and hopefully, trusted.

Obviously, the mechanism for developing this dialogue will vary from plant to plant. Successful techniques have included community newsletters, plant neighborhood nights, specialized hotlines to answer questions, advance communication of unusual activities such as noises and/or flaring, and regular meetings with a committee of community representatives to discuss what is going on at the plant as well as what is going on in the minds of the neighbors. Many plants use periodic newsletters as an effective vehicle in dealing with specific plant issues and more general questions having to do with the relationship between cancer and chemicals, and so on.

In the United States, the SARA 313 emissions report has become a focal point for such dialogue. Originally, many plants simply submitted the data to the appropriate regulatory agencies and hoped for a quiet response. While there was some delay before the first year's data were available in a form that could be easily analyzed by environmental groups, the environmentalists wasted no time in publicizing their interpretation of the significance of the large amounts of emissions that were made public. Some companies seized the moment to make commitments for future reductions. Monsanto was perhaps the most visible of these and gained wide recognition for its proactive approach. From the practical standpoint, Monsanto simply painted its own view of its own responsible actions at the time the data were released. This gave Monsanto the advantage of having its interpretation before the public much sooner than the more critical view of environmental groups. The value of this initiative is clearly appreciated by most companies these days, but the focus of the environmental-group criticisms is changing, so each company must still assess the attitude of its own neighborhood's groups before following blindly in any other company's successful footsteps.

The risk of negative reaction from the public cannot be defined out of existence. It does appear, however, that well-thought-out, open discussion of each plant situation can do much to create an atmosphere of trust in the community *provided* real progress is being made in reducing the risks imposed on the community by an industrial facility.

Risk Communication

The purpose of this section is not to provide all the information needed to develop and implement a good risk communication program but rather to convince managers that risk communication is a tricky business and needs to be carefully prepared for and well implemented.

Since the requirements imposed by the SARA 313 law in 1986 to disclose emissions, much has been written about how to communicate to the public about risk. The CMA publishes many handbooks and has presented seminars on the subject, based on the research of V. T. Covello, P. M. Sandman et al. (*Risk Communication, Risk Statistics and Risk Comparisons,* CMA, Washington, D.C., 1988), and others. The major thrust of this work is that the psychological aspects communicating risk to a community or group are probably more important than the technical.

A corollary is that there is not a lot of good data on the health consequences of any exposure, especially chronic exposure to small concentrations.

Covello summarizes eighteen factors derived from his research and that of others that influence the public acceptance of risk. These make the simplicity and intuitive appeal of risk comparisons extremely difficult to carry out successfully. These factors are:

Factor	*Conditions associated with increased public concern vs. decreased public concern*
• Catastrophic potential	Fatalities, injuries grouped in time and space vs. scattered
• Familiarity	Unfamiliar vs. familiar
• Understanding	Mechanisms or process not understood vs. understood
• Uncertainty	Risks scientifically unknown vs. known
• Personal controllability	Uncontrollable vs. controllable
• Voluntariness of exposure	Involuntary vs. voluntary
• Effects on children	Children specifically at risk vs. not
• Effects manifestation	Delayed effects vs. immediate
• Effects on future generations	Risk to future generations vs. none
• Victim identity	Identifiable victims vs. faceless statistics
• Dread	Effects dreaded vs. not
• Trust in institutions	Lack of trust in responsible institutions vs. trust
• Media attention	Much media attention vs. little
• Accident history	Major and sometimes minor accidents noted in past vs. no prior publicity
• Equity	Inequitable distribution of risks and benefits vs. equitable
• Benefits	Unclear benefits vs. obvious ones
• Reversibility	Effects irreversible vs. repairable
• Origin	Caused by human actions or failures vs. those of nature

Thus, comparing the risk of cancer from plant emissions to the risk from smoking fails on a number of issues. Such a comparison matches a voluntary vs. an involuntary action, one with clear benefits (to the smoker) vs. probably unknown benefits, one that might affect the children of the receptor vs. one that will not, familiar risk to an unknown, and so on. Since this comparison differs in a number of qualitative dimensions, it is likely to be meaningless and often counterproductive.

Many examples of bad comparisons are discussed in the CMA manual. It suffices to say that one's instinctive approach is probably not the best approach. Significant research should be done before approaching even friendly audiences. In addition, the process must be ongoing. The objective is to become familiar to the community, to build trust against the day when hard times arrive.

The story about the Exxon dock study in New Jersey discussed in Chapter 1 helps to illustrate this point. Of its own volition, Exxon shut down marine operations (and ultimately the refinery, for lack of crude oil) until a major marine safety study was done. The results of the study were then shared with the agencies, local government, and the press. The docks were reopened and the agencies invited to watch the initial operations. The results of the study were then shared with other marine operators in the area.

Shortly after all of this, there was another barge accident at the Exxon docks. Two non-Exxon barges collided causing a leak of about 40,000 gallons. Although Exxon was not responsible, the press was invited for a briefing and live TV was permitted in the dock area. Reportage characterized the incident as an accident, showed the emergency response to be excellent, and was delighted at the open approach. The incident ended as fast as it had begun. In contrast to the drawn-out period of investigation and allegation associated with the previous incidents, the treatment of this one was a major reward for the program of communication and trust-building described here.

Summary

Managers must recognize that the prevention of any incident is the only acceptable standard these days. They must also recognize that their efforts to achieve this objective need to be communicated to their employees, their host communities, and to state and local agencies as required by law in an effective way. Management's highest priority in conducting its day-to-day business must be the protection of its employees, its customers, and its neighbors. Failure to accomplish this truly threatens the viability of the enterprise.

Management must be visible, accountable, and credible on the issue of incident prevention. It must set objectives, measure accomplishments against them, and explain any failure. Further, the company must "walk its talk" from start to finish to be credible.

More importantly, management must provide leadership. This means having a vision of excellence and taking prudent risks to achieve it. Leadership starts with defining excellence and then providing the resources and motivation to achieve it. Excellence can no longer be a private matter. Each milestone needs to be shared with employees and neighbors and a continuing dialogue must be maintained. All of this requires new skills and approaches from managers. The price of procrastination in learning and practicing them could well be disastrous.

CHAPTER 3

Hazard Identification

Introduction

The purpose of this chapter is to discuss methods for identifying hazards and how to to assess the importance of each hazard relative to the specific plant location. In short, what hazards exist and how bad are they?

Our focus will be on acute situations (i.e., potential accidents) and how the "system" responds. Some analysis can be done very early in the development of a process or project, including study of the choice of chemicals and system inventories needed to implement the process; decisions on other topics require much more information about the details of the design, site location, instrumentation, and so on.

The impact of some hazards is obvious. For example, if large volumes of chlorine are used, it is intuitively clear that many precautions must be taken to ensure maximum risk control. Less-obvious hazards often require more sophisticated analysis to evaluate.

This invariably involves judgment and, therefore, differences of opinion. Even in cases where detailed calculations of concentrations in the air are made, questions can be raised by concerned citizens or workers: Is the "worst case" scenario credible? Is the concentration judged safe by the plant or state agency really safe enough for the community? Thus, one needs to be prepared for logical questions about how important each hazard is.

Chronic hazards, i.e., those that cause an effect through the cumulative impact of continued small exposures, are also important, especially since fugitive emissions are reported under SARA III. Not much is known about the health effects of small concentrations of many chemicals. In the United States, OSHA sets workplace standards for many chemicals. Given reasonable separation from the community, meeting these standards often provides a good margin of safety for the community, even though some of its inhabitants may be in more fragile health than the average industrial worker. Various associations of industrial hygienists endorse standards for acceptable long- and short-term exposures in many countries, and so do some state and federal governments. These provide useful reference points in communicating the importance of hazards in specific situations.

What is needed is the realization that at one stage or another in the process of hazard identification, judgments are made about how serious specific hazards are. These judgments and their underlying assumptions must be scrutinized initially and reevaluated periodically as plant and community situations evolve.

Types of Hazards

As was noted in Chapter 1, there are three fundamental types of hazards. These include the energy content of a material, the toxicity of a material, and the potential for a material to react uncontrollably with other materials either inside or outside the system being studied or managed.

Energy

Energy-related hazards cover the broadest range of possibilities. The obvious hazard is the ignition of a material with high energy content after it has escaped from containment. This could result in a pool fire, the ignition of a vaporizing liquid, or a vapor cloud explosion. Most *controlled* releases of high-energy materials occur through flare systems where ignition is purposely done at a safe place, but some safety valves still release directly to the atmosphere, and small vents (such as on-line analyzer vents) and leaks provide the potential for uncontrolled combustible mixtures being produced outside of the intended containment system. While vapor leaks of five tons or less are unlikely to cause vapor clouds, one event (e.g., a pool fire) may cause another (tank rupture), cascading into a major risk for both the plant and community. The possibility of cascade effects must also be considered in evaluating hazards.

Many high-energy chemicals also pose the hazard of *causing* loss of containment because of their own physical properties. This can happen if a highly volatile material is heated to the point where its vapor pressure exceeds the maximum working pressure of its containment. It can also occur as a result of another accident, for example, an external fire, or because of the intrusion of enough oxidizer (e.g., air) to form a combustible mixture inside the containment, which then encounters a source of ignition (e.g., a static electricity charge).

Static accumulators are materials that build up static electricity charges while flowing through a pipe or splashing on a liquid surface. They pose special hazards during transfer operations because, of themselves, they provide two legs of the fire triangle—the fuel and the source of ignition.

We know it takes all three components of the fire triangle, that is: fuel, ignition source, and enough oxygen to form an explosive (flammable) mixture to result in a fire or explosion. A number of industrial processes deliberately operate in the explosive range by carefully excluding a source of ignition. Often major precautions are taken to deal with the possibility of explosion even then. The oxidation of *o*-xylene to phthalic anhydride is a good example of such a process and compatible equipment design.

Much less exotic operations, such as the loading of gasoline barges and tanks, also have the potential to operate in the explosive range if *procedures* (rather than hardware) are not scrupulously followed. In the spring of 1990, a barge being loaded with heating oil exploded at a New Jersey commercial terminal. The cause was thought to be residual gasoline vapors from the previous cargo. Loading procedures and facilities were proper for the less volatile heating oil, but cleaning of the barge between cargoes may not have been appropriately complete. Human error or incompatible materials and procedures can cause an accident even though the knowledge is ostensibly available to prevent the incident.

Over the years, I have become much less sanguine about relying on eliminating only one leg of the fire triangle to control energy hazards. Surely if one is to rely for safety on the absence of only one leg, he/she must fully understand all of the reaction characteristics of the system (more discussion in the section on reactivity hazards). There have been too many explosions in which the source of ignition was never clearly identified for one not to be concerned.

In 1970, the reactor section of a caprolactam plant in Spain exploded. (This was a very different situation from the Flixborough incident in Britain, which has been the subject of much study.) The reactor consisted of a large vessel containing cyclohexane and catalyst at a pressure of about ten atmospheres. Vapor from the vessel was drawn off by recycle compressors and air was introduced to provide the O_2 for the oxidation reaction, all under carefully controlled conditions. One day the recycle compressors shut down suddenly but the interlock to shut down the air supply did not operate. At least part of the vapor recycle line thus was put in the explosive range. When a recycle compressor was restarted, the recycle system exploded, creating a twenty-inch-diameter blowtorch until the reservoir of cyclohexane was depleted.

There was no obvious reason why starting the compressor should have caused a spark. The motor was explosion proof and there was no obvious friction in the small machines. A great deal of time and energy was spent examining the chemistry of the system to see whether an unstable peroxide might have been formed inside the system in the unusually high O_2 content, but no definitive conclusion was ever reached. Was it the machinery, was it the chemistry, was it an unusually high level of some impurity in the air? We will never know, but

I, for one, like to test designs against the contingency of a source of ignition being present unless the design has been thoroughly developed with the intention of preventing one.

As can be seen, energy hazards are really a group of related hazards that stem from the physical and chemical properties of the materials in question.

The size and design of the physical systems in which a potentially hazardous material is contained can also play a role in the judgment of whether a hazard can result in a need to reduce the risk it imposes on people both inside and outside the plant. If the contained volumes are small, the magnitude of the hazard is obviously reduced (more about this approach in Chapter 4). If critical vessels are well separated from sources of ignition and other flammables, if they are a significant distance from the plant fence, if they are protected from external heat (fireproofed), and if they are designed to withstand maximum credible pressures and temperatures, the risk imposed by energy hazards may well be negligible to the community except in the case of a major loss of containment. However, the possibility of the material itself causing a leak or internal explosion cannot be ignored in the process of identifying these hazards.

Toxicity

Over the years, the words "toxic" and "hazardous" have assumed new meanings. Almost everything is hazardous (by law) and the word "toxic" screams at us daily from the headlines. Industrial hygienists will tell you that anything (including water) can be toxic or hazardous at the wrong time or in the wrong amount. There was one case where a batch of baby formula contained salt instead of sugar and made children ill. Too little salt causes muscle cramps in athletes on hot days. Too much vitamin C can be excreted, but too much of other vitamins in the body may cause just as many problems as too little.

One of the great unknowns in evaluating hazards is what the acceptable range of any chemical is for good health. In many cases, the health professionals do not know. Clearly, they continue to learn, and generally, the acceptable concentration in any chemical exposure becomes more restrictive with time: for example, the OSHA workplace limits for benzene have just been reduced from ten parts per million (ppm) in the air to one part per million. Yet we can measure some concentrations down to one part per trillion and many down to a part per billion (ppb—equivalent to one second in thirty-two years of time), and some feel that even these low levels can have health effects.

As a result, the definition of what is toxic and hazardous is often left to state regulations or judgment (and salesmanship). Clearly a conservative and informed posture is needed in any industrial enterprise, as is an effective way to communicate that to the agencies and the public. Given all of that, there will (not may) still be risks (real or perceived) remaining to deal with. Since Material Safety Data Sheets (MSDS) are often very conservatively written, each company is often left to find its own position in the gray areas. Even so, as I shall hope to demonstrate shortly, it is still possible to develop reasonable definitions of toxicity hazards for each plant.

Reactivity

Since Bhopal, reactivity hazards have assumed a new importance. The range of possible situations that may pose a hazard has grown, perhaps even to include sabotage. In response, the AIChE and others have developed a great deal of technology aimed at identifying and quantifying reactivity hazards. In addition, the potential for introducing extraneous materials is being considered much more frequently than before.

One can classify reactivity hazards into various groups based on the initiating factor. The first class can be attributed to the *instability* inherent in the molecule itself.

Perhaps the most frequent cause of instability is overheating the material in question. Raw materials, products, and intermediates must be considered in this analysis. In many cases, the impurities, co-products, and so on present in crude products, reaction masses, and the like can contribute to instabilities not present in the final product.

It is essential to know the complete envelope of temperatures at which every mixture in a process is thermally stable and the margin of safety to the lowest decomposition temperature. This implies that the variation of the decomposition temperature is known. It is also important to know whether the beginning of decomposition causes the formation of a product that catalyzes the decomposition reaction. Thus, it is necessary in many cases to get data about the decomposition reactions as well. As will be discussed later, this is essential to design safety systems adequately.

In defining the safety envelope, it is also important to consider abnormal operating conditions and *all* sources of heat. A temperature-sensitive material may have no obvious heat source connected to it, but if the tank is agitated, the mechanical energy input from the agitator needs to be considered, especially if the liquid level is lower than normal. Another concern associated with low liquid level is whether the temperature sensor being relied upon to signal a decomposition is still submerged in the liquid. If not, it will continue to register a safe temperature even if the bulk temperature begins to rise. These are but two examples of how the interaction of the material in question and the processing facilities and procedures may combine to escalate a relatively small hazard into a potential incident requiring risk reduction.

Chemical sources of heat also have to be considered. The obvious source would be the addition of another material to the brew at a temperature higher than called for by the process. Its sensible heat may bring the bulk of the material or a segment of the material to decomposition temperature. Slow oxidation may also cause heat-up. The air in an agitated vessel has occasionally reacted with the contents to cause a slow release of heat. This can become a serious hazard, especially at low levels or long agitation times.

This leads to the consideration of *impurities/catalysts* as causes of instability. Clearly, air needs to be excluded from a vessel that contains a highly flammable material. However, there are many other equally reactive materials that may cause a hazard in your system. Remember, Bhopal was caused by the introduction of water to a place where it did not belong. Impurities in raw materials are a common cause of decompositions, as is the failure to remove by-products that catalyze unwanted reactions. Obviously, the introduction of chemicals that are known to be hazardous in the mixture merits special preventive measures.

In some cases, even introducing reactants in the wrong order or the wrong physical state can be hazardous. For example, gaseous chlorine may react quite controllably with an organic, but liquid chlorine can react quite violently in the same equipment. Similarly, reacting *A* with *C* before introducing *B* may cause a catastrophic alteration in the chemistry desired. The possibilities for mischarging in order, amount, temperature, or material need to be carefully considered in evaluating the potential impact of even an apparently minor hazard.

The impact of *external chemicals* already has been discussed at some length. However, this hazard deserves separate consideration as well. The reason is that such chemicals may be introduced by routes thought to be secure. Usually this occurs when there has been a monumental foul-up in the supply system for the desired reactant or utility. For example, a truck driver may inadvertantly unload chemical *B* into the storage tank for chemical *A*. A plant nitrogen system may become contaminated with oil or air, or perhaps the plant air system is backed up with nitrogen. (The latter is why no one *ever* relies on plant air for breathing air in confined spaces.) In all these cases, something quite unexpected comes out of the pipe. For this reason, potable water supplies are always protected from contamination by a physical disconnect in the flow system (an absent length of pipe that will absolutely prevent backflow), and similar precautions are sometimes needed in chemical or utility systems.

The contamination of a recycle solvent is a common occurrence. In many cases, a solvent is evaporated from an intermediate material, condensed, and recycled. If the condenser uses water as coolant, the solvent could become contaminated by a leaking condenser tube. Process designers often take care of this by planning for the water pressure to be lower than the solvent pressure. Such subleties are sometimes lost on plant operators. They may turn on the spare water pump because more pressure is needed or close a valve downstream because of a problem. The message is simple: If the introduction of water would be catastrophic, it must be inherently impossible to contaminate the solvent.

In general, recycle processes provide opportunities for contamination. Impurities collected or formed in tiny quantities in each pass of the recycled material through the system build up with time and can become hazardous. As discussed, they may react in the system. Materials may also accumulate in the piping system, causing problems later when the system is opened for maintenance or chemically cleaned. Iron sulfide, which is pyrophoric in air, often forms in refinery systems handling vapors rich in H_2S. In many cases, it is necessary to clean these systems chemically before they can safely be opened for maintenance or inspection. A good understanding of what might accumulate in recycle systems is very helpful in identifying some unusual hazards.

Another source of unwanted hazards is *leaks*. Some materials react violently with chemicals commonly encountered in the environment. The leak of a hydrocarbon to the air above its autoignition temperature causes an instant fire; spilling molten sodium into a typical sewer causes an instant explosion; and leaking acid causes instant corrosion, perhaps setting off a chain of unwanted events. These are only some of the reactivity hazards. Vapor cloud explosions and toxic clouds may seem much more serious, but even small incidents can have serious impacts and need to be prevented.

To conclude this section, we can say that a broad range of possible reactions need to be considered before declaring a reactivity hazard insignificant. If *A* can react with *B* to cause an incident, then any way *A* and *B* could get together must be considered and protected against before concluding that *A* does not represent a hazard.

Identifying Hazards

The first step in identifying hazards in a process plant involves developing a data base on *all* materials used (even the janitors' supplies) and evaluating the potential hazards posed by each individually and in credible combinations with other chemicals, the environment, and the system in which they are contained. The data base provides the necessary technical information, but the evaluation of this information will require judgment and, therefore, debate pertinent to the specific plant situation.

For example, one ton of chlorine in the middle of a 1,000-acre petrochemical facility may pose no problem to the community because of distance, whereas the same quantity at a municipal water-treating plant may pose a major threat to neighbors.

There are a number of approaches to doing such a chemical screening study. The methodology described below is one that was accepted by the New Jersey DEP as the basis for an Environmental Accident Risk Assessment Study at Exxon's Bayway facility.

One-Chemical Screening Study Methodology

The first step in the screening process is to compile a complete listing of all chemicals used, produced, stored, or disposed of at the site. The list will include all raw materials, finished products, intermediate products, by-products, catalysts, solvents, cleaning agents, waste streams, and so on.

In the second step of the screening process, a data bank is developed for each chemical. The data include a description of its physical state as well as the typical and maximum amount produced or handled. Also included in the data bank are the typical and maximum quantity stored in inventory (including material within process areas), its location, and the type of storage container.

The data bank will also include various properties of each chemical and its reactions to the extent that these properties are available from reference materials.

In the third step, a screening analysis is made to classify the chemicals. Each chemical identified is classified based on its physical and chemical properties, and the other data are assembled. Special attention is given to vapor pressure, thermal stability, chemical stability, chemical reactivity, combustion products (flammability), and toxicity.

References used for classifying chemicals typically include the following:

1. Material Safety Data Sheets from the manufacturer of the material
2. *Patty's Industrial Hygiene and Toxicology,* Clayton, G. D. and Clayton, F. E., Wiley Interscience, New York 1978
3. *ACGIH Documentation of Threshold Limit Values,* American Conference of Governmental Industrial Hygienists, 6500 Glenway Avenue, Building D-7, Cincinnati, OH, 1986
4. CRC *Handbook of Chemistry and Physics,* The Chemical Rubber Co., Press Cleveland, OH (1988 edition)
5. *Chemical Hazards of the Work Place,* Proctor, N.H., Hughes, J.P., J. B. Lippincott Co., Philadelphia, PA, 1978
6. *Clinical Toxicology of Commercial Products,* Gosselin, B. E., Smith, R. P., Hodge, H. C., Williams & Wilkins, Baltimore, MD (1984)
7. *Casarett & Doulls Toxicology,* Doull, J., Klassen, C. D., Amdur, M. O., MacMillan Publishing, New York, 1980
8. *Handbook of Toxic and Hazardous Chemicals,* Sittig, M., Noyes Publication, Park Ridge, NJ, 1981

Each chemical can be classified in one of the following categories. Please remember that any chemical can be a hazard under certain circumstances. One may drown in pure water or be burned by steam. It is the job of Risk Management to recognize all hazards, eliminate as many as possible, and then control the remaining risks to an acceptable level.

Classification	Description
A	Stable solids with little or no potential threat of injury to public health or welfare, but that might be an in-plant hazard
B	Stable liquids with little or no potential threat of injury to public health or welfare, but that might be an in-plant hazard
C	Gases with little or no potential threat of injury to public health or welfare, but that might be an in-plant hazard
D	Solids that could pose a potential threat of injury to public health or welfare
E	Liquids that could pose a potential threat of injury to public health or welfare and have low volatility (vapor pressure [V.P.] $\leq$50 mmHg at 25°C)
F	Liquids that could pose a potential threat of injury to public health or welfare and have a high volatility (V.P. > 50 mmHg at 25°C)
G	Gases that could pose a potential threat of injury to public health or welfare

Chemicals classified as A, B, or C can be excluded from further consideration as a potential threat of injury to public health or welfare. We need to recognize that the reasons for making these judgments must take a form that can be used to explain them to the public, if needed. There are few universally black-and-white decisions. In addition, the magnitude, if any, of the hazard to in-plant workers must be assessed and the associated risks managed.

The reasons for classifying a chemical A, B, or C might involve its properties, the quantity of the material present, its container, its location in the plant, or the procedures used to manage the risks. It is important that the reason be clearly identified so that *the safeguard can be maintained,* that is, can be recognized as *safety critical.*

Consider Table 3.1, the chemical screening data sheet for sulfur. The material is virtually nontoxic, has low vapor pressure, high flash point relative to storage temperature, and is chemically stable. It is present in process vessels and in a storage pit at temperatures high enough to cause severe burns to workers, and it is combustible. However, it is stored below its flash point and snuffing steam is provided in the unlikely event of fire.

The likelihood of this material causing public injury is essentially zero, but workers must be protected from splashing or falling into the pit and adequate

Table 3.1. Chemical Screening Data Sheet—Sulfur

Unit	*Product Type*	*Product Name*
CAPH	Raw Material	Sulfur

Screening classification: B	
Amount used typically:	X klb/yr
Maximum:	Y klb/yr
Inventory typical:	56.0 klb
Maximum:	112.0 klb
Location:	Block 62
Container:	Pit
	Process vessel
Physical state:	Liquid
Vapor pressure:	0.105 mmHg at 284°F
Flash point:	370°F
NFPA rating:	1
Storage/usage conditions	
Pressure (psia):	34.70–14.70
Temperature (°F):	300–250
Thermally stable:	Yes
Combustion products:	Oxides of S
Chemically stable:	Yes
OSHA toxicity (PEL):	No limit set (NLS)
OSHA ceiling:	NLS
ACGIH toxicity (TWA):	10.000 mg/m^3
ACGIH toxicity (STEL):	NLS
AIHA (WEEL):	NLS
Comments:	Sulfur is virtually nontoxic

fire protection must be provided. Thus, a B classification is assigned and the critical steps for in-plant protection are identified:

- Fire protection
- Inspection of containment to ensure its integrity
- Railings and the like to prevent falling into pit
- Proper protective equipment and procedures for handling
- Training of operators and mechanics to use them

For each chemical classified as D, E, F, or G, additional analysis must be made to assess potential threat of injury to public health or welfare in addition to the in-plant hazard each poses. In determining the maximum potential threat to the public for the selected chemical, the equipment associated with the chemical is examined at its maximum production rates and at its maximum inventory. The judgments in these categories become more subjective.

Before using more sophisticated methods of assessing the potential impact on the community, we will explore some examples of the evaluation of several hazardous substances that can be classified more simply.

The properties of a caustic solution are listed in Table 3.2. This is a 20% solution used in pollution control scrubbers with recirculation pumps. The material is nonvolatile, water soluble, and very corrosive to organic matter. It reacts violently with acidic materials.

Clearly, the only way this material could escape and impact the community would be via a spill into a public waterway. If there is secondary containment around tanks and process areas—and this includes a wastewater treatment system—the hazard to the public is essentially zero.

Hazards inside the plant are substantial, however. Any spills must be kept from splashing on workers and acidic materials must be segregated both inside and outside process vessels. This should be part of normal plant precautions, but periodic reassessment is important. Fire is not a consideration with caustic.

As a result of the above considerations, a classification of B—a liquid posing no threat to the community is justified—and sustainable under public scrutiny.

The properties of phenol (monohydroxy benzene) (see Table 3.3) present both similar and different considerations. Phenol is a solid at normal ambient conditions. It is usually handled as a molten liquid, however, where it has a medium-low flash point of 175°F. It is exceedingly hazardous to workers and overheating can cause fumes and mist that are corrosive.

The phenol is received in a tank car, stored in a tank, and pumped from there to batch reactors. Workers must be protected from spills, but otherwise spilled material will rapidly solidify and present no hazard. If some does reach a process sewer, some reaction may occur, and any direct discharge to public waters would cause a pollution incident. Thus, it is similar to caustic in some ways.

Where it differs from caustic is in its storage temperature and flammability. Overheating or exposure to fire could cause fumes that could impact the community. These issues are dealt with in one plant as follows: The phenol is stored at a maximum temperature of 180°F. At higher temperatures, it is

Table 3.2. Chemical Screening Data Sheet—Sodium Hydroxide 20% Solution

Unit	*Product Type*	*Product Name*
VIS-J	Raw Material	Caustic

Screening classification: B	
Amount used typically:	1890.0 klb/yr
Maximum:	2000.0 klb/yr
Inventory typical:	6.0 klb
Maximum:	15.0 klb
Location:	Block 2
Container:	Process vessel
Physical state:	Liquid
Vapor pressure:	9.751 mm Hg at 68°F
Flash point:	°F
NFPA rating:	0
Storage/Usage Conditions	
Pressure (PSIA):	19.70–14.70
Temperature (°F):	120–40
Thermally stable:	Yes
Combustion products:	Does not burn
Chemically stable:	Yes, but reacts with acids violently and generates heat when reacting with water
OSHA toxicity (PEL):	
OSHA ceiling:	
ACGIH toxicity (TWA):	2.000 mg/m^3
ACGIH toxicity (STEL):	
AIHA (WEEL):	
Comments:	Corrosive; can cause severe chemical burns

contained only in process vessels. The phenol storage tank typically contains 800,000 lb. The tank itself is contained in a separate diked area. The closest adjacent hydrocarbon storage is in tanks located twenty-five feet from the phenol tank. All of the tanks adjacent to the phenol tank contain high-flashpoint materials (~300°F) in which lube oil is the most volatile component. Each of these adjacent tanks is located within dikes separate from the phenol tank. Thus, there is a low probability of a fire in the tanks impinging on or engulfing the phenol tank. If there were to be a fire in the adjacent area, there are sufficient hydrants and monitors both to contain the fire and to deluge the phenol tank.

Thus, considering the properties of phenol, its usage and storage conditions, and the low probability of a direct fire, it has little or no potential to cause an offsite impact. Therefore, this material has been classified as B, a stable liquid with little or no potential threat of injury to public health or welfare. Note that part of the rationale for this conclusion is the arrangement of facilities to prevent involvement of phenol in external hazards and, therefore, reducing the risk to the public to a level *judged* to be negligible.

Turning to a more volatile material, the properties of sulfur dichloride are

Table 3.3. Chemical Screening Data Sheet—Monohydroxy Benzene

Unit	*Product Type*	*Product Name*
NP/DDP	Raw Material	Phenol

Screening Classification: B	
Amount used typically:	X klb/yr
Maximum:	Y klb/yr
Inventory typical:	847.0 klb
Maximum:	1090.0 klb
Location:	Block 62
Container:	Tank, tank/car, process vessel
Physical state:	Liquid at 220°F, solid at ambient temperatures
Vapor pressure:	1.000 mm Hg at 104°F
Flash point:	175°F
NFPA rating:	2
Storage/Usage Conditions	
Pressure (PSIA):	49.70–14.70
Temperature (°F):	275–68
Thermally stable:	Yes
Combustion products:	Fumes, smoke, CO
Chemically stable:	Yes, but reacts with many chemicals
OSHA toxicity (PEL):	NA
OSHA ceiling:	NA
ACGIH toxicity (TWA):	NA
ACGIH toxicity (STEL):	NA
AIHA (WEEL):	NA
Comments:	Vapors and mist are locally corrosive; over exposure to high concentrations can cause collapse, convulsions, circulatory failure, and death

shown in Table 3.4. The material is received in tank cars and pressured by low-pressure nitrogen directly into a reactor, where it is used to sulfurize an organic material. A small fraction of the material is unreacted and goes to a neutralization unit before being discharged to the sewer.

Sulfur dichloride is obviously a reactive chemical. In addition, there is an equilibrium between the di- and monochlorides that will evolve chlorine gas. At higher temperatures, the equilibrium shifts toward the monochloride and chlorine. The material also reacts with water to form SO_2, HCl, and H_2S. Thus, spills to the ground are potentially hazardous and cannot simply be washed down the sewer.

Clearly, this material has the potential to impact the community. If the quantities used were small and relatively far from the property line, one might conclude it would be worthwhile to make calculations about the concentrations in the community and their potential health effects before deciding to do a risk assessment for this material.

Table 3.4. Chemical Screening Data Sheet—Sulfur Mono/Dichloride

Unit	*Product Type*	*Product Name*
NPS	Raw Material	Sulfur Dichloride

Screening Classification: F	
Amount used typically:	X klb/yr Confidential
Maximum:	Y klb/yr Confidential
Inventory typical:	301.0 klb
Maximum:	465.0 klb
Location:	Block 62
Container:	Tank/car, process vessel
Physical state:	Liquid
Vapor pressure:	7.501 mm Hg at –9°F
Flash point:	N/A °F
NFPA rating:	0
Storage/Usage Conditions	
Pressure (PSIA):	54.70–14.70
Temperature (°F):	285–68
Thermally stable:	No
Combustion products:	Does not burn
Chemically stable:	Yes of itself, but reacts with water and organic materials to give off HCl and generate heat
OSHA toxicity (PEL):	
OSHA ceiling:	
ACGIH toxicity (TWA):	1.000 mg/m^3
ACGIH toxicity (STEL):	
AIHA (WEEL):	
Comments:	Acid gas; irritating, corrosive to eyes, nose, and respiratory tract

With tank-car quantities present, this makes no sense. A thorough risk assessment study must be done.

Air-Dispersion Modeling

Once the powers of logic to develop simple answers have been exhausted, it is necessary to do air-dispersion-model calculations to predict community concentrations of chemicals released to the air. These, in turn, may be used to evaluate health risks more quantitatively on both an acute and chronic basis. This sounds simple enough, but conclusions are strongly affected by the assumptions used in the calculations and the perceptions of the community about what (if any) concentration is safe to breathe.

In some locations, state, federal, and local governments may have established concentrations judged to be safe for community health. In the absence of any such standards, the plant may develop its own standards or draw upon those of other states or countries.

Many computer models are available with which to calculate the concentrations in a plume of gas released from a plant or evaporating from a volatile liquid spill and the resultant ground-level concentrations at any point downward. Some models deal only with gases that are released at densities lighter than that of ambient air. Others deal with gases that are heavier than air at release conditions. Obviously, the correct model for the case must be chosen. Some governments specify which models are acceptable to them for permit applications or for certain technical problems. In some cases, a company (or consultant) may need to use a more sophisticated model to maximize its own understanding of a problem, but it may also be required by a government to use another model to get official approval of the results of the study.

The major variables used in dispersion models cover a wide range of factors. In addition to the properties and physical conditions of the gas released, the height, direction, and velocity of the release are also important. The plume is markedly affected by the meteorological conditions such as wind direction and velocity and atmospheric stability. (There is generally much more vertical mixing of gases released on a sunny day than at night or during a temperature inversion in the atmosphere.) The geography around the point of release is also important. This includes both natural obstacles such as hills, trees, and rivers, and also man-made facilities such as buildings, towers, and the like. The various models deal with all of these variables in more or less elegant and sophisticated ways, making the choice of model the subject of some expertise rather than automatic.

Released gases form what is termed a "plume," which moves away from the release point at the behest of the wind and begins to disperse vertically and horizontally depending on a number of variables.

The plume starts to move in the direction of the prevailing wind *at the altitude of the release.* Wind directions are generally given as the direction *from which* the wind is blowing. Consequently, if the wind is (from the) south, the plume will start to move north.

The plume will rise or sink from the point of release, depending on its density, release direction and velocity, and the geographic obstacles it encounters. As wind directions may be different at different altitudes, the plume may change directions as it moves away from the release point. Figure 3.1 pictures such behavior schematically.

The plume will travel at the wind velocity, if constant, so that if the wind is ten miles per hour, the leading edge of the plume should be ten miles downwind in one hour.

The width and depth of the plume will evolve with distance from the source depending on a number of factors. Chief among these are wind speed, terrain, temperature, and specific gravity of the gas.

Generally, plumes are wider and deeper at lower wind speeds and narrower at higher speeds. At twenty miles per hour, the plume could be half as thick as at five miles per hour. Rough terrain—that is, hills, city buildings, and so on, will cause the plume to be larger and more quickly diluted. Flat terrain and large bodies of water permit it to become more narrow and symmetrical in shape. The sketches in Figure 3.1 summarize these general trends.

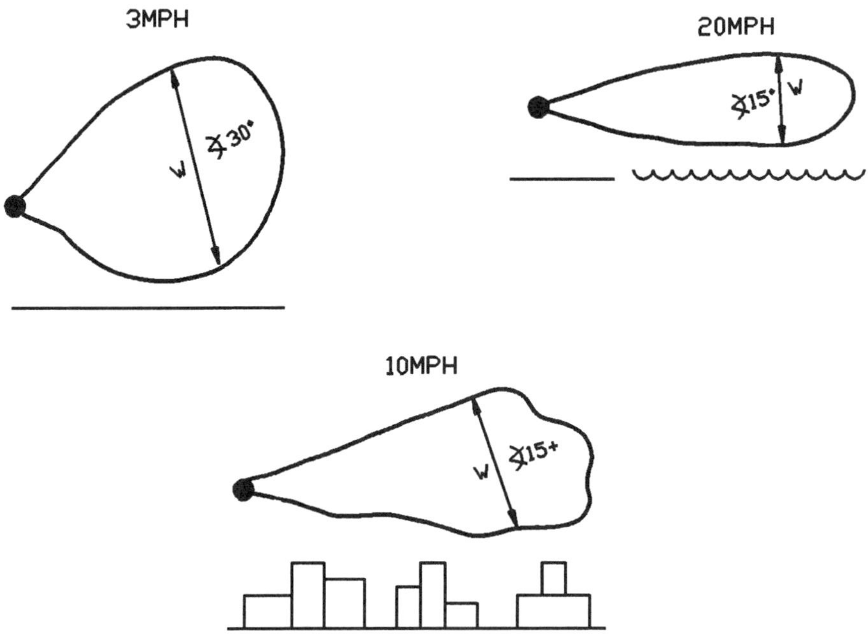

Figure 3.1. Impact of variables on plume shape.

The concentration of the toxic plume generally decreases as it extends outward from its source. The pattern of concentration is not always symmetrical or evenly distributed and decreasing. A concentration can be heavy close to the release point, drop off as the plume moves away, and then increase again.

The causes for the uneven distribution of plume concentration are dependent upon several factors. These include:

- Vapor density
- Temperature of the vessel, air, ground, and chemical released
- Height of release (ground, elevated stack)
- Areas of release and velocity vector
- Starting concentration of chemical in vessel or pipe
- Duration of release
- Wind speed
- Weather (rain, night, humid)
- Perturbations

Atmospheric stability (stability is a function of the rate of radiant energy input and the air-density distribution with altitude, among other things) is also a major factor in the rate at which the concentrations in plumes change. Various stability classes have been established to characterize the atmospheric conditions for dispersion models. Nighttime is generally more stable than daytime because of the mixing induced by the heating of the sun during the day. Table 3.5

Table 3.5. Stability Class

Stability Class	Surface Winds, mph	Cloud Condition	Time of Day
A	<5	Clear	Day
A/B	<5	Broken middle	Day
	5–7	Clear	Day
B	<5	Broken low	Day
	5–7	Broken middle	Day
	7–11	Clear	Day
B/C	7–11	Broken middle	Day
C	>11	Clear	Day
	5–11	Broken low	Day
C/D	11–13	Broken middle	Day
D	>13	Broken middle	Day
	>11	Broken low	Day
		Overcast	Day
		Overcast	Night
	>7	Broken	Night
	>11	Clear–scattered	Night

correlates these stability classes with common observation of atmospheric conditions. As shown at a given wind velocity, cloud cover can affect the stability class somewhat and nighttime generally produces the most stable conditions at common wind velocities.

Another factor that influences the calculated concentrations at any point in the plume is the averaging time used in the computer algorithm. For some models, this is a variable; for some, it is fixed by the intended purpose of the algorithm (long-term, steady state vs. short-term, acute). Generally, shorter averaging times will give higher concentrations in transient situations. Thus, in calculating the maximum ground-level concentration of a chemical in a short-term release, the averaging time may be important. For continuous, steady-state release, longer averaging times may give values that are more indicative of cumulative exposures.

Normally, plume concentrations are calculated at a number of combinations of meteorological variables for the site in question and for each of the specific accidents thought to be credible. The worst case for each accident is then plotted on a map to facilitate an assessment of the impact on the community. The maps should show:

- Streets, roads, and rivers as well as large bodies of water
- Critical facilities such as hospitals, schools, shopping centers, relocation facilities, and so on
- Industrial facilities with identification of their business and materials present (i.e., fuel storage, chemical processing, etc.)
- Major traffic areas such as airports, turnpikes, bridges, ports, and railroads
- Any sampling or other air-monitoring points already established

For a given set of atmospheric conditions and a particular release, the dispersion model will calculate the locus of points with the same concentration of the chemical released. These lines of constant concentrations are called isopleths and are generally overlaid on the maps mentioned here. In this way, the anticipated concentration at any location can be seen and its potential impact can be assessed.

An example of calculations made for steady-state emissions of NH_3 from multiple sources is shown in Figure 3.2. This is a somewhat artificial example

Figure 3.2. Annual average ammonia concentrations (micrograms/cubic meter)

because it shows annual averages, not a time-dependent incident plume. It is meant only to show what a group of isopleths looks like and the kind of information such a plot makes available. At any point on the map, the NH_3 concentration can be interpolated. In an incident scenario, the locations of the isopleths would be changing with time, as would the extent of the plume. We shall discuss some examples of that situation.

An example involving only flammability is instructive. In a plant turnaround situation, its 200-foot-high flare was to be taken out of service. Since some isobutylene would remain in storage on-site, an alternate relief point for the safety valve on the storage in the event of overpressure or fire was required. An existing vent point some seventy-five feet above grade and 300 horizontal feet from a major highway was available. A number of questions were raised about the safety of using this vent lest a flammable plume intersect the highway or the heat from an ignited vent cause damage to passing vehicles.

Thus, two problems had to be analyzed: the hazard from an unignited vent and the heat release from an ignited vent. The worst case for both—the highest flow rate and the worst atmospheric conditions—needed to be considered.

Figure 3.3 shows the calculated plume profile for the worst release. It shows that the plume will be well below the flammability limit before it reaches the road some 300 feet from the release point. In addition, its elevation will probably still be well above the road as the diluted gas cloud goes by. We can conclude that sufficient margins of safety exist—that is, the plume is at less than twenty-five percent of the lower flammability limit (LFL): LFL is the lowest concentration of a given hydrocarbon in air that will ignite)—such that it is reasonable to feel secure that some car or truck will not ignite a flammable mass on the roadway.

However, there is a major volume of the release that will remain in the flammable range near the vent point. If it ignites, what will be the impact of the heat released on the roadway and on plant workers? (Recognize that this represents a double contingency: There must be a release *and* something to ignite it.) Radiant-heat-release calculations can be made using standard procedures to define the heat fluxes at various points near the vent should it ignite. Figure 3.4 shows the results of these calculations. At the base of the stack, the heat flux is 3,000 BTU/hr ft^2. At the road, it will be 374 BTU/hr ft.

Those are the facts, but we cannot tell what they mean without some additional data. How much heat can a person stand and still be safe? Fortunately, these data are available. Solar radiation in a hot climate can be as high as 300 BTU/hr ft^2. The maximum heat flux allowed for continuous exposure of workers in an industrial setting is often set at 500 BTU/hr ft^2, but exposure to 3,000 BTU/hr ft^2 at the base of the stack is safe only for immediate escape (six seconds). Workers in the immediate area need to know escape routes and be trained to keep them clear. With these data, it seems reasonable to conclude that no hazard would exist to passing motorists if the vent ignited. Even motorists caught in a traffic jam (or stopping to look) would be safe, albeit uncomfortable, if the vent were to ignite. Equipment—for example, a

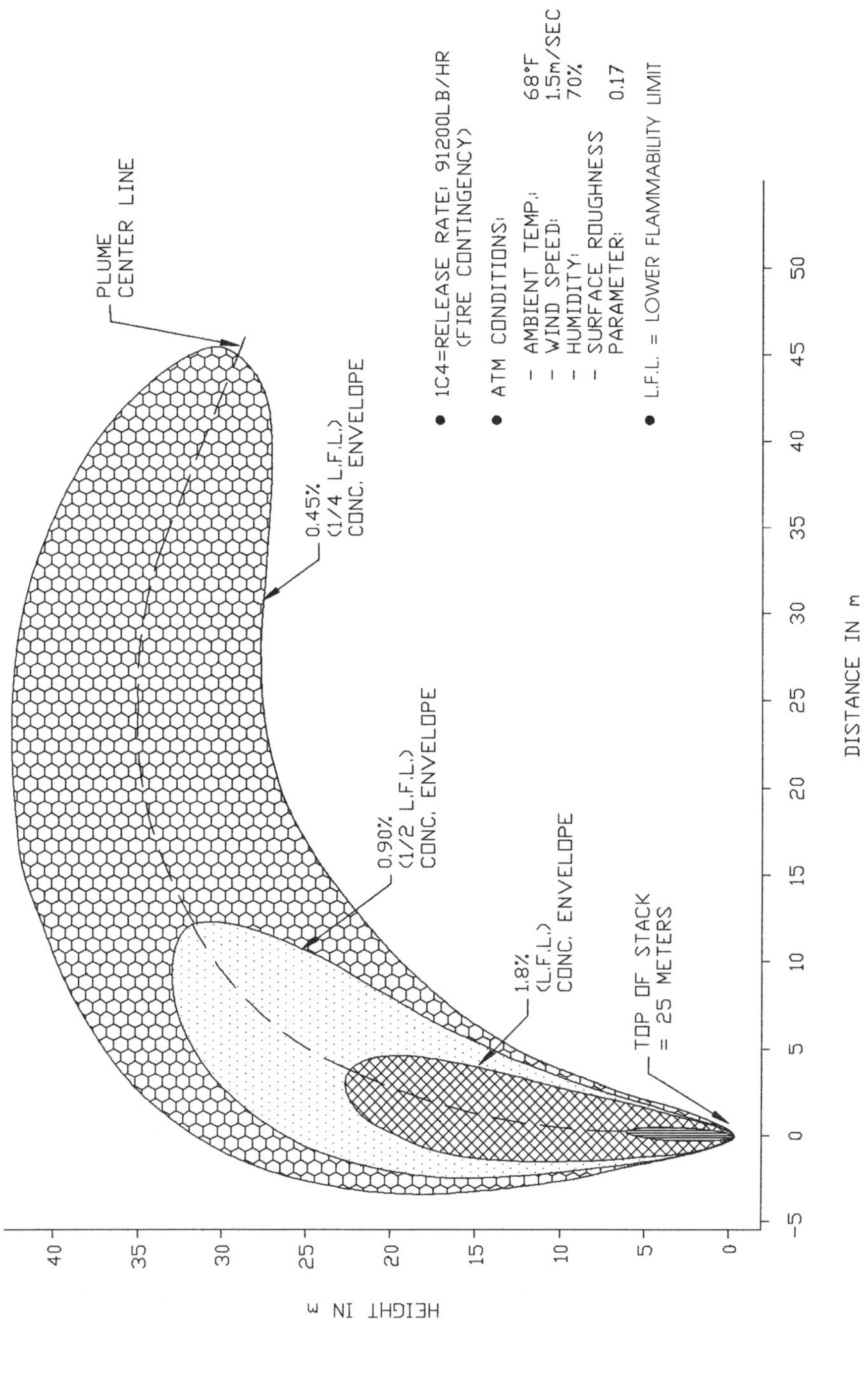

Figure 3.3. Isobutylene dispersion profile.

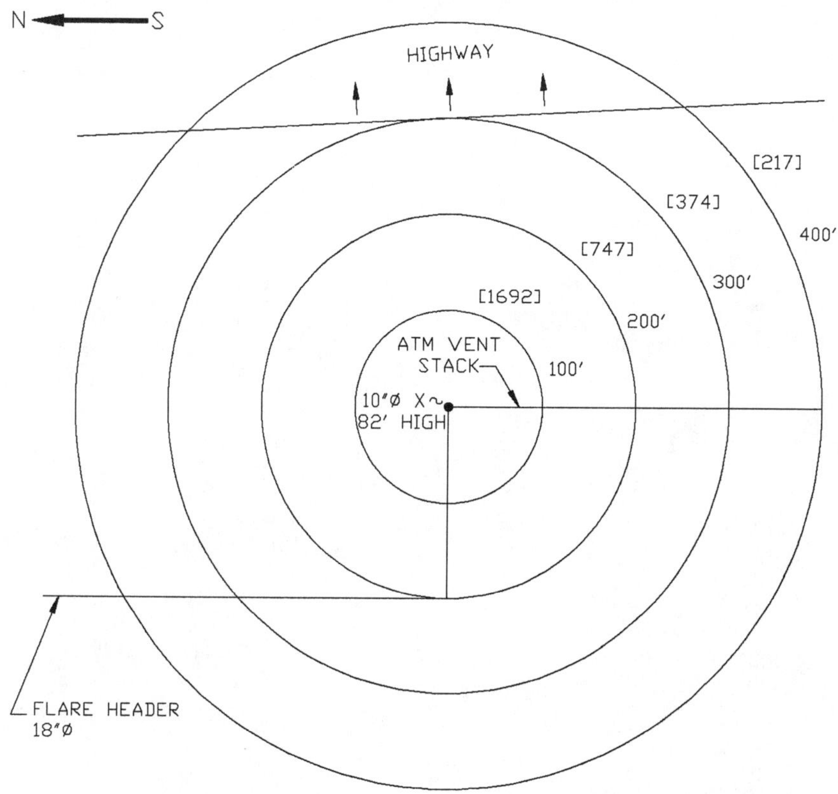

[X] ; X = RADIANT HEAT DENSITY, BTU/HR/FT^2, AT GRADE

LOCATION	DISTANCE FROM STACK,FT	HT. ABOVE GRADE,FT	RADIANT HT. DENS,BTU/HR/FT^2
HIGHWAY	300	25	392
WORK PLACE	250	6	518
VENT STACK	0	6	3044

Figure 3.4. Radiant heat density levels.

passing gasoline tanker—can usually tolerate higher heat fluxes than people (unless made of low-melting-point materials—(e.g., PVC), so people should be the controlling case.

Since the contingency that would have caused the release was itself remote (fire), it seems reasonable to conclude that using the atmospheric vent for the period of the maintenance outage would be a responsible thing to do.

Let me point out that this decision is not free of risk. It is an engineering decision based on engineering calculations. There are many in the public who distrust industry and its engineers to the point of automatically rejecting such a conclusion. More importantly, some driver may panic at the sight or sound of the fire and have an accident. Another may abandon his or her car in fear and run into the path of another vehicle. In short, the perceptions of the public and the irrationality of possible human response have not been dealt with.

The hazard is low, perhaps negligible, but a risk of adverse results still exists.

Assuming the release conditions are immutable, are there ways to reduce the risk? What extra mile might this plant go to prove its good citizenship?

Some technical steps may be possible. Every step to reduce the duration of any release should make sense. It might be possible to reduce the cross section of the vent to increase the upward velocity and reduce the heat flux on the road. It might be possible to build a wall or plant trees to shade the road, and so on. This kind of risk mitigation should be considered for any significant risk and will be discussed later.

Another approach might be to share the risk by having in-depth discussions with the fire department and the highway police to see if they feel the situation is reasonable. Perhaps in a few minutes they could be at the scene to provide traffic direction, cooling water sprays, and so on to avoid some of the scenarios described. Perhaps they would assess the risk as unacceptable and provide management with incentives to do better. Thus talking to the authorities in hopes of reducing "outrage" reactions is also a risk.

This demonstrates that even when the facts are available and accepted by the professionals in the area, some subjective aspects of the problem will remain in the health assessment area. When chemical (toxic) effects are considered, the data are less clear and the judgments more subjective.

Dispersion Models and Chemical Health Assessments

Consider a study made on the community impact of a possible safety valve release of ammonia vapor within a large petrochemical complex in New Jersey. Dispersion-model calculations showed that at the maximum ammonia release rate and the worst meteorological conditions, the maximum ground-level concentration occurred at a distance about 800 meters from the release point. This concentration would be about 100 volume parts per million, rising rapidly from zero at the release point, which was 137 feet above grade, and falling rapidly to a level of twenty parts per million at a distance of 3,000 meters from the release point. The plume itself was calculated to be about 200 meters wide and over 1,000 meters long before concentrations at ground level again approached zero. The details of the ground-level concentrations as a function of distance are shown in Figure 3.5.

The question is whether these concentrations would pose a threat to the community. Table 3.6 provides the chemical screening data sheet for the ammonia stored in the plant. Large quantities of material are present in both process equipment and a single storage drum. The OSHA workplace limits are set at fifty parts per million, while that of the ACGIH is one-half that, or

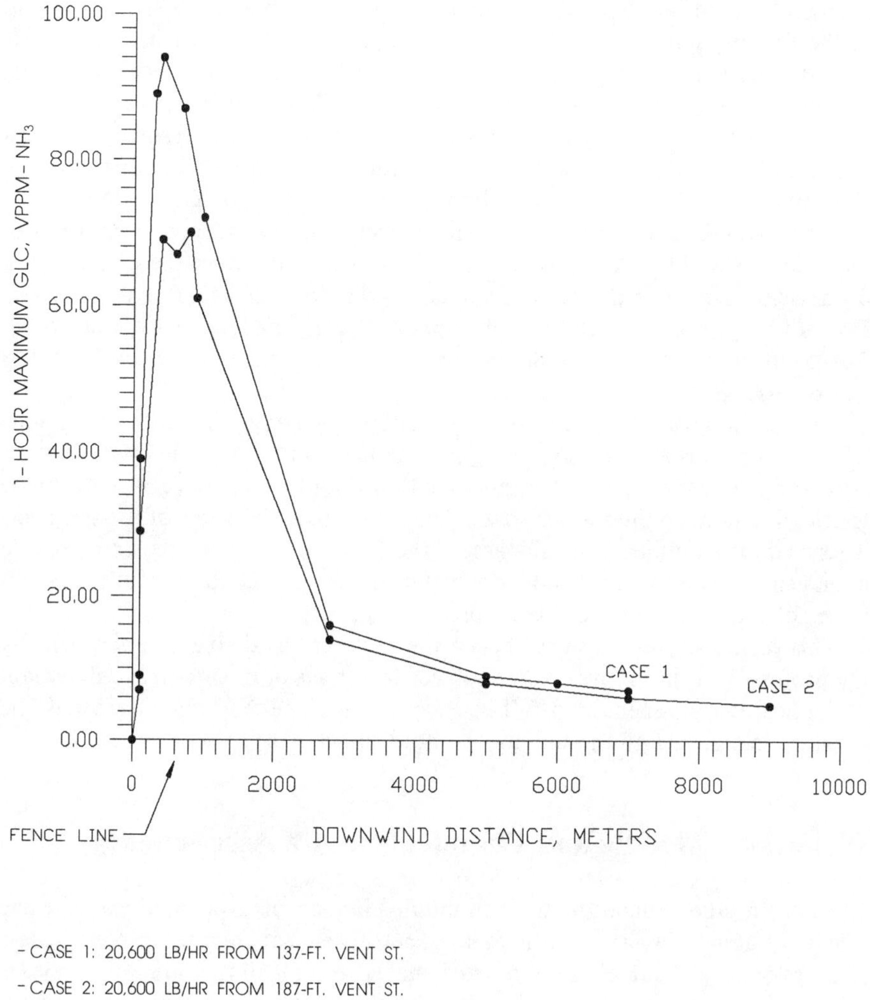

Figure 3.5. Ammonia release modeling 20,600-lb/hr releases from 137- and 187-foot stack.

twenty-five parts per million. Thus, we can conclude that our calculated maximum ground-level concentration outside the fence exceeds the safe work-place concentration, for a forty-hour work week, by a factor of from two to four. We are still left with the question, "What does that mean for a relatively short-term exposure to the general public?"

The toxicology of ammonia has been carefully described both in laboratory animals and in humans. Some industry associations may have published guidelines for safe concentrations in air. In addition, the New Jersey Department of Environmental Protection has set a community hazard standard of 423 ppm.

Table 3.6. Chemical Screening Data Sheet—Ammonia Anhydrous

Unit	*Product Type*	*Product Name*
VIS-J	Raw Material	Ammonia

Screening classification: G	
Amount used typically:	155.0 klb/yr
Maximum:	160.0 klb/yr
Inventory typical:	50.0 klb
Maximum:	80.0 klb
Location:	Block 2
Container:	Process vessel
	Process vessel
Physical state:	Liquid/gas
Vapor pressure:	2,507.501 mm Hg at 19° F
Flash point (°F):	
NFPA rating:	0
Storage/Usage Conditions	
Pressure (PSIA):	234.70–14.70
Temperature (°F):	110 to −30
Thermally stable:	Yes
Combustion products:	Oxides of N_2 (toxic)
Chemically stable:	Yes
OSHA toxicity (PEL):	50. ppm
OSHA ceiling:	NA
ACGIH toxicity (TWA):	25. ppm
ACGIH toxicity (STEL):	35. ppm
AIHA (WEEL):	NA
Comments:	Ammonia gas is very irritating to the eyes, nose, and respiratory tract

This number is for short-term exposure and is significantly higher than the forty-hour-per-work-week standard. In some studies for the Fertilizer Institute, researchers found that 100 ppm were tolerated by healthy acclimatized workers for periods of up to six hours per day without apparent effect. Concentrations of 400 ppm were tolerated by healthy acclimatized workers for no more than one hour per day with considerable discomfort. Other studies have found that some subjects exposed to 500 ppm for thirty minutes suffered severe upper respiratory irritation that lasted for up to twenty-four hours after exposure, but no long-lasting impairments were noted. Concentrations of 2,500 to 4,500 ppm were fatal to otherwise healthy adults within thirty minutes. On the other end of the spectrum, exposures to concentrations of ten parts per million or more tended to produce slight-to-moderate irritation to the eyes and and nose in a general population. Volunteers exposed to thirty and fifty parts per million of ammonia for ten minutes all suffered irritation but none described their reactions as painful.

Thus, human toxicity data would seem to indicate that exposures in the 400-ppm range would produce severe effects on healthy humans but might not necessarily leave any lasting impact. The NJDEP acute toxicity concentration seems to be supported by these findings. On the other hand, significant irritation had been noted at significantly lower concentrations, and we can obtain little guidance as to what constitutes a satisfactory, short-term community health limit from the data presented so far.

The Chemical Manufacturers Association (CMA) has prepared a manual that gives several ways to establish standards that have been found acceptable for chronic (i.e., long-term) exposures. In some places, the simplest and perhaps more widely used approach is to take the accepted workplace standard (fifty or twenty-five parts per million, depending on source) and divide by a safety factor. These safety factors can be constant or variable depending on the nature of the toxic effect and the sensitivity of the surrounding population. Fixed or constant factors generally have two components. The first is 4.2 to account for the forty-hour work week vs. the 168-hour community exposure week. This might be appropriate for low-level, continuous emissions but it comes into question for short-term exposure. The second factor is an outright safety factor to acknowledge the quality of the data used to develop the workplace limit and the severity and reversibility of the toxic effect anticipated. This factor ranges from ten to 100. Thus, the final long-term community exposure standard might range from the workplace standard divided by forty-two to the workplace standard divided by 420. For ammonia, this would leave us with a chronic exposure level of from one to 0.1 ppm. This is much lower than the thirty to fifty parts per million that volunteers described as "irritating" and does not address the question of acute exposures that is appropriate for a short-term release.

There are several undesirable characteristics of the safety-factor approach described above even for chronic effects. It is often criticized as not founded in scientific fact and subject to serious credibility deficits. The practice also often results in unnecessary and extremely expensive regulation and conservatism. The inflexibility and lack of thoughtful judgment that arbitrarily adjusted factors reflect ensure that some data will be discarded, disregarded, or disbelieved entirely. Lastly, the safety-factor approach removes the driving force to carry out state-of-the-art scientific analysis to characterize health impacts more precisely, because these more costly experiments are not justified if large safety factors are going to be arbitrarily applied. Thus, the approach undermines obtaining the data that are so urgently needed to form a rational basis for risk assessment.

Louis et al. (S. C. Lewis, J. R. Lunch, A. I. Nikiforov, "A New Approach to Deriving Community Exposure Guidelines From No Observed Adverse Effect Levels," *Regulatory Toxicology and Pharmocology 11,* 314–330, 1990) recently published a paper exhorting a more scientific approach to the setting of both acute and chronic community exposure guidelines. Their approach is based on the analysis of a great deal of data on how such guidelines are established and tries to address the known weaknesses in current practice. It attempts to take into account the following factors:

- Known differences between laboratory animals and humans and between experimental conditions and the real world
- The sensitivity of exposed human populations
- The strengths of the evidence that the chemical in question posseses a real hazard to human health
- The general quality of the experimental data base
- The uncertainty of extrapolating from laboratory animals to humans
- The potency of the toxic agent
- The type and severity of the adverse effect

They propose a formula that calculates the human "no adverse effect limit" (NAELHuman) from the "no observable adverse effect level" (NOAEL) detailed in animal studies and a series of corrections and safety factors developed by a consensus of scientists from the data and from knowledge of the exposed population.

The formula is as follows:

$$\text{NAEL}_{\text{human}} = \frac{NOAEL_{animal}}{(I)(R)(Q_1Q_2Q_3)(U)(C)}$$

Definitions of the correction values are as follows:

- *S* is a factor to account for the known quantitative differences between species. It may take virtually any positive value. The default value is 1, indicating that animals and humans are judged to be equivalent.
- *I* is an adjustment factor for greater susceptibility among members of the test animal population (intraspecies variability). It may take values from 1 to 10 with the most likely range being from 1 to 3 when the animal data are derived from a state-of-the-science study.
- *R* is an adjustment factor acknowledging anticipated differences in susceptibility between humans and laboratory animals. *R* can take virtually any value but a range from less than 1 to 10 is common. The default for this factor is 10, indicating a conclusion that humans are much more susceptible than animals.
- Q_1 reflects the data evaluators' certainty that the agent actually causes the specific effect in humans. It make take values from 0.1 to 1.
- Q_2 is employed when extrapolating data from subchronic studies to estimate risk from lifelong exposures. It may take values from 1 to 10, but will more commonly be in the range of from 1 to 3. The larger values indicate that the chemical tends to accumulate in the body or to cause repeated and cumulative injuries.
- Q_3 is employed when extrapolating LOAEL (low observed effects) to NOAEL. (no observed effects) From a state-of-the science study, a Q_3 value of 2 is usually adequate.
- *U* is used to acknowledge residual uncertainty in the estimates of *(S) (I)* and *(R)*. An example is data from one route of exposure used to estimate a

NOAEL by another route of exposure. *U* may take values from 1 to 10.
- *C* is a nonscientific judgmental safety factor that is a social or political value judgment. *C* may take values from 1 to 10 with typical values of 3 or less.

The net result of these ranges of possibilities results in a range of 250 to 100,000 in the denominator of the equation.

All of these factors would be derived by developing a consensus among a group of toxicologists and other health officials who could assess the susceptibility of a given population. Presumably, this complex and obviously costly approach would be appropriate for chronic exposures to slightly toxic chemicals, or those that might be suspected of being carcinogenic.

For a short-term exposure to chemicals for which simple irritation is the health-effective concern, the article goes on to suggest a short-term (thirty-minute) community exposure guideline equal to the highest level at which irritation does not occur during an eight-hour workday exposure (for ammonia twenty-five to fifty parts per million). The article goes on to point out, however, that "expert judgment will always be required to ensure that this simple approach does not permit a seriously hazardous substance to be undercontrolled." In short, it is very difficult and complex to develop community exposure guidelines either because there are insufficient human toxicity data or because the sensitivity of the population relative to that of the group that provided the human toxicity data is unknown.

To return to our ammonia example, if we make the assumption that plant workers are properly protected, is the plant likely to perpetrate an "atrocity" on its neighbors should its safety valve relieve under the conditions described in Figure 3.5? From the toxicity data, one might conclude that there is low probability of inflicting any serious damage on anyone in the community. Even if this is true, there may well be other factors to consider.

In a relatively small community in Texas, a safety valve on a ammonia refrigeration system relieved and created a plume that entered a nearby community. The plant in question had worked hard to establish communications with the community and responded immediately to calls indicating a severe odor in the neighborhood. Plant personnel toured the community assessing the strength of the odor and querying the neighbors about any adverse effects. For the most part, there were none, but one little boy felt nauseous. Plant people took him and his parents to the hospital, where he was examined and found perfectly healthy. However, upon return to his home, which was now free of any odor, the symptoms returned. Further medical examinations were carried out with similar results, but it took a day or so until the boy reported he felt well and for his mother's anxiety to be assuaged. Was the child cunningly getting endorsement for a day out of school, or was he genuinely affected? We shall never know.

For the most part, the efforts of plant personnel received a positive response from the neighbors. They were seen as caring and concerned and willing to deploy resources to address the adverse impact of the incident. I think the point is that the acceptable level of exposure in a community is likely to be much less than that which might be evaluated to cause an adverse health impact. If anyone were indeed to get sick as a result of a plant incident, extreme negative reactions in the community would be almost inevitable. The question for management must be, "Can we tolerate *any* health impact if we have to deal with the perception of the community that if noticeable impacts on community health are indeed observed, silent and mysterious impacts might also be taking place?"

A standard to be considered might be that in addition to imposing no known health impact on the community, there should be no *perceived effect* of any sort on the community. In some cases, odor might be a rational indication of what an acceptable community guideline might be. In the case of ammonia, the odor threshhold is believed to be about five to ten parts per million, or about one-fifth to one-tenth the workplace exposure limitation. For a chemical leak of hydrogen sulfide, which has a workplace concentration limit of ten parts per million, the odor threshhold is approximately one to five parts per billion, or the workplace concentration divided by several thousand. Again, a very wide range of potential impact is possible. The cost of reducing the concentration that results from an incident to below what might be perceived in the community needs to be balanced against the potential impact of community reaction on the total cost of plant operations. In short, each management must establish principles for the types and extent of community risk it can live with. The answer is trending toward zero, but once a limit is set, I would recommend explaining the remaining risk to community leaders in detail, and perhaps establishing a communication system to demonstrate the measurements which prove safe concentrations. Figure 3.5 also provides some data on a possible way to reduce community concentrations. For ammonia, the second case presented is for a fifty-foot-higher release point. The maximum concentration occurs at about the same location but is about seventy parts per million vs. the ninety-five parts per million of case 1. Thus, it is likely that an extension to the stack could lower the ground-level concentration experienced by the community to perhaps fifty parts per million, the OSHA workplace limit. It would have been prudent to have explored this and any other obvious mitigation techniques in developing the background information for any risk management decision. Remember that this decision will likely have to be justified to someone outside the company sometime. For the case being discussed, a risk mitigation program seems to be indicated even though state guidelines have been met.

Summary

The first step in answering the question, "What hazards exist in this plant and how bad are they?" involves identifying all of the chemicals in the plant and tabulating their properties. The nature, amounts, and location of these chemicals in the plant, combined with a list of possible scenarios that would cause exposure either to the workers or to the community, set the basis for determining the level of risk present. These need to be evaluated simply, if possible, and a conclusion needs to be reached about whether the risks are acceptable. Presumably, workers will be properly protected, and such protection may also reduce community exposure. In some cases, a more detailed evaluation involving air-dispersion modeling and health assessment may be required. In any event, judgment by responsible plant people about what is acceptable will undoubtedly be required. It will also be necessary to provide appropriate documentation on the basis for this judgment because it is likely to be necessary to defend it in a public forum, if an incident occurs.

Obviously, these judgments will be affected by the corporate culture and its tolerance for real or perceived risk in the community. It is necessary to determine what risks the community will tolerate. Assumptions made by plant managers are invariably flawed in this arena, and it is almost always necessary to establish communications with the most sensitive community groups to provide the necessary data for informed judgments on acceptable risks.

CHAPTER

4

Is This Hazard Necessary?

Introduction

What you don't have cannot leak or hurt you or your neighbor. Such is the philosophy of those who recognize that inherent safety is far better than the extrinsic, or added on, kind. The contrast to be drawn is between that which cannot cause an accident and that which is prevented from doing so by adding safeguards.

Thus, a plant that has no gasoline storage tank is truly safe from a gasoline explosion. One with a five-gallon tank is "inherently safer" than one with a ten-gallon tank regardless of the number of alarms and cutoffs on the tank, the extent of the fire protection system, and the level of training of the employees. All of the latter may be employed to reduce the risk of having the gasoline tank, but the only way to *eliminate the hazard* is to do away with the tank.

Trevor Kletz (See list at the end of this chapter), a retired ICI (Imperial Chemical Industries) employee now teaching at a British university, has been the most vocal exponent of inherently safe design, and he points out that the concept is much more broadly applicable. Application begins with the fundamental decisions of business managers, the selection of plant location, the establishment of plant design philosophy and mechanical design standards, and extends on through the distribution of the product, the education of customers in how to use it safely, and the safe disposal of any residues. Preferably, any product would be used in a way that did not produce any waste. In short, the concept applies to the whole of the enterprise.

Business Aspects

The fundamental business question is whether the profit potential for making or using a given chemical justifies the risk undertaken to make or use it. In most cases, this question is not addressed explicitly. There are many possible reasons for this aside from gross negligence on the part of venture managers.

The most important reason is that the data are simply not available to make any reasonable assessment. At the early stages of development, the technology to be used may not even be established. Choices may remain about whether to develop or buy the technology. Perhaps an acquisition is contemplated and must be kept shrouded in secrecy. Laboratory synthesis may be at a very preliminary stage with many alternates still to be considered. Yet someone has developed a "worst case" scenario to be able to understand how the material would be manufactured on a commercial scale. On this basis estimates have been developed for the cost to manufacture and for the venture economics.

This would be a fine time to stop and assess the likely risks associated with the venture. If top management asks the question, an answer will be generated in hopes of getting on with project development. This can be evaluated.

Another approach is to establish guidelines such as those set up in the "Seveso Directives" (rules adopted by European countries to help reduce community risks after an incident at Sevesso, Italy) that force more detailed consideration of process routes that involve more than specified amounts of certain types of substances. Thus, if a chemist knows he is likely to need a halogen for his reaction chemistry, he knows he must deal with a highly toxic, volatile liquid or gas. Perhaps he will need a light olefin. Is it likely that on-site inventory will exceed the threshold quantities set by governments to trigger special safety analysis and procedures? If so, the company can initiate those analyses at this very early stage to begin the process of understanding the risks it may be undertaking, even if the early basis is partly (or even largely) educated guesswork.

One list of threshold quantities is given in Table 4.1. For commercial operations, some of these quantities are very low and serve to emphasize the magnitude of the hazard these chemicals introduce. Many of these trigger quantities for specific chemicals are based on about one percent of the triggers used by the European Economic Community to require a "Safety Report" under the "Seveso Directives." The purpose of setting such a low percentage as a trigger is to make the risk issue visible very early in the development of a business venture. Clearly, other approaches to setting trigger quantities are

Table 4.1. Triggers for Early Management Risk Review

Substance	Quantity
Pressurized hydrocarbons	5 tons
Atmospheric storage <2 psig	500 tons
High flash storage	10,000 tons
Chlorine	250 kg
Ammonia	5 tons
Sulphur dioxide	2.5 tons
Sulphur trioxide	750 kg
H_2S	500 kg
Ethylene oxide	500 kg
Propylene oxide	500 kg
Acrolein	2 tons
Bromine	5 tons
Sulphur dichloride	10 kg
HCl	2.5 tons
Other material labeled "very toxic"	0.2 tons
Material that self-accelerates to decomposition	100 kg

equally valid. The key is to get early management recognition of risk factors. While qualitative, such insights may give pause to the headlong development of a marginal project. For truly attractive ventures, they should catalyze the analysis needed to make correct and timely risk management input to venture analysis.

Site Selection

Once the decision has been made to go forward with a project, a location must be chosen. Clearly, the risk imposed by a facility on its host community can be reduced by siting it in a location remote from people or property that might be affected by an incident. Just as clearly, this is also a business decision because many other costs are also affected by it. Spacing can be obtained by buying (or using) more land, an obvious economic impact. Locating a hazardous facility in a remote part of the country (or state or the world) also means it is farther from markets, raw material supplies, and skilled workers. All of the above and many other considerations impact the profit potential of a new enterprise.

The impact of an incident is also a serious business consideration. One needs only to look at the $2 billion spent by Exxon on the *Valdez* cleanup (before any legal penalties) to realize this. While there was little Exxon could do about where the oil was located, one wonders if there should not have been second thoughts many years before about where the port was located, or whether transferring the crude to a ship at all was a reasonable business decision. Would a pipeline across Canada have been more sensible given the wisdom of hindsight?

Unfortunately, the cost of credible incidents is rarely considered in siting decisions or venture economics. Mostly this is because the data are not available, but sometimes it is because no one cares to ask. There are many far-less-monstrous examples than Bhophal that should inspire business management to consider what might go wrong in a venture before becoming too far committed to change their minds. Indeed, some companies pledge in their corporate policies not to manufacture or use chemicals for which it is not possible to control the associated risks.

There are many computer programs now available which, when coupled with a preliminary idea of the technology, can provide a rough estimate of the possible impact of credible plant incidents on the surrounding area. Most of these will provide simple models of the impact of a release, explosion, or other accident on the adjacent community. This can provide considerable insight about the risks associated with any plant location and about the incentives for various technology choices as project development continues.

Consider the case of the monoethylamine plant study presented by D. C. Hendershot at the August 20, 1990, meeting of the American Institute of Chemical Engineers. Preliminary community impact areas were estimated for two technology alternatives at one possible plant site. These impact areas, which were calculated from an air-dispersion model, are shown in Figure 4.1 as circles delimiting a critical atmospheric concentration of monoethylamine.

The direction of the plume will, of course, depend on the direction of the wind, as discussed earlier, so the generalized impact area is shown as a circle.

The circle labeled “A” represents the distance from the release point (the failure of a one-inch pipe) at which the critical concentration will be measured during summer maximum ambient conditions. Inside the circle concentrations will be higher. Clearly a large portion of an adjacent residential area would be affected by the release from the failure of even this small-size pipe. Also, the location of the source within the site is of no importance because the diameter of the circle is so large. Unless the wind is *always* from the east, locating the monoethyamine plant as shown involves a risk to the community.

The circle labeled “B” in Figure 4.1 represents the impact of a different technology. In this case, the amine is stored refrigerated at its atmospheric boiling point. When a comparable failure occurs, much less vaporization takes place and the circle encompassing areas at concentration higher than the critical is much smaller. Since these calculations would be carried out for the worst meteorological conditions, if it were not for the highway, one might conclude the risk was tolerable. Also, it might be possible to move the exact location of the amine to the west of the site, avoiding an impact on populated areas altogether.

Please note that the inventory of amine has not been a consideration so far. The inventory in a one-inch pipe is small. The assumption has been that only a small amount is needed to create the plume. Under these circumstances, the mere presence of the hazard on the site leads to the risk because of the proximity of the site to the community. Extrinsic safety devices may be added to reduce the risk, but inherently it cannot be eliminated.

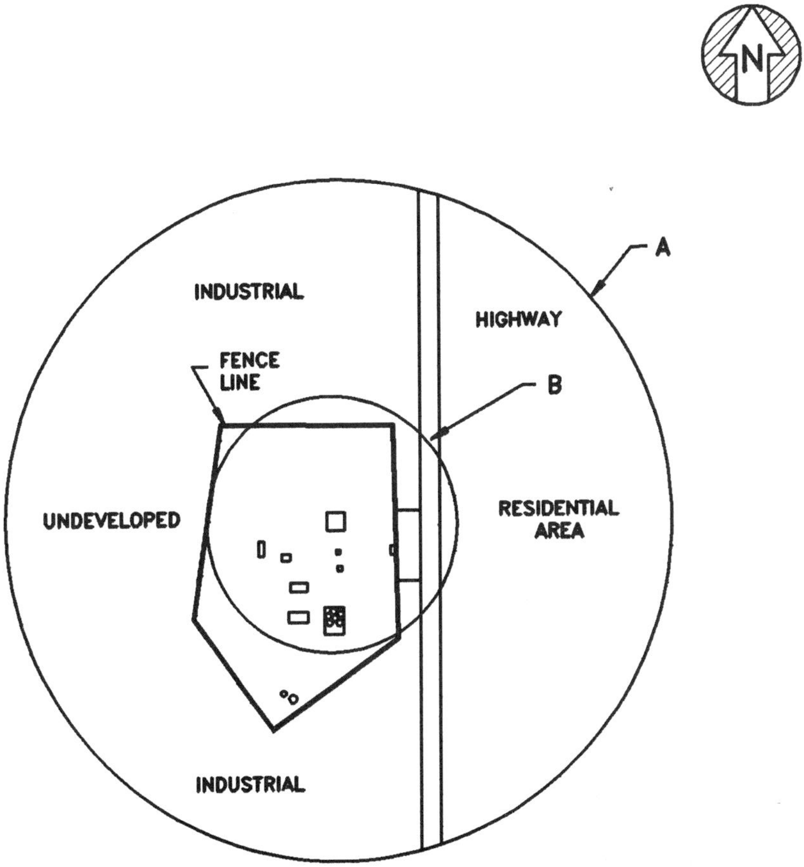

Figure 4.1. Relative hazard zones for monomethylamine releases—areas exceeding specified atmospheric concentration, one-inch liquid pipe rupture.
A, At summer ambient temperature.
B, Refrigerated to its atmospheric-pressure boiling point.

The impact of a different technology approach is shown in Figure 4.2. The circles depict the same critical concentration level as in Figure 4.1, and the circle marked "A" is exactly the same. It represents the above-critical concentration area for a leak of anhydrous monoethylamine from a broken one-inch pipe. The circle marked "C" shows the impact area for a similar-sized leak of an aqueous solution of amine. The diluent reduces the vaporization of amine to a greater extent than refrigeration did in the previous figure. The critical concentration is essentially limited to the plant boundaries and provides a degree of inherent safety not provided by any anhydrous case.

These studies can be used to impose restrictions on the technology chosen for the project at this location as well as to suggest possibilities for relocating

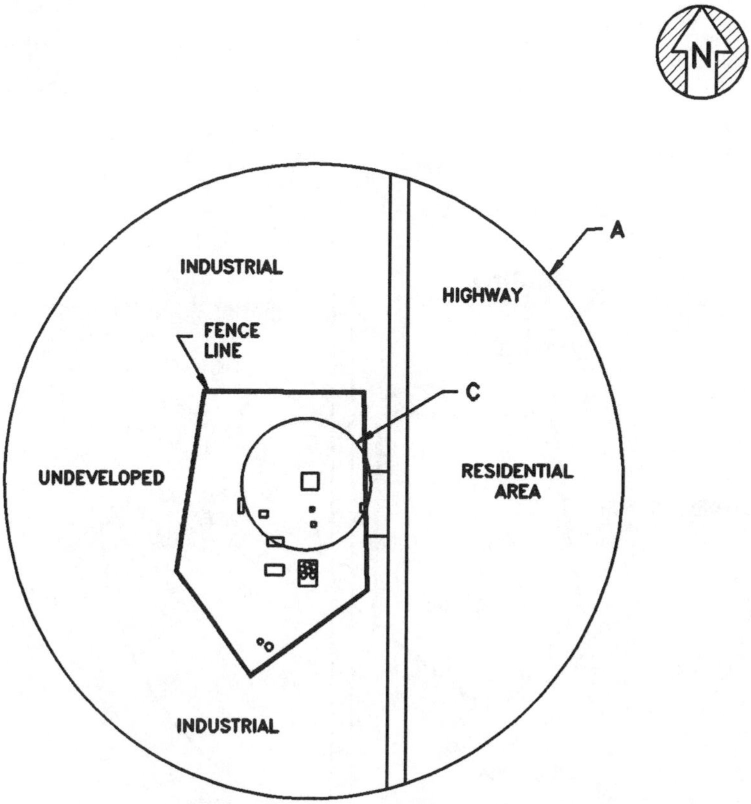

Figure 4.2. Relative hazard zones for monomethylamine releases—areas exceeding critical concentration, one-inch liquid pipe rupture at summer ambient temperature.
A, Anhydrous monomethylamine.
C, Aqueous monomethylamine.

the hazard on the site. The interaction between the site and the technology has been considered at an early stage of project development. The conclusions reached are twofold:

- With certain technology, the site appears inherently acceptable from a community risk standpoint.
- Certain technically feasible technology choices have been shown to be less desirable.

This is not to say that these technology choices are unacceptable. Using aqueous amine solutions will likely require larger storage tanks and greater transportation costs (if amines are not produced on the site) and perhaps greater processing costs to reach the desired product purity. On the other hand, the cost of preventing or repairing potential community impact can also be estimated and included in venture economics for any technology.

The American Institute of Chemical Engineers, in its handbook, *Guidelines for Safe Storage and Handling of High Toxic Hazard Materials* (1988), discusses a number of other considerations for siting plants that will use or store highly hazardous materials. While these guidelines are primarily aimed at toxic materials, they provide prudent guidance for the siting of all facilities that will employ hazardous materials of any kind. These are largely based on common sense, but the breadth of their purview is helpful in bringing some not-so-obvious potential hazards under scrutiny.

A brief compilation of items pertinent to site selection follows:

- Access roads to the site should provide adequate capacity for emergency service vehicles and provide no unusual risks for truck transporation of hazardous materials in or out of the plant.
- Soil loading capability should be adequate for major vessels and equipment and not in active earthquake fault zones.
- There should be adequate protection from vehicular impact and spacing from aircraft takeoff-and-landing patterns.
- Major leaks of hazardous materials should not flow toward populated areas unless adequate potential for diversion can be provided in the design.
- Similarly, any such leaks should not contaminate local waters, particularly drinking water supplies.
- The site should have access to reliable utility supplies (power and water) even in major emergencies to allow adequate response.
- Emergency response resources needed from the community should be available or be able to be developed and trained in the time needed.
- The various local regulations and environmental permit requirements should be manageable and future trends compatible with the business to be established.
- There should be potential to provide adequate security against vandals, sabateurs, and so on.

Other considerations will be pertinent at some sites and some of the above will not apply at all sites. There is no substitute for a specific risk analysis for each potential site. Clearly, the scope of the economic choice can be faced early in a project. This avoids ignoring the risk-induced costs, making the wrong choice of either location or technology and then paying for numerous extrinsic safety devices later, in an attempt to correct the risk situation after it is too late to deal with it more fundamentally.

Isolation Zones

A further consideration is the size of the land parcel purchased or allocated for the project. In our example, the possibility that the undeveloped land and/or some of the industrial land could be acquired to provide both more flexibility in locating the hazard and a bigger buffer zone around it can be explored. This, too, can be fit into the economic equation.

There is a caveat that goes with buffer zones. Unless controlled carefully, they tend to fill up with people. We are all aware of the tendency of airport workers to build their homes at the end of the runway and then complain about the noise. At Pemex in Mexico, a very large liquified petroleum gas (LPG) facility was built well away from any community. As time went on, more and more squatters set up "temporary" housekeeping in the space around the plant, many of them on slopes looking down on it. On the day that Pemex blew up, hundreds of these folks were literally broiled alive by the heat from the fireballs. Others were killed by flying metal from the explosions. The erosion of the buffer zone resulted in truly tragic consequences in this case.

In recognition of this, some process companies are trying to reestablish safety zones around their existing plants. A *New York Times* article dated November 21, 1990, reported on a number of attempts by major chemical companies in Louisiana, Oklahoma, and Texas to buy out neighboring homes and relocate whole neighborhoods to safer distances. They are finding that simply paying good prices for the land is not enough. There is great difficulty in providing satisfaction to those relocated because of the "emotional attachment" to the land, their homes, and the existing social structure in the community. Even when a location is provided for the entire community to move to, many do not go there, even if they do sell their homes to the company.

A spokesperson for a New York City relocation company was quoted as saying that a growing number of companies were considering this route to establishing a safety zone. "From a business sense, it is economically sound. . . . It makes sense in putting a program together instead of waiting for an accident and lives to be lost." Plants located near sensitive areas have an even greater interest in having an appropriate safety zone. Locating a plant that uses large quantities of a respiratory irritant next to a nursing home that specializes in dealing with emphysema patients (or vice versa) makes no sense at all. Neither does locating large quantities of LPG on a site prone to earthquakes, nor placing tall towers near the glide path to an airport or in an area subject to frequent hurricanes.

These days, terrorism must also be considered. The recent war in the Persian Gulf made many companies reevaluate their security procedures and strengthen perimeter patrols even in relatively well-established and secure areas. Even so, some people managed to plant bombs on some chemical tanks near Norfolk, Virginia, albeit for purely monetary reasons. A year or so earlier, Greenpeace breached a DuPont plant in southern New Jersey and carried out a demonstration for a couple of days. If this kind of incident can occur in major industrialized areas and at well-organized sites, it implies that security needs significant consideration at new sites and "greener" areas.

Much-less-dramatic factors also deserve consideration in site selection. Normal weather can create hazards as well. Northern plants must deal with freezing weather and icy conditions. Southern plants must deal with the impact of higher temperatures on process and storage containers. If a material begins to decompose at 150°F, I would much rather store it in New Jersey than in Texas. The opposite is true for materials that freeze at prevailing ambient temperatures. Both of these issues can be dealt with extrinsically via equipment design and instrumentation—at a cost in money and in peace of mind.

This author believes that accidents are *not inevitable.* However, it does make sense to provide as much inherent safety as possible. If economics persuade management that dealing with a particular hazard has high incentive *and* that the risks of dealing with it are manageable, then it makes sense to begin the risk management at an early stage. Early recognition of hazards can make any project *inherently safer,* if not *inherently safe,* and do so at lower cost than adding on extrinsic safety devices.

Process Research Phase

It is in the process research phase that chemists select the sequence of chemical reactions that will result in the desired product. With some input from process engineers at an early stage of process research, it is possible to guide the selection of process chemistry to less hazardous paths. This amounts to moving the concepts of inherent safety forward from the process design/development phase into the process research phase. Or, if you will, telescoping these related phases together to achieve the minimum hazard process route. The key element in this effort is to practice continuously the steps of hazard identification and elimination. It is not harmless to let a chemist develop a process using chlorine (or other toxic chemical) because it is convenient, if alternate chemicals or processes are available. The use of the toxic route will soon be "essential" for the new process because of the pressures of time, or uncertainties about product quality, or some other "overriding" reason.

Substitution

Process research is the stage of development where there is time to practice the principle of substitution. This means using something nontoxic in place of something toxic (e.g., sodium hypochlorite in place of chlorine). Other steps involve the use of a catalyst in place of large volumes at high temperatures and pressures, using purer raw materials to avoid making a hazardous by-product, using water as a solvent in place of hydrocarbon or an inert gas in place of a flammable one, or using a diluted hazardous material in place of a pure one. For example, many polymerization catalysts are pyrophoric in pure state. Some polymerization units form the catalysts in situ while many others buy theirs diluted (in a compatible solvent) to a point at which the mixture is no longer pyrophoric when exposed to air.

There are, of course, many other ways to reduce process hazards by substitution. These can be identified by having an early hazard review as the chemistry for the process is developed and by repeatedly asking the question, "Is the hazard necessary?" Once the process has been purged of all but the minimum number of hazards, process development and design can proceed on the basis of minimizing the extent of even these hazards by applying other techniques (intensification, simplification, and mitigation) to be discussed later, and additional reviews.

It is important to note that the process research phase is often the only possible phase in which substitution can be applied in a practical way. The progress of any venture or project tends to accelerate markedly once it enters process design, where equipment is sized and laid out, control schemes are developed, and specifications are written. Time pressures, market demands, plant tie-in and environmental permit requirements, and so on force the fixing of many aspects of a design very early in project development. Often there are good economic reasons that prevent going back to change a fundamental chemistry route once the process is off the lab bench.

Perhaps the airship *Hindenburg* is a good example of the point. It was designed for and ran on hydrogen, the lightest element and a very flammable gas. Since the explosion which destroyed the *Hindenburg,* lighter-than-air ships have been designed to use helium, twice as heavy, but inert. Could helium have been substituted for hydrogen in the *Hindenburg* once the airframe had been designed?

Perhaps, but the net lift from the heavier gas would have been less and the performance of the airship would have surely been affected. In commercial terms, its capacity would have been less and, therefore, its profitability. Also, helium would have cost more than hydrogen, and may not have been as readily available, causing an additional debit to the ship's economics.

Somewhere the original design would likely have to have been reworked, causing a delay in the building of the ship and increasing its cost. These factors are relatively generic when retrofits or redesigns are considered in any commercial project. We are repeatedly left to live with original decisions, adding this safety device or that to compensate for an inherent weakness in the system. The intelligent alternative is to consider the safety of any process early in the lab phase to eliminate as many hazards as possible.

Table 4.2 gives a general indication of the cost of change at various stages of the development of a manufacturing process. The cost of change increases geometrically as the project progresses.

Often the identification of inherently safer alternatives is not the easiest thing to do even when the practice of hazard reviews is accepted in process research. Foresight and experience are required, as is a great deal of chemistry expertise

Table 4.2. Cost of Change

	Cost, K	Engineering Time
At process basis stage	1	10 manhours
At design specification stage	10	100 manhours
Late in mechanical design stage	100	1000 manhours
Well into construction stage	1,000	? manhours

and the patience to consider alternative approaches that may not be well established. However, the incentives for a better process are often economics as well as risk reduction.

Support Innovation

In addition, support for innovators must be provided. Much more than lip service is required on this point. Consider the following different ideas aimed at dealing with various chemical hazards. What would your reaction be?

1. Use ultraviolent light as a catalyst in place of a hazardous chemical.
2. Replace combustion for hazardous waste destruction with a low-temperature electrolytic process to convert waste to carbon dioxide and water, producing an off gas that contains no impurities.
3. Use coal for energy without producing CO_2 to avoid the "greenhouse" effect. This would involve hydrogenating the coal to produce methane, followed by decomposition of the methane to carbon and hydrogen. Some hydrogen is recycled to the first step of the process; the excess is burned for fuel. The carbon is buried against the day the "greenhouse" effect turns out to be a myth or other technology for using CO_2 productively is developed.

Each of these ideas represents another step on the support-for-innovation scale. Not many would go as far as step 3, and few would tolerate step 2. To what extent does your company have the resources and patience to pursue innovative approaches?

Data For Development and Design

A process development phase usually forms a bridge between research and design. A key element in process research is to get the necessary data to allow downstream process developers and designers to exercise their creative risk-reducing capabilities. In effect, this means, "Do it right the first time." Those readers familiar with *quality* principles will recognize this as the first rule espoused by Deming, Crosby, Conway, Juran, et al. If we are to prevent waste, recycle, and eliminate other activities that add no value and no profit to our enterprise, process research must be well planned and executed effectively.

Understanding the process kinetics is often a critical step in many organic chemical reactions. It is not unusual for bench chemists to have limited knowledge of commercial-reactor design options. In a laboratory environment one can often use the nearest beaker and stirrer to carry out reactions. Having successfully produced a desirable product using pure laboratory reagents, the chemist sometimes is not given the time to explore the nuances of the chemistry he has employed. He turns his results over to an engineer to run in a pilot plant of one type or another. The chemist's reaction rates and conversions may not have been analyzed to identify the kinetic parameters for the desirable and competing reactions. In addition, in the absence of any reagent impurities, there may be no data at all on the production of unwanted and/or hazardous by-products.

Even though it may still be early in the formal process development stage, the pilot-plant engineer may have little choice but to scale up the chemist's beaker. Indeed, available pilot-plant equipment may also force him in that direction. To make an attempt to test another type of reactor vessel, the engineer must have some basis. To campaign for a high-intensity mixer, he must have some data to show that mixing is a limitation in laboratory kinetics. To demand time and money for a vapor-phase reactor study simply because it is inherently safer may indicate a career death wish. Without some kinetic data on the main and competing reactions, the pilot-plant engineer's hands are often tied, and again, improvement is defeated. There is no substitute for a well-planned and evaluated process research phase to avoid such limitations on process development.

In the area of by-products resulting from impurities in raw materials, the pilot-plant engineer can bring more to bear. He can find out what impurities are typically present in commercially available feedstocks. At the least, he can synthesize such a feed for his pilot unit and measure impurities in the product. All this assumes, however, that the analytical methods have been developed to find the impurities, and further, to evaluate their impact on product performance.

At this point, product purification alternatives can be hypothesized and another safety review can be carried out. This marks the beginning of more serious process development activity and perhaps the last chance to go back to the laboratory bench for further data on reaction kinetics.

It probably also represents the first opportunity to begin the accumulation of data on the hazards associated with intermediate and waste products and on the stability limitations of streams that are likely to be present at some point in the commercial plant. This information is essential for the process design phase in many areas. A few of the possible questions that should be answered are listed here as examples, but many other questions can also come up:

- Temperature limitations for safe operation
- Possible decomposition reactions and rates
- Basis for sizing safety relief devices
- Inert blanketing requirements
- Corrosion problems
- Possible odor problems
- Discharge points/systems for vents and safety relief devices
- Waste disposal problems
- Possible pyrophoricity or other undesirable reactions of leaks/spills
- Design pressures and temperatures for equipment.
- Purge requirements
- Limitations on utility characteristics
- Fouling tendencies

Thus, the process research phase must overlap the more traditional process development phase to provide information for dealing with those process hazards remaining as the commercialization process continues.

Reactive Chemicals Screening

Special expertise may be required to screen for chemical reactivity hazards and to define their extent. Various calorimetry techniques have been developed for these purposes. These are quite effective but require some appreciable expertise to operate effectively and interpret intelligently. Some companies (Dow, Merck, et al.) have established a separate process safety laboratory in which reactivity/stability tests are run by a small group of experts using sophisticated equipment. Most companies cannot justify the investment in equipment and highly trained personnel and use outside laboratories to provide these services.

A variety of calorimeters and tests are available. Although modifications and improvements have been made to some of the apparatus, three fundamentally useful types are:

- Differential Scanning Calorimetry (DSC)
- Accelerating Rate Calorimetry (ARC)
- Vent Sizing Package (VSP)

In many cases, a combination of techniques may be required to characterize any hazards completely.

The DSC approach is useful for screening the stability limits of a chemical or mixture to determine if the potential for energetic decomposition or reaction exists. DSC is relatively inexpensive and can detect both endothermic and exothermic reactions. The calorimeter measures the difference in heat input required to maintain a sample and a reference substance at the same temperature during a programmed heat-up cycle. As temperatures increase, any differences in heat input between the two samples can be interpreted to show the type of reactions taking place in the substance being studied. Information on the type and heat of reaction can be obtained directly but the temperature at which a reaction started must be determined by graphical extrapolation.

The ARC is an adiabatic calorimeter and can detect only exothermic effects. It has the advantage of measuring the initial reaction temperature directly, but is more expensive to use. Often the best heat-up program for running the ARC needs to be determined by using the DSC data. The ARC can also give data on the maximum pressure reached during any reaction. ARC measurements for the temperature at the onset of any exothermic reaction are often lower than those extrapolated from DSC data.

An example of data from an ARC run (Fig. 4.3), taken from S. M. England, "The Design and Operation of Inherently Safer Chemical Plants" (AIChE meeting presentation 8/20/90 San Diego, California), shows the type of information that can be derived. The technique is to heat the sample externally by an increment and wait to see if an exothermic reaction results. Generally, much

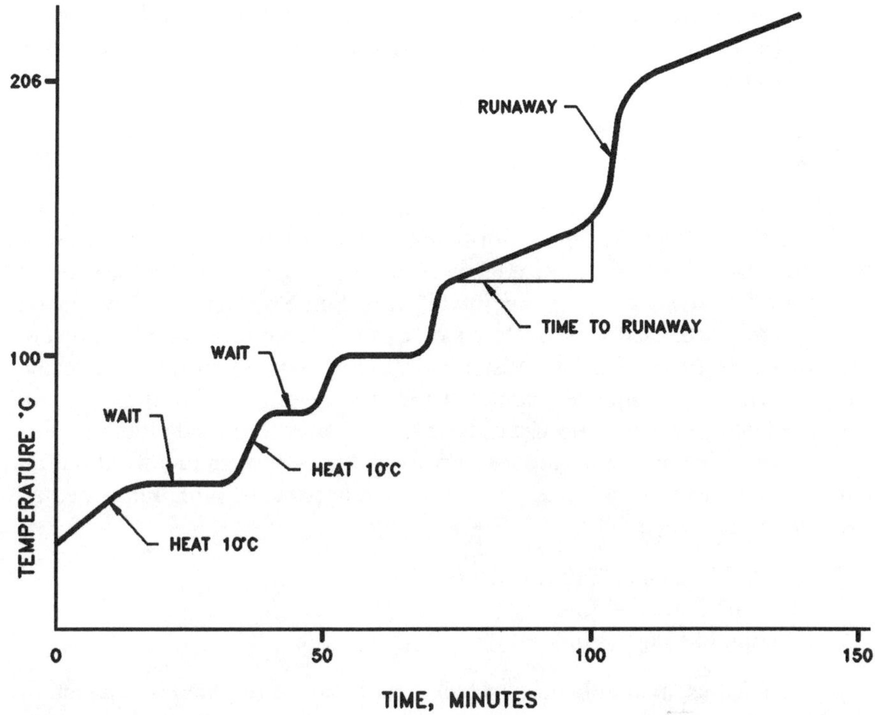

Figure 4.3. Example of ARC data.

smaller increments than those shown are used to pinpoint the initiation temperature of the exothermic reaction. The time taken by the sample to reach uncontrolled runaway conditions can also be measured directly. From the amount and heat capacity of the sample, the heat of reaction can be calculated.

Obviously, the impact of different impurities and catalysts can be evaluated in multiple runs. All of the data are extremely useful in determining the safe envelope for operations needed by designers and plant operators. The data can also be used to provide the basis for relief vent sizing.

The VSP is a specialized calorimeter that provides data needed for the calculation of how big a relief device is required to prevent overpressure of a process vessel. It requires less engineering expertise to do the actual vent sizing than does the ARC. It is very sophisticated, complicated, and expensive to run, but is the best system for obtaining such data in reactive and multiphase systems. The VSP method allows direct and safe extrapolations to full-size process vessels once they are designed. If the simpler calorimeters indicate reactivity problems, then process design can best be completed using VSP data. Data from smaller calorimeters are often necessary to show that the relatively large samples (100 ml) used in the VSP are safe to heat. There are presently few laboratories and companies around the world able to make VSP runs. Figure 4.4 shows a typical

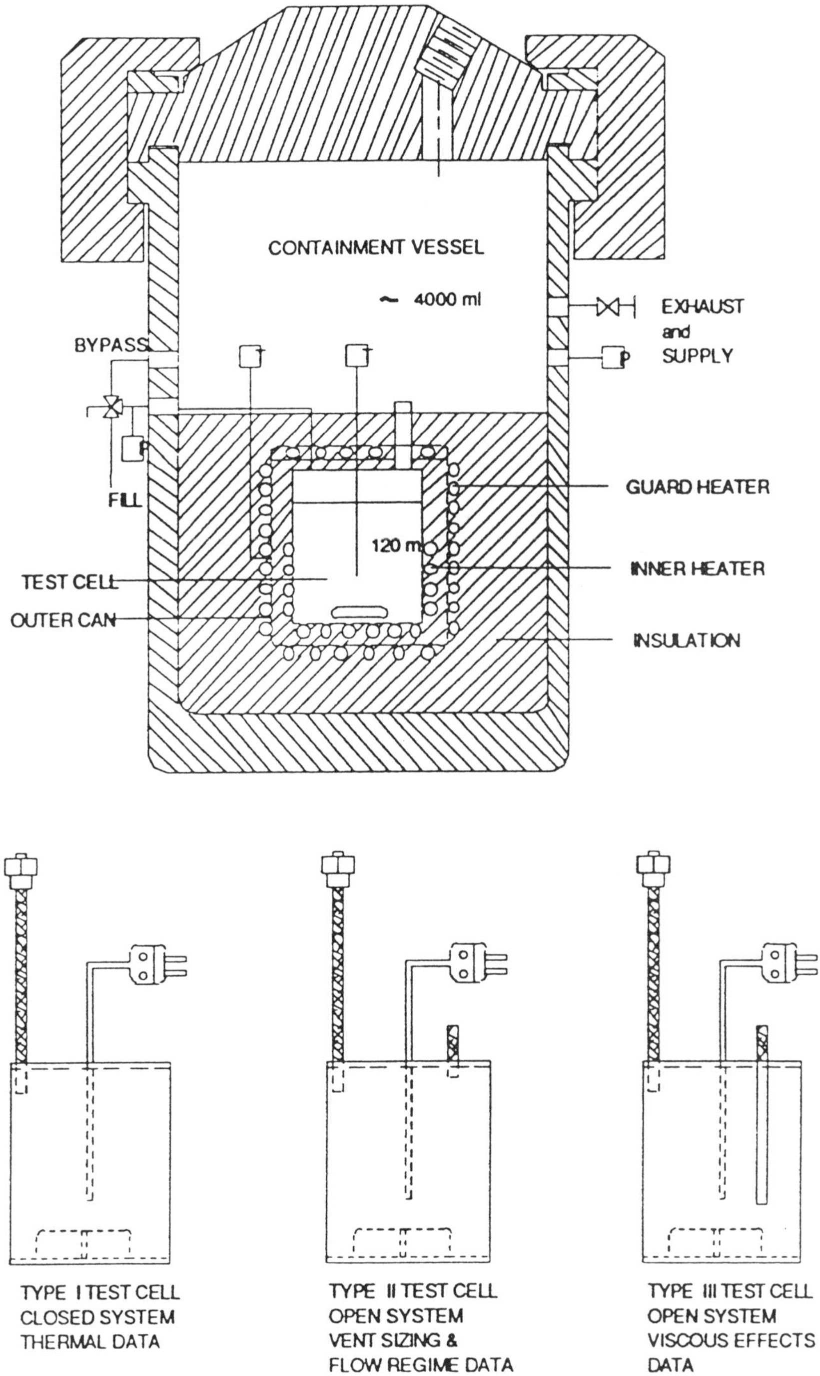

Figure 4.4. The VSP and configurations.

VSP and the configurations possible in order to obtain various types of data.

A simpler modification of VSP has emerged that is cheaper to purchase and use, but still requires professional expertise. It is called the Reactive System Screening Tool (RSST). In recent years, many more of these have been purchased than VSP units, but the number is still not large.

This area of technology is rapidly developing. The relief vent sizing calculations that had previously been endorsed by API and major technical organizations have only recently been found to give inadequate results for certain multiphase reactive systems.

Further developments can be anticipated as more expertise in this area becomes available.

Even more sophisticated general-purpose reaction calorimetry systems are in use in some research departments. Their use can result in a much more complete understanding of all of the reactions that may be going on in a particular system. From the *quality* standpoint, investing an appropriate amount of time and money in exploring the boundaries of the anticipated operating envelope during the process research phase may pay off handsomely, not only in safety, but also in optimum reactor design and in future process improvements.

While the above discussion has been framed largely in terms of a new plant, the same needs exist for a new or modified product being proposed for existing plant facilities. The time frames and equipment limitations in this case are even more stringent. Time pressures are often such that fundamental equipment changes are precluded, and sometimes interim facilities are used up to bridge the gap between the current and the ultimate facilities for the new product or process.

There is tremendous risk in bypassing the fundamental data needs for a safe process design that must also be constrained by what equipment exists or that which can be quickly modified. Many plants have found themselves with inadequate safety facilities, exorbitant waste-disposal costs, and labor/industrial hygiene problems as a result. The inevitable reliance on high-cost, extrinsic "Band-Aids" can undermine the long-term viability of the enterprise.

A last thought: Early safety reviews are often met with the argument that not enough data are available to do a meaningful analysis.

I believe responsible management cannot accept this answer. They should instead require that a chemist wishing to use any quantity of a hazardous material prove to management that using or making it is essential (not convenient, not merely economic) to the success of the chemistry. This done, we have done our best to identify the potential problem.

Process Development and Design Phase

As indicated earlier, the cost to change a process and/or equipment design escalates rapidly as project development advances. Once process development work is done, and this usually overlaps appreciably with the process design, there is often little opportunity for significant change for any reason.

In the early stages of the process development phase, there are many opportunities to have a major impact on the inherent safety of a process plant. The objective of this phase is to develop a process flow plan with the related flow rates, pressures, temperatures, and equipment configurations. During this phase, many decisions are made that restrict the flexibility for downstream changes, and these should be made while recognizing the principles of an inherently safer process.

Inherent Safety Principles

As previously mentioned, these principles have been advanced by Kletz and expanded upon by others. Those that apply particularly to process development (just as substitution applied particularly to process research) include:

- *Intensification*—that is, do the job with the smallest equipment, applying whatever external input required to accomplish this end—for example, maximum reactor conversion, mixing, catalysts, heat exchange, minimum inventory, and so on.
- *Simplify*—that is, *Simple* configurations: Is all that flexibility and/or spare equipment necessary?
- *Mitigation*
  + Minimize potential for leakage—that is, lowest pressure, vapor-phase reactors vs. liquid phase, lowest inventory, nonflashing conditions.
  + Safer physical states—that is, dilute or refrigerate hazardous materials, use vapor-phase reactors.
  + Segregate reactions—that is, provide a reactor specifically designed for each reaction rather than carry out multiple reactions in a single vessel.
  + Select nonhazardous or minimum hazard solvents, diluents, and heat exchange media—that is, water vs. hydrocarbon, tempered water vs. steam, noncorrosive vs. corrosive.
  + Limit the potential effects of an upset—that is, prevent overheating by limiting steam pressure; ensure that one upset does not trigger another (Kletz's "knock on effect"); use self-limiting reactor configurations.

Many of these principles bear on the selection of reaction and other process conditions. Also to be considered are the methods of heat transfer, mixing, product purification, and providing the necessary available energy to drive the process.

The reactor section provides a number of possibilities to improve inherent safety. Large, high-pressure and -temperature, batch reactors are at the risky end of the scale, and small, tubular, continuous reactors are at the other, more inherently safe, end. Highly volatile, flammable solvents/diluents with complex product recovery systems are at the risky end, and aqueous, simple recovery systems are at the other. A single large reactor containing toxic materials, carrying out multiple reactor steps at different conditions requiring different utility supplies and control systems is more risky than several smaller reactors each designed for a specific reaction step with a single utility and control system.

Reactor Examples

Clearly, life can become complex quickly, and the potential for conflicting indications is great. Let us consider an example from Hendershot's paper (see reading list) in an attempt to clarify the situation. The challenge is to conceptualize the least-hazardous reactor system for a complex synthesis of a chemical.

The process is a multistep synthesis involving the following steps:

1. A highly exothermic reaction of *A* and *B* in solvent *C* at low temperature requiring refrigerated brine cooling and a large rupture disk for emergency relief.
2. A less-energetic reaction of the product of the first step with a highly toxic and volatile material *D* at higher temperature. The reaction mixture would freeze at refrigerated brine temperature.
3. Stripping to remove solvent *C,* using steam heating. The product is redissolved in solvent *E* before shipping.

The single reactor design of Figure 4.5 is possible, but includes many potential hazardous interactions, such as:

- Accidentally applying steam during the highly exothermic first reaction step
- Accidentally applying brine, freezing the reaction mixture, during the second reaction step
- Failure of the large rupture disk during the second reaction step, releasing the highly toxic and volatile raw material *D* to the environment
- Numerous potential errors in the order of addition of raw materials or contamination of a process step with materials intended to be charged in a different step

Kletz would go on to point out that a continuous tubular reactor would be ideal for the exothermic step since it would facilitate heat transfer and minimize holdup. Such a configuration would also be advantageous for step 2 to minimize holdup of the toxic material.

To carry the continuous analogy to its end, solvent *C* could perhaps be stripped in a tower and mixed in line with final solvent *E* if viscosities permit.

Another approach would be to split the batch reactions into several separate steps. Separation of the process steps into three vessels, as shown in Figure 4.6, greatly reduces or eliminates the potential for many of these hazardous interactions. The design of each vessel can be optimized for its particular process step; each vessel is only connected to the raw materials, utilities, and auxiliary equipment required to carry out that process step; each vessel is sized for the

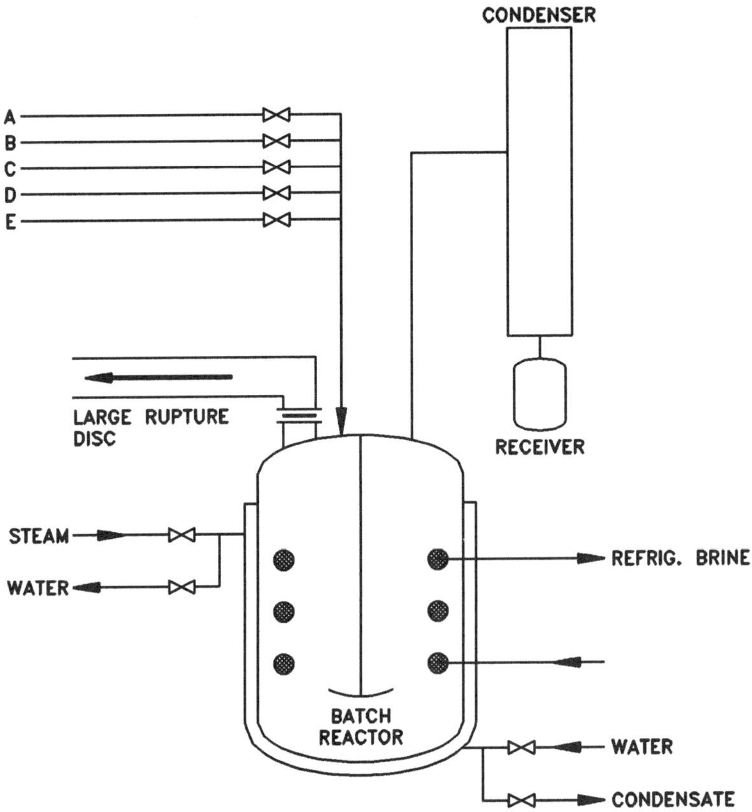

Figure 4.5. A single reactor design for multistep batch chemical process.

residence time needed for the specific step; and a specific control system is applied to each.

Many of the potential interaction hazards are eliminated by this configuration. In addition, the residence time in each vessel can be optimized for the reaction of immediate importance only, rather than be constrained by the requirements of the other reaction steps. Metallurgy may also be a consideration since a single vessel would have to be fabricated of materials to withstand the most corrosive conditions in any reaction step. On the other hand, multiple transfers present multiple opportunities for piping leaks, and three vessels will cost more than one. Thus we continue to face the need for trade-offs.

In the same paper by Englund (see reading list, item 9) another example of an inherently safer reactor configuration is presented. He refers to the production of specialty synthetic rubber latex from styrene and butadiene that cannot be made by a continuous process.

In conventional batch processes for making latex, all the reactants are added

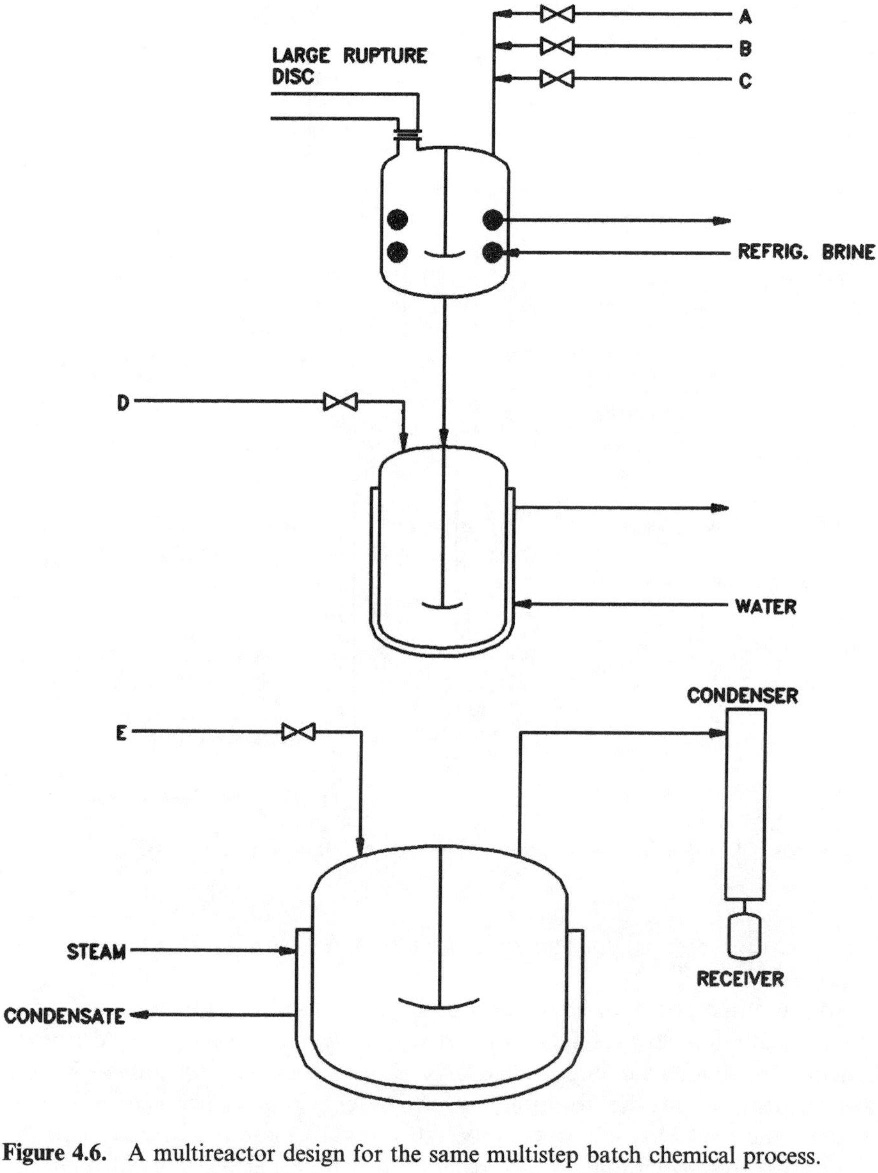

Figure 4.6. A multireactor design for the same multistep batch chemical process.

at the beginning of the reaction. There is a considerable amount of hazardous, flammable material present in the reactor at the beginning of the reaction, and many scenarios can be visualized that could cause safety problems, such as leaks, agitation failure, loss of cooling, and runaway polymerization. A typical reactor configuration is shown in Figure 4.7.

Methods have been developed for improving batch process productivity in

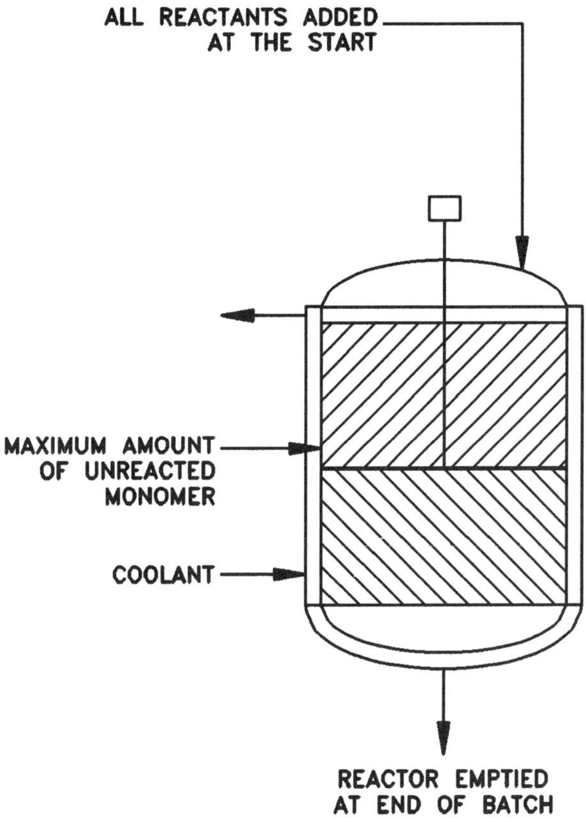

Figure 4.7. Conventional process.

the manufacture of latex by the continuous addition of reactants so the reaction takes place as the reactor is being filled (Englund 1982). These are not continuous processes even though the reactants are added continuously during most of a batch cycle. The net result is that reactants can be added about as fast as heat can be removed. There is relatively little hazardous material in the reactor at any time because the reactants, which are flammable or combustible, are converted to nonhazardous and nonvolatile polymer almost as fast as they are added. See Figure 4.8.

Such an approach is given in U.S. patent 3,563,946 (1971). In this example, styrene-butadiene latex is prepared by simultaneously feeding two separate streams (one a mixture of styrene and butadiene monomers, the other an aqueous solution of sodium laurayl sulfate as emulsifier and sodium persulfate as initiator) to a reactor that has an initial charge of water and emulsifier. The reactants are added at about 25°C and the reaction occurs at 90°C. The configuration also makes it easier to have good temperature control, which is good for product quality. Productivity is significantly better than conventional

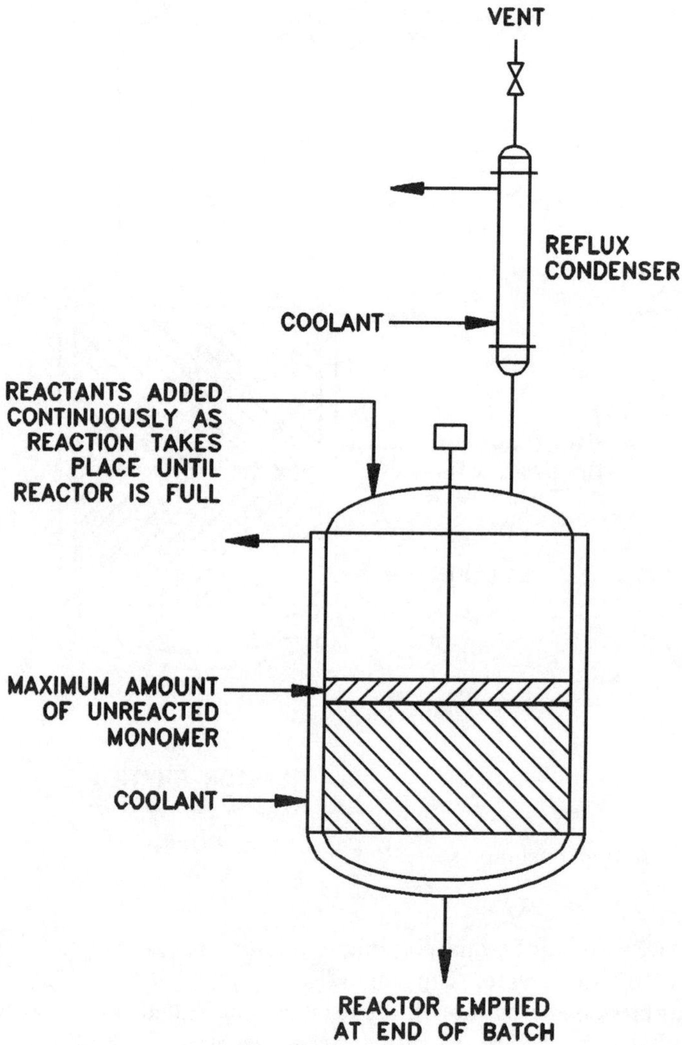

Figure 4.8. Revised process.

batch processes since the reactor is polymerizing at its maximum rate most of the time, normally limited only by heat transfer capability. Heat transfer capability can be increased by adding a reflux condenser (U.S. patent 4,024,329, 1971).

What begins to emerge at this point is the potential for simultaneously achieving enhanced productivity, product quality, and inherent safety through the use of the same concept. Given the time frame of this particular improvement, the driving force was almost certainly based on the economics of the process improvements and not on the safety issues. The possibility of achieving both economic *and* inherent safety benefits is not trivial and more will be said about it later.

Inventory Considerations

As the process develops toward the process design stage it becomes possible to address the inventory and conditions in the system downstream of the reactors. This does not merely refer to the off-site storage tanks used for raw materials, intermediates, and products. It also includes process inventories and conditions, and interacts with the layout of facilities within a given site. The inherent safety of many process facilities can be significantly enhanced through a design strategy that minimizes hazardous material inventories at the plant. Some features to be considered are:

- Locating process equipment handling the hazardous material such that transfer piping length is minimized.
- Minimizing or eliminating in-process surge tanks for hazardous materials.
- Minimizing line sizes for hazardous material transfer piping, using pipe no larger than required to provide the desired flow rate.
- Transferring hazardous materials in the vapor state rather than as a liquid.
- Minimizing piping complexity.
- Storing and processing liquified gases at low temperatures and pressures (use nonhazardous refrigerants). This may not apply to shipping toxic materials.
- Selecting low-inventory-process equipment types and control systems.
- Minimizing potential leak sources and pressure of equipment.

These principles are typically applied in an organized review early in the design phase. At this point, such a review gives direction to the designers and identifies most of the final alternate studies to be considered. The designers can then select equipment consistent with this overall strategy.

One Exxon study for a steam cracker plant, which, unfortunately, was never built, gave some remarkable results. It started from a preliminary contractor process design that was consistent with current best industry practice. Major changes to the process sequence—for example, demethanizer first vs. deethanizer first separation sequence, were not considered due to time constraints and contractor expertise in the sequence proposed.

The final configuration, including only proven technology, reduced flammable inventories by fifty percent, eased a congested plot plan, reduced the size of potential spills/leaks, and produced a plant that was judged easier to control *and to be of lower investment and operating cost*! (Again, we see the confluence of economic benefits and improved inherent safety.)

As an example, consider the flow/equipment diagram for the ethylene splitter (which is a large refrigerated tower separating ethylene from ethane) shown in Figure 4.9. The diagram shows process flow rates, pressures, and temperatures, but more importantly, it also shows equipment configurations and inventories. The total inventory in the system is 225 tons, excluding piping and the tube

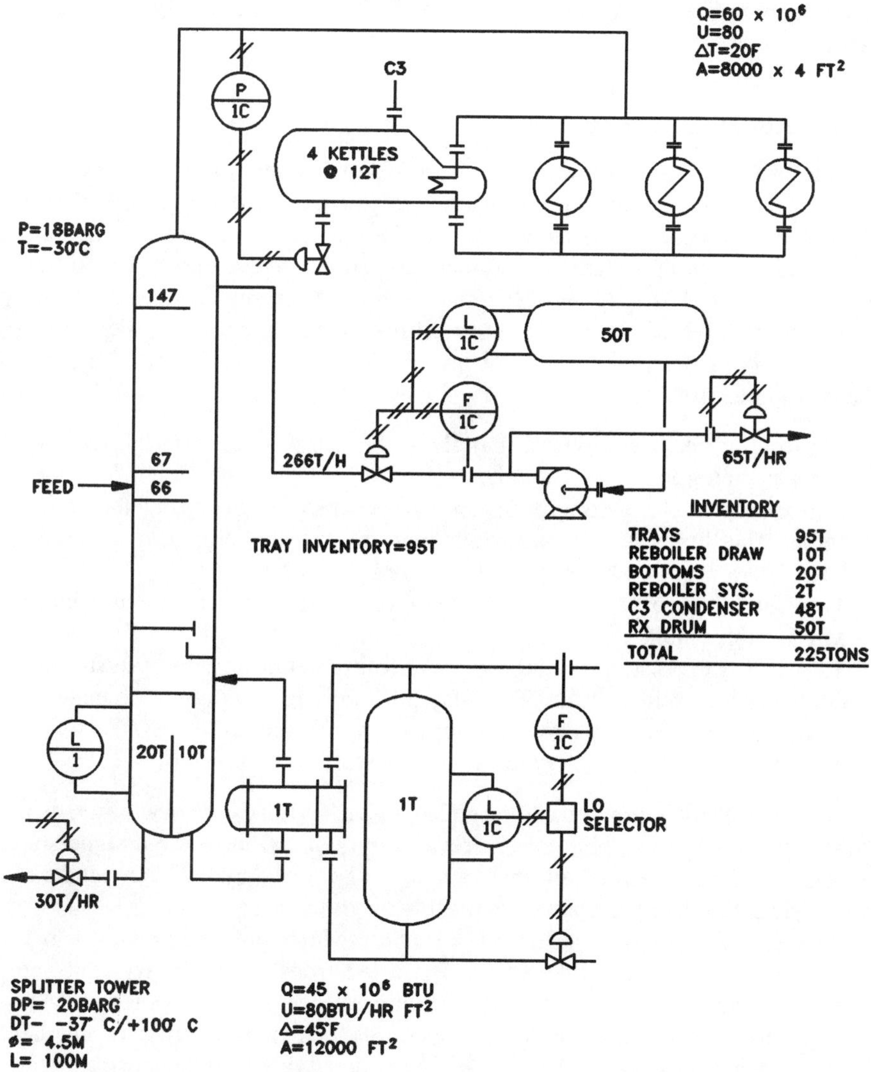

Figure 4.9. C Splitter example.

side of heat exchangers. The question to be addressed is whether this configuration represents the most inherently safe one—that is, the one with the lowest practical inventory of flammables. Please note that the condenser is cooled by propylene refrigerant and, therefore, contributes significantly to flammable inventory.

To comment on this problem, one needs to know something about alternate equipment configurations and process possibilities. For the moment, let us consider equipment aspects only.

In initiating our inherent safety review aimed at inventory, we will be asking questions related to the principles outlined earlier:

- What is the basis for the flammable inventory hold-up chosen?
- Can it be eliminated?
- What are the trade-offs for reducing it?
- Does it have to be stored at this pressure?
- Is all the complexity shown necessary?

The more fundamental questions like, "Can a substitute material be used?" and "Are these process conditions necessary?" will be addressed later.

The inventory in the overhead system (98 tons) comprises almost half the total. Assuming that the total heat transfer surface area listed is required, kettle reboilers provide it at about the maximum inventory. Their eccentric design requires a large inventory of boiling liquid on the shell to maintain the bundle flooded. Thermosiphon heat exchangers require less inventory, and plate fin exchangers even less. Consideration of these types of heat exchangers rather than the kettles seems warrented.

A second question to explore if the kettle approach were essential for some reason would be, "Why were four chosen?" Perhaps two would serve to provide the necessary surface area at lower inventory. In connection with this, perhaps the use of high flux (enhanced boiling surface) tubes might serve to make it possible to provide the necessary surface area in two shells even if kettles are chosen.

Many light hydrocarbon condensers can be operated without reflux drums. The total flow through the drum in this scheme is about 330 tons/hr, giving a hold up of about ten minutes. This is not an excessive control inventory if a drum is required. The dynamics of the control system and both start-up and shutdown procedures need to be examined before deciding to eliminate this drum, but many successful designs have done so.

The potential for reducing inventory in the overhead system would seem to be large. The use of plate fins plus a reflux drum could eliminate about fifty tons. The use of two kettles and no drum holds the same potential. Eliminating both the kettles and the drum might be possible as well. Except for the last idea, there is plenty of practical data against which to evaluate the current design. Significant reduction should be possible in the overhead system.

Additional savings may be possible in other parts of the refrigeration system if on-site refrigerant inventories could be reduced. It is typical to design refrigerant system hold-up drums to allow draining of the system inventory during a shutdown. Less on-site inventory means a smaller refrigerant hold-up drum and lower inventory during shutdown periods as well.

Notice that lower inventory means less hardware. Again, better inherent safety may mean lower capital cost as well.

Let us now turn attention to the inventory in the tower. More than half of the total is contained in this single piece of equipment. The split bottom section provides forty minutes hold-up for the bottom's product. This is excessive with even older control systems. We cannot evaluate the hold-up requirements of the reboiler feed compartment because no flow rates are given. It is possible to reduce the diameter of the tower bottom to reduce inventory. About five tons for bottom's product would be a more appropriate figure based on earlier data.

The tray inventory in the tower itself is ninety-five tons. Two types of questions are appropriate. The first is whether the number of trays could be reduced and the reflux rate be increased to give the same separation efficiency. The second is the basis for the hold-up on each tray—whether it could be reduced on most, if not all, trays.

A further, more radical, question can also be asked. With today's advances in packing technology, trays can be abandoned altogether in many clean services and be replaced with dumped or structured packing. Today, this is possible at significant inventory reductions. This is an example of intensification of the process.

Thus, several serious questions need to be explored before even the tower itself, a critical device for accomplishing a critical process step, can be declared as inherently safe as possible. Clearly, process risks must be evaluated more critically for the tower than for the condenser system because no corrections afterward are likely to be practical if the design does not measure up.

I think that a reasonable conclusion is that the inherent safety of the tower system could be improved through equipment changes that would result in lower inventories. In general, conventional process equipment design, no matter how well proven, needs to be reevaluated periodically even if no radical departures in concept are possible. Many technology developments have taken place and continuous improvement is essential, not only in business operations, but also in safety.

Layout

Earlier, I mentioned the importance of astute layout in minimizing inventories of hazardous materials. This obviously has implications in the mitigation of any incident as well.

In the discussion of site selection, I made some comments on the location of particular equipment inside a site relative to susceptible facilities or populations outside the fence. Clearly, major inventories of hazardous materials should be located as far as possible from plant boundaries. Moreover, they should be located apart from other hazards that may be involved in the site, such as sources of ignition and so on.

Englund (1990) suggests a triangular approach with loading/unloading at one apex, storage at the second, and the process unit at the third. (See Fig. 4.10). No facilities involving people, significant equipment, or buildings would normally be allowed inside the triangle. Water facilities, effluent treatment ponds, and so on might be appropriate inside the triangle, but boilers, control

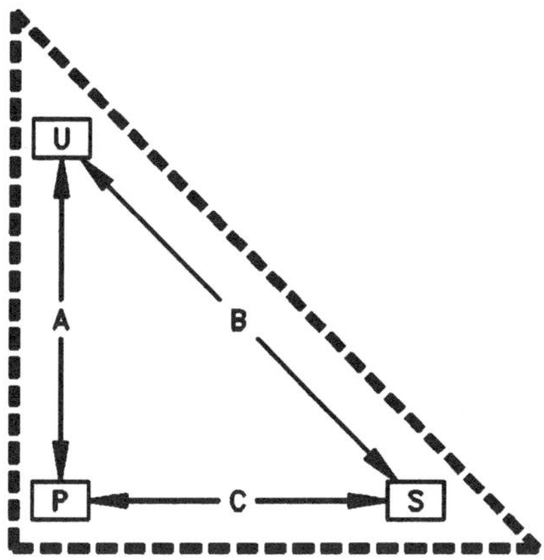

U = LOADING AND UNLOADING OF TANK CARS AND TANK TRUCKS OF HAZARDOUS MATERIALS

P = PROCESS PLANT

S = STORAGE OF HAZARDOUS MATERIALS

Figure 4.10. Triangular concept of facility layout.

rooms, offices, and so on would not. Such an arrangement provides reasonable protection from what Kletz calls "knock on" or "domino" effects in that hazards are separated, allowing for better isolation and mitigation in an incident.

Obviously, structures should be appropriate for the equipment contained. With flammable materials, walls and roofs are usually avoided to promote dispersion of any leaks. Walls are usually eschewed even for machinery bays. Such structures are also often used for equipment handling toxic materials. However, lately there has been a trend toward total containment with scrubbing for really toxic materials like chlorine. Sometimes water or steam curtains can be used effectively in mitigating a leak of toxic material from an open structure.

Another layout consideration is preventing leaks of flammable materials from accumulating under process equipment or surrounding storage tanks. This is a matter of grading the ground appropriately and providing appropriate containment around storage vessels. There are a number of codes related to such questions and insurance inspections often prescribe standards that must be followed. The principle of getting any spills away from facilities containing hazardous materials to prevent domino effects is also an important consideration.

Definitive Process and Mechanical Design Phase

I have tried to picture a project evolving from conceptual to more and more specific decisions. The process is a continuum, and different companies and, indeed, different individuals mark the phase boundaries at various points along it. Thus, one process developer will say that his or her work ends only when the final heat and material balance is done, while another will stop as soon as the process sequence is finalized, leaving the process designer to explore more alternatives and combinations as he or she develops the final heat and material balance. Similarly, some process designers wish to specify the details of heat exchange trains while others are content to specify only their duties and temperature ranges and leave the number of shells, type of exchanger, and so on to the mechanical designers. Wherever the boundaries are drawn, the same thinking must be done. Thus, the principles discussed below might be applied by an in-house process-oriented person or by an engineering contractor.

A rather long checklist could be written at this point, but it could not be comprehensive. Try as I might, I could not address every opportunity that might exist across such a broad spectrum as the process industries. Others have written down some principles, and using these as a base, I shall try to flesh them out with some discussion and examples.

Again, I am indebted to the authors mentioned earlier, to various practitioners at several process companies, and to others who have reduced some of these principles to practice, such as the AIChE.

Once the process sequence has been chosen, designers need to choose equipment, control schemes, and relief systems that will minimize the probability of a release and minimize its consequence, all without inordinate cost. Some principles to achieve a more inherently safe plant design are:

- Incorporate planning into process design.
- Use proven and robust equipment.
- Avoid catastrophic failure of any containment equipment by using conservative materials and provide for any possibilities like brittle fracture, stress corrosion cracking, fatigue, and so on where potential exists—for example, auto refrigeration.
- Minimize hazardous inventories on-site and off-site; optimize size and number of containers.
- Use less-severe process and storage conditions: refrigerate volatile materials, slurry dust hazards.
- Use passive rather than active protection systems—that is, apply the lowest exergy level needed to drive the process—for example, heating media that cannot overheat a sensitive material.
- Simplify piping and equipment. Keep lines containing hazardous materials short. Minimize installed spares, flanges.
- Incorporate fail-safe process features—for example, process conditions that have a negative temperature coefficient on reaction rate.

- Provide adequate equipment spacing to prevent cascading of any incident into other process or storage areas.
- Make sure safety relief systems will function in both the "worst case" and the "most likely" case and properly account for two-phase flow and ultimate disposal needs.
- Provide for the ability to withstand extremes in weather conditions, earthquakes, floods, and so on.
- Use seal-less pumps, all welded pipe to minimize leakage points.
- Design to prevent incorrect assembly.
- Provide rapid isolation systems for large hazardous inventories.
- Design so that the status of any piece of equipment is clear—for example, quarter-turn valve handles in line with pipe when open, rising stem valves.
- Minimize sample points.
- Minimize small piping connections especially in areas where physical damage may occur.
- Provide equipment and controls that are tolerant of mal-operation or poor maintenance—for example, expansion loops vs. bellows joints.
- Provide redundant/voting control sensors on critical safety devices.
- Provide slow response systems so that the operator has time to analyze the problem and choose the proper corrective action.
- Provide maximum practical separation of the operating conditions from those at which the process may run away.
- Provide double containment where prudent.
- Provide software that signals obviously when not functioning properly.
- Provide unequivocal standards for design—for example, those familiar to the owner and designer (One can use API 620 in the United States, but must remember that it is not a European standard).

I promised you a short list but found it difficult to stop. Perhaps some of the analogies Kletz presented in his July 1989 article in *Chemical Engineering Progress* magazine can capture the essence of what the designer should be trying to do. Kletz says that if the meat of lions tastes only as good as the meat of lambs, why raise lions for food? If a tricycle will get you there, why use a bicycle? If we have the space to build what we need on ground level, why build a two-story structure and have to construct stairways, inspect and maintain them, train people to use the handrails, and keep them clear of clutter. In short, design a plant that cannot or is less likely to cause a specific incident even if things go wrong.

Some Additional Examples

Figure 4.11 illustrates how equipment design can reduce the inherent potential for operating errors in a system. The vessel shown is a feed tank for a batch reactor. The amount of the charge is critical, and the feed tank has been sized to hold exactly one batch charge. To ensure that material cannot be transferred

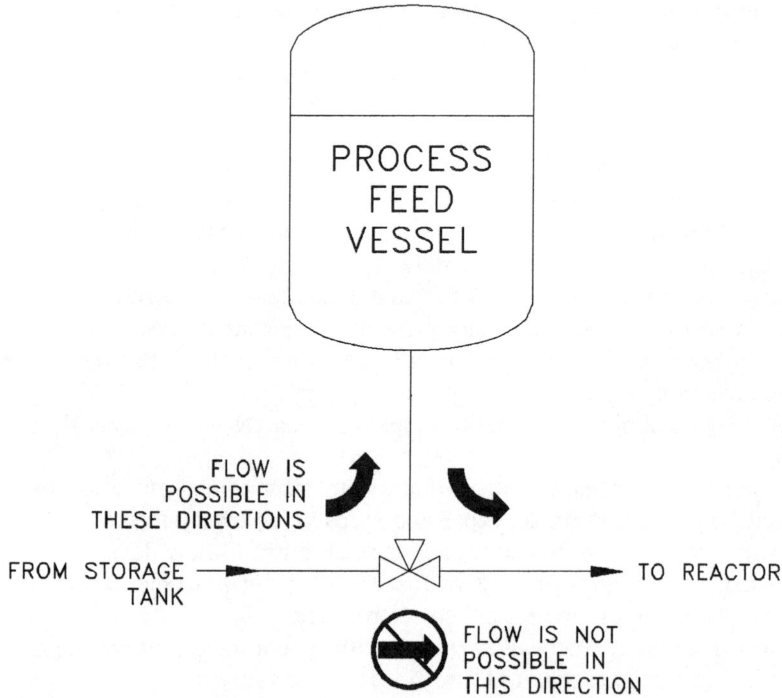

Figure 4.11. Mistake-proof measuring vessel system (from Hendershot 1990).

directly from the storage tank to the reactor, and to ensure that it is not possible to add accidentally more feed to the charge tank while it is emptying to the reactor, the piping system shown is used. This three-way valve does not allow flow directly from the storage tank to the reactor, even if it is left in an intermediate position. Note that the three-way valve must be specified to function in this manner.

In another example, consider the choices we must make in the way in which the established minimum inventory of any hazardous material is provided. Several smaller tanks would limit the amount discharged in any single vessel failure. However, a large portion of the risk in any storage facility depends on the number of nozzles, pipes, and valves associated with it. Thus, a single large tank may offer greater safety because there are fewer associated nozzles, valves, and so on. This depends on the assumption that piping sizes are still set by the flow rates required in the process and not arbitrarily increased because of tank size. In addition, a single larger tank will be more robust and resistant to external damage than several small ones, and will have less external surface area. Again, trade-offs must be evaluated.

The handling of ethyl aluminum dichloride (EADC) illustrates application of inherent safety in product configuration. Pure EADC is pyrophoric and reacts violently with water. To avoid violent reactions to small spills in customers' plants, the manufacturer ships the product diluted in a compatible and relatively inert solvent. Small spills result in some fuming, but this problem can be easily controlled.

Procedure also plays a role in inherent safety. Bottom loading of tank trucks is generally considered safer in many respects than top-loading. All normal oprations are done from the ground, eliminating the need for personnel on top of the truck. Fire risk is reduced because splashing and vapor emissions are minimized; a high-level shutoff avoids overflow. However, static electricity may be a greater problem in bottom loading than in top loading. This is especially true if a truck that has just been carrying gasoline, or another volatile product, is reloaded with a less-volatile product (jet fuel) that is a greater static-charge accumulator. This is often called switch loading. Emptying the truck of the gasoline introduces air that forms a flammable mixture. Reloading with the heavier liquid from the ground rarely involves opening the dome hatch to purge the vapors. Thus, the elimination of static charge generators (e.g., filters in the pipe) and spark promoters (e.g., a gauge rod) is essential in a safe bottom-loading system, but elimination of switch loading is more fundamentally sound.

ARCO's public briefing on their July 1990 explosion in Texas provides an interesting example both of how an inherently safer design might have prevented the disaster and of the high cost of fixing the issue after the fact. A 900,000-gallon storage tank exploded when a flammable mixture (unknown to the operators) was sucked into a compressor that was being restarted.

The waste storage tank was designed to serve several purposes in processing waste water from the propylene oxide–styrene monomer (PO/SM) unit prior to disposal of the treated wastewater in the deep disposal well on-site. Those purposes included:

- Temporary storage of waste water
- Separation and temporary storage of light hydrocarbon liquids contained in the waste water
- Decomposition of light hydrocarbon peroxides produced in the PO/SM process and purged from the unit in the waste
- Separation of the resulting free oxygen and purging of the gaseous stream

The tank was designed as an atmospheric pressure tank with dual conservation vents to relieve pressure buildup and to break any vacuum resulting from normal operations.

Because gaseous oxygen was liberated in the tank by light peroxide decomposition, the tank was equipped with an inert nitrogen sweep system to purge the oxygen and maintain an oxygen concentration within safe limits. Due to the presence of light hydrocarbon liquids, the nitrogen sweep leaving the tank also contained hydrocarbon vapors. These hydrocarbon vapors were scrubbed from the tank vapor purge stream by a pressurized recovery tower located in the PO/SM unit. A compressor was used to transfer the vapor from the atmospheric-pressure storage tank to the pressurized recover tower.

The tank vapor space was sampled by an oxygen analyzer, providing a continuous readout in the control room of the tank oxygen concentration and controlling the N_2 purge rate.

The underlying reason for the explosion was found to be an accumulation of oxygen in the vapor space of the tank, which was not recognized because of an apparent failure of the oxygen analyzer.

The oxygen buildup was caused by a reduction in the nitrogen sweep rate. For a period of thirty-four hours before the explosion, the nitrogen sweep was shut off completely because the compressor failed and was removed for repair. Since the oxygen analyzer continued to show an extremely low—and safe—oxygen level, the nitrogen sweep was allowed to remain off until just a few minutes before the compressor restart and the explosion.

In short, we had hydrocarbon, oxygen, and energy from the machinery (the source of ignition). Normal operations on an abnormal process mixture (which resulted from the failure of a single analyzer) resulted in the explosion.

The repair and corrective action cost ARCO $20 million. One feature of the corrective action is changes that will result in an inherently safer design from several viewpoints.

- The hydrocarbon will be stored in a separate vessel from the water/hydrocarbon separator, meaning a much smaller vessel in hydrocarbon service (minimum inventory).
- Redundant O_2 analyzers and N_2 supplies (no flammable mixture).
- A continuous N_2 flow at a minimum level for a safe purge O_2 concentration (no flammable mixture).
- Elimination of the compressor by the direct connection of the hydrocarbon vapors to a dedicated flare (no source of ignition).

With the wisdom of hindsight, this configuration is free of a number of risks inherent in the original design. The trick is to do this through foresight.

Exergy Change, Energy Efficiency, and Risk

The end of a chapter on improving the inherent safety of systems seems a good place to philosophize a little on fundamentals. I have written extensively (*Energy Conservation in the Process Industries,* Academic Press, New York, 1984) on the benefits of exergy analysis in helping improve the fundamental energy efficiency of processes, and in the course of so doing, have presented some data to show that the capital cost of inherently more efficient systems is generally lower. The logic is simple: If you use less exergy, you need less equipment to transfer it to the process and less to generate it, so your plant costs less. There is some parallel with inherent safety.

Consider the propane/propylene splitter tower shown in Figure 4.12. A feed containing sixty-five percent propylene is purified in a superfractionator (200 trays) to a 99.9% pure overhead product. At the operating pressure shown, only low-level heat is needed to drive the reboiler, but a refrigerated condenser is needed to remove the heat overhead. The refrigerant is also propylene. The

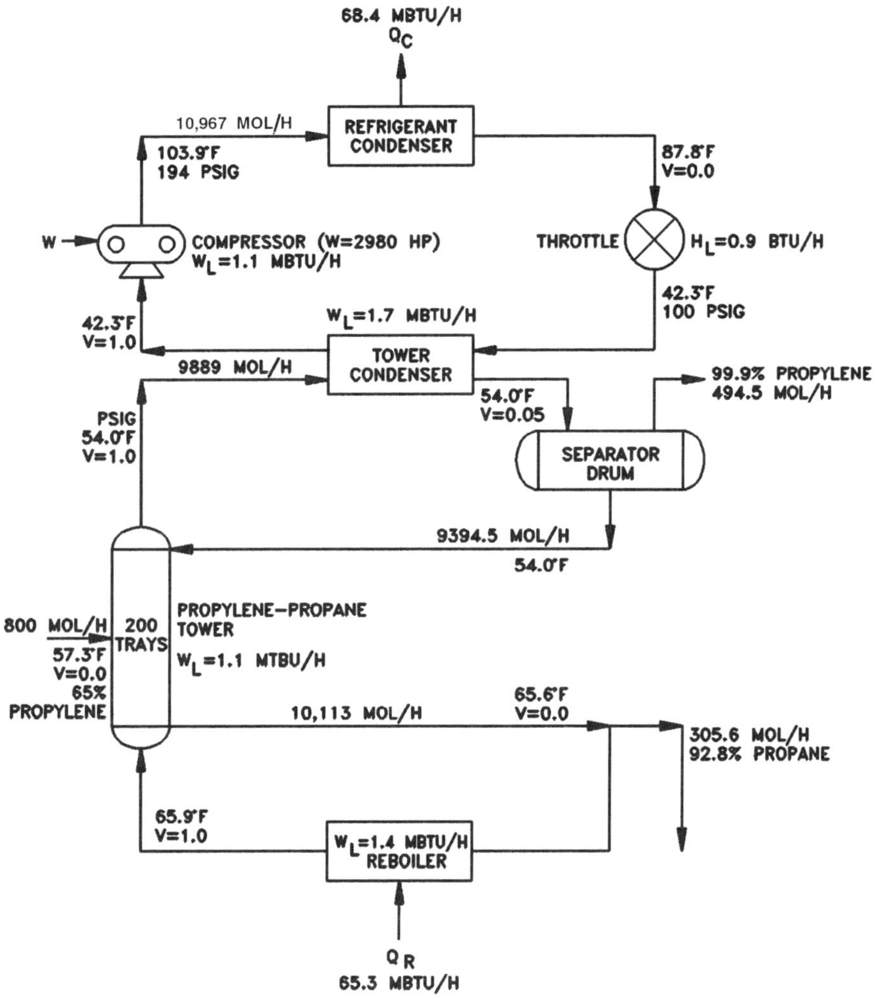

Figure 4.12. Energy flow for a refrigerated fractionation tower.

system consists of three major heat exchangers, the tower, a drum, the refrigerant compressor, and two piping systems.

The theoretical work needed to accomplish the separation is 0.3 MBTU/hr. This represents the thermodynamic minimum exergy input needed to drive the process. The actual *exergy* consumption for the process shown is calculated to be 7.6 MBTU/hr. You will note that this is much lower than the total *energy* input to the process (about 70 MBTU/hr heat to the reboiler and compressor horsepower) because the *exergy* content of the low-temperature heat used in the reboiler is very low.

An alternative process configuration to do the same separation would involve a vapor recompression (heat pump) system. In this system, the vapors from the top of the tower are compressed to a point at which they can be condensed in the reboiler. One alternative for a heat pump configuration is shown in Figure 4.13. It reduces the size of the compressor, eliminates two major heat exchangers, and adds two tiny, trim coolers.

As shown in Table 4.3, the total work required by the alternative process is 5.1 MBTU/hr even though the theoretical work required remains the same. Thus, the wasted work (exergy) in the process is reduced from 7.8 to 5.1 MBTU/hr (35%).

In addition to the energy savings in operation, the installed cost of the system will obviously be less. There is less hardware, the compressor (and its power supply) is smaller, there is only one piping system (and, therefore, only one control system), there is no large cooling water system for the now nonexistent refrigerant condenser, and so on.

There are commensurate benefits from the risk viewpoint as well. The system

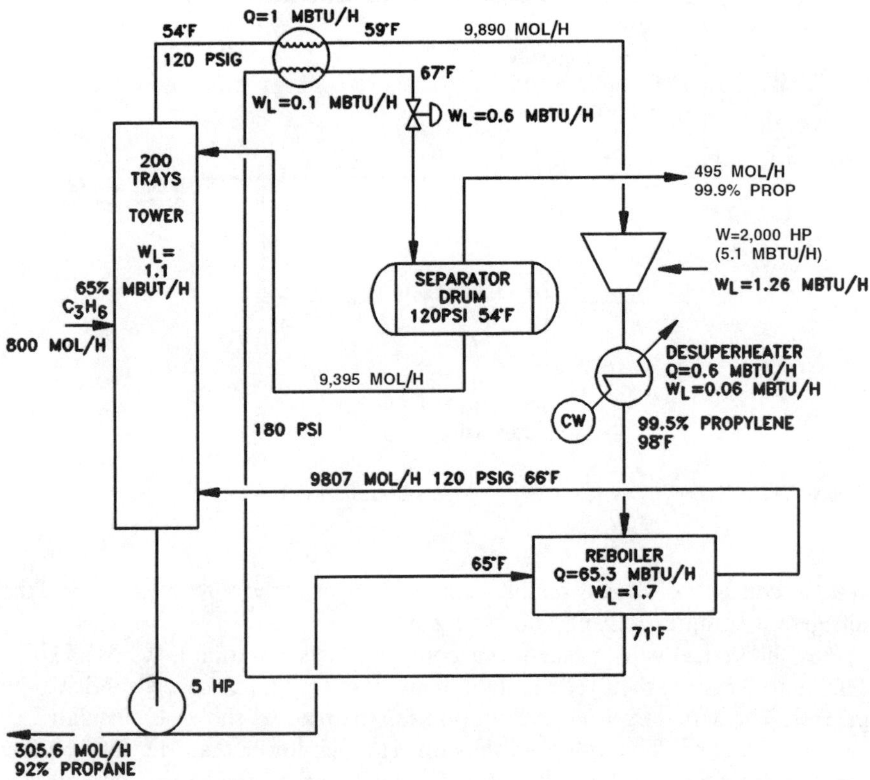

Figure 4.13. Energy flow diagram for a heat pump system in the same fractionation process.

Table 4.3. Thermodynamic Analysis for the Propylene—Propane Separation Process

Component	Lost Work (Exergy) (M BTU/H) Refrigeration Option	Heat Pump Option
Tower	1.1	1.1
Tower condenser	1.7	—
Tower reboiler	1.4	1.7
Compressor	1.1	1.26
Desuperheater	—	0.06
Refrigerant condenser	1.3	—
Throttle valve	0.9	0.6
Superheater	—	0.1
Total lost work	7.5	4.82
Theoretical work	0.3	0.3
Actual work	7.6	5.1

is simpler, the hold-up is less, as is the amount of piping. As such, there will be fewer places to leak, less material to leak, and a less-complex control system, so the potential for human error is reduced. These benefits also involve a lower capital cost and higher energy efficiency. While one example does not prove the general relationship, the intuitive logic does much to support the hypothesis that a thermodynamically better process may also be inherently safer. For example:

- Low driving forces for heat exchange not only minimize exergy input and maximize exergy recovery but also reduce the accident potential of a system. If we use low-pressure steam in place of high pressure, there is less chance of a leak and its consequences are less. If we use steam to heat a process in place of a furnace, we do not face the possibility of a fire or explosion in the furnace.
- If we eliminate a process step or storage tank (simplification), we not only save its energy usage but also eliminate its risks.
- If we maximize yields in reactors (intensification), we minimize separation steps, recycles, and raw material waste and the attendant risks.
- If we carry out separations at optimum conditions, we have smaller equipment (inventory), less reflux (recycle) and smaller utility systems.
- If we minimize air leaks into a furnace, we save fuel and also reduce the risk of secondary combustion in the convection section.
- If we have an efficient and flexible steam/power system, we minimize the amount of fuel burned in the boilers (and perhaps can shut one down), eliminate venting of excess steam (and the attendant risk and noise), recover condensate (and avoid the risks associated with putting hot water in the sewer or splashing it on the ground), and so on.
- If we combine many dispersed sources of exergy loss into a larger single unit,

not only can we economically provide a more efficient system but it can perhaps also be simpler and, therefore, safer.

While much of the above is qualitative, my instinct says the relationship is real. I would suggest that the fundamental thermodynamic analysis of processes may provide a basis for establishing the most inherently safe configuration as well.

Summary

The elimination of hazards in a process plant can be much more successful if such efforts are begun very early in the conceptualization of new processes or facilities. There are many more opportunities to eliminate hazards during the process research stage than in the process or mechanical design stage. Multiple "inherent safety" reviews are essential as a project progresses from research through start-up if management is to maximize its responsibility to protect people, property, and the environment in its profit-making enterprise.

The principles against which both new and existing processes should be compared can be summarized briefly as follows:

- Review raw materials and intermediates to see if less hazardous chemicals can be used (substitution).
- Evaluate storage and handing requirements to see if hazardous materials operations can occur under less-severe conditions (e.g., pressures and temperatures nearer to ambient—mitigation).
- Size systems to reduce hazardous material inventories (intensification).
- Eliminate of nonessential complexity (simplification).
- Choose a site and plan the site layout based on the estimated extent of credible accidental releases of hazardous materials.

After the planning stages, final system design should be based on minimizing the likelihood of hazardous material releases or, should a release occur, limiting the amount of release. Finally, consideration should be given to effective emergency response that can reduce the impacts of a release.

Table 4.4 lists four stages of project development and the types of decisions likely to be encountered in each that can affect inherent safety. Clearly, the more fundamental opportunities exist at the left of the table, but meaningful opportunities still exist in the right-hand column. Indeed, inherent safety is not to be ignored even after a unit is built and operating. Many kinds of design items can be retrofitted to an operating unit and may even be more effective with the benefit of some operating experience.

As a final thought, fundamental thermodynamic analysis of process options may also contribute to defining the most inherently safe configurations for the plant system.

Table 4.4. Some Inherent Safety Questions In Project Progression

Business Decision	Process Research	Process Development & Design	Detailed Mechanical Design
• Make, buy, or forego opportunity + Are potential benefits worth the risk? • Plant location + Available land + Safeguards/spacing + External hazards + Impact on process technology	• Raw material choices + Specifications • Reactor type • Solvents, diluents, catalysts • Process design optimization data • Minimize downstream problems + Eg., pollution from waste products • Support innovation	• Intensification + Mixing + Reaction hardware + Separation/ purification sequences + Inventories + Heat exchange • Limit available energy input • Equipment selection + Seal-less pumps + Self-limiting configurations • Selection of operating conditions • Layout/spacing	Clear mechanical STD's • Small pipe integrity • Simpler designs • Spiral wound gaskets, no asbestos • Minimize sample points • Minimize maintenance needs • Consider attenuation + Storage conditions + Containment • No "domino effect" from single failure • Tolerant of bad installation and mal-operation • Control systems + Flat response curve

Inherent Safety Reading List

1. A. Barrell, *The Chemical Engineer,* No. 451, Aug. 1988, p. 3.
2. A. Borning, Commun. *ACM,* 30, Feb. 1987.
3. P. A. Carson, C. J. Mumford, R. B. Ward, *Loss Prevention Bulletin,* No. 065, *Loss Prev. Bull,* Oct. 1985, p. 1; No. 070, Aug. 1986, p. 15.
4. A. D. Craven, "Fire and Explosion Investigations on Chemical Plants and Oil Refineries," in *Safety and Accident Investigations in Chemical Operations,* edited by H. H. Fawcett and W. S. Wood, 2nd ed., Wiley, New York, 1982, p. 659.
5. W. S. Davies, *Essays on the First Hundred Years of Anesthesia,* Longmans, New York, 1982, pp. 3–5.
6. H. A. Duxbury, R. D. Turney, "Techniques for the Analysis and Assessment of Hazards in the Process Industries," Seminar of Safety and Hazards Evaluation, New Mexico Technical Research Center for Energetic Materials, April 1989.
7. S. M. Englund, "The Design and Operation of Inherently Safer Chemical Plants," AIChE Meeting, San Diego, August 1990.
8. S. M. Englund, "Opportunities in the Design of Inherently Safer Chemical Plants," *Advances in Chemical Engineering,* Vol. 15, Academic Press, Inc., 1990.
9. S.M. Englund, "Chemical Processes, Batch or Continuous?" *J. Chem. Educ.* No. 9, 1982, p. 59.
10. D. Farmer, *Health and Safety at Work,* No. 8, Nov. 1986, p. 54.
11. R. Fowler, *The Chemical Engineer,* No. 456, Jan. 1989, p. 35.
12. N. Franklin, *The Chemical Engineer,* No. 430, 1986, p. 17.
13. H.G. Gerrisen, C. M. Van't Land, *I&EC Proc. Des. Dev.,* 24, p. 893, 1985.
14. J. L. Hawksley, *Plant/Operations Progress,* Vol. 7, No. 4, Oct. 1988, p. 265.
15. J. L. Hawksley, *The Safety Practitioner,* Oct. 1987, p. 10.
16. D. C. Hendershot, "Design of Inherently Safer Chemical Storage and Processing Facilities," AIChE Meeting, August 1990, San Diego.
17. J. E. Huff, *Institute of Chemical Engineers Symposium Series,* No. 85, p. 109, 1984.
18. James E. Huff, "Emergency Venting Requirements for Gassy Reactions from Closed System Tests," *Plant/Operations Progress,* Vol. 3, No. 0.1, Jan. 1984, pp. 50–59.
19. R. W. King, *Industrial & Engineering Chemistry,* Vol. 61, No. 9, Sept. 1969, p. 66.
20. C. J. Kirkland, *The Channel Tunnel: Engineering Under Private Finance—Innovation or Frustration,* Fellowship of Engineering, London, 1989, p. 8.
21. T. Kletz, "Friendly Plants," *Chemical Engineering Progress,* July 1989, pp. 18–26.
22. T. A. Kletz, *Cheaper, Safer Plants—Notes on Inherently Safer and Simpler Plants,* Chap. 4, 2nd ed., Institution of Chemical Engineers, Rugby, United Kingdom, 1984.
23. T. A. Kletz, *An Engineer's View of Human Error,* Institution of Chemical Engineers, Rugby, United Kingdom, 1985.
24. T. A. Kletz, *Safety and Health at Work: Report of the Committee 1970–1972 (the Robens Report),* para. 261, Her Majesty's Stationery Office, London, 1972.
25. T. A. Kletz, *Chemical Industry,* 6 May, p. 287, 1978.
26. T. A. Kletz, *Institute of Chemical Engineers Symposium Series,* 100, p. 153, 1987.
27. T. A. Kletz, "What Went Wrong?" sections 13.5 and 17.1, Gulf Publishing Co., 1985.
28. T. A. Kletz, *Learning From Accidents in Industry,* Butterworths, Tonbridge, UK, 1988, Chapter 8.
29. T. A. Kletz, *Plant Design for Safety—A User-Friendly Approach,* Hemisphere, New York, (1991).
30. Trevor Kletz, *Cheaper, Safer Plants or Wealth and Safety at Work,* Institution of Chemical Engineers, Rugby, England, 1983.
31. Frank Lees, *Loss Prevention in the Chemical Industries,* Butterworths, London, 1980, and appendices.
32. L. M. Litz, *Chemical Engineering Progress,* Vol. 81, No. 11, Nov. 1985, p. 36.

33. M. Mackley, *The Chemical Engineer,* No. 433, Feb. 1987, p. 18.
34. J. C. Middleton, B. K. Revel, "The Intensification of Chemical Reactors for Fluids, Institution of Chemical Engineers Research Meeting, Manchester, UK, April 1983.
35. A. Nienow, *The Chemical Engineer,* No. 459, April 1989, p. 73.
36. Parker, R. J. (Chairman), "The Flixborough Disaster: Report of the Court of Inquiry," Her Majesty's Stationery Office, London, 1975.
37. Ramshaw, C., Private communication.
38. C. Ramshaw, *The Chemical Engineer,* No. 389, Feb. 1983, p. 13.
39. C. Ramshaw, *The Chemical Engineer,* No. 416, July/August 1985, p. 25–30.
40. C. Ramshaw, *The Chemical Engineer,* No. 437, June 1987, p. 17.
41. C. Ramshaw, "Problems of the Innovator in a Large Company," Symposium on Management of Change in the Process Industries, Institution of Chemical Engineers North West Branch, Manchester, UK, 1989, p. 5.1.
42. D. F. Steele, *Atom,* No. 393, July 1989, p. 10.
43. R. D. Turney, "Designing Plants for 1990 and Beyond," International Conference on Safety and Loss Prevention in the Chemical and Oil Processing Industries, Singapore, Oct. 1989, Paper H2.
44. D. E. Wade, D. C. Hendershot, D. C. Caputo, S E Dale, Papers in *Proceedings of the International Symposium on Preventing Major Chemical Accidents,* edited by J. L. Woodwar, AIChE, 1987.

Copies of papers presented at AIChE meetings are available from the institute, 345 E. 47th Street, New York, NY 10017.

CHAPTER

5

Risk Identification

Introduction

At this point, it is probably appropriate to recapitulate for a moment. In previous chapters, we have discussed the importance of managing process risks and some definitions of important factors in that process. We have spent appreciable time defining management's role and discussing techniques for identifying the inherent process hazards in our plants. At this point, we took what might seem like a diversion to explore the elimination of unnecessary hazards in some depth. This focused primarily on process hazards because these are the core of the risks we must control.

Each of the hazards that remain pose some risk to the workers and the community in the operation of the plant. These risks need to be identified specifically and evaluated. However, we must also recognize that the remaining hazards and their attendant risks do not all come from the process hazards upon which so much time has been spent. They may well come from the equipment used to implement the process and/or from the utilities needed to drive it. All of the resultant risks must be controlled/minimized in the plant and the impact of incidents prepared for in the event plant controls fail.

Risks are controlled in several ways and these indicate other sources of risk. In many cases, facilities are put in place to prevent a bad incident from occurring. These facilities must be maintained and inspected periodically to ensure that they are able to function as planned. In addition, their operation must be monitored appropriately. Thus, another source of risk is plant procedures and the people who implement them. To take full advantage of this type of preventive control, it is obvious that the necessary information must be on hand and imparted to the people.

A second approach to managing the remaining risks is to provide systems to mitigate the impact of an unwanted event through either containment or safe disposal procedures and effective emergency response planning. Here again, both facilities and procedures play a part.

The final approach to controlling risks is to prevent the evolution of new ones by ensuring that changes are controlled and analyzed from the risk standpoint. Potentially risk-inducing changes can come from myriad sources. They could include changes in facilities, procedures, operating conditions, and the people involved with a given unit.

This chapter will discuss ways of identifying the remaining risks in a process plant. Subsequent chapters will deal with minimizing and controlling these risks and with the mitigation of incidents and the management of change.

Risk Identification

To identify risks in a given plant configuration, a great deal of fundamental information is required. This starts with the identification of hazards and continues on through the complete description of the hardware and utilities systems needed to implement the process in question. The identification of process hazards has been discussed at some length. To the extent that these hazards have not been eliminated, there remain residual risks introduced by the process itself.

The hazards introduced by the utilities systems used to drive the process and handle its wastes are fairly obvious. The hazards and control of electricity are well known. The potential for burns from hot utilities such as steam and their ability to overheat process material and induce process risks are well known. The use of high pressures in steam and other utilities presents another series of problems that are relatively straightforward. These include overpressure, too much flow, and the possibility of reverse flow, all of which could cause an incident in the plant. Sometimes, chemicals are used like utilities. In many refineries, there is an extensive caustic distribution system, leaks from which might cause personnel injury and/or introduce unwanted reactions and/or external corrosion. In some plants, acids are also part of the utility system and used for neutralization of wastes or other caustic streams. In many utility systems, plugging from dirt, scale, or precipitants from treating chemicals represents a significant risk. Plugging of fire water systems can be very painful if discovered when a hydrant is needed to put out a fire. Similarly, plugging in the cooling water system to an exothermic reaction is usually painful to discover as the reaction begins to run away. Nitrogen and other inert gases pose hazards of their own, particularly if they are used to back up an instrument air system. Here, the hazard is asphyxiation of workers who open a valve in a line that they believe to contain air only to find nitrogen instead.

The equipment used in the plant confers a series of risks specific to its configuration. These risks include its susceptibility to corrosion, the risks associated with machinery used in the process, the risk of structural failure due to any overload, and the risks of failure of sealing devices that might release hazardous materials to unwanted places.

Even water brings a series of hazards and risks of its own to a number of process units. As mentioned earlier in this book, it was the introduction of water to a place where it did not belong that caused the Bhopal disaster. Aside from the reaction potential of water, its physical properties must also be considered. The introduction of water to a hot hydrocarbon tank in a refinery has been known to blow the roof off the tank and cause explosions. The freezing of water in a process line has caused the failure of many pipes with subsequent releases of hazardous materials.

Thus, the analysis of the residual risks in any process plant requires a considerable amount of information and a broad exploratory viewpoint. Typically, the information required includes the following:

- Process description, including materials, conditions, hazards, flow rates, and material balance
- A full description of utility systems, operating and design pressures, and temperatures
- A complete description of all equipment in the plant, including materials and design standards
- Complete process and instrument flow diagrams (P&IDs) including not only conventional instrumentation but also the computer control system description and alarm and trip settings
- Written operating and mechanical procedures
- A visual survey of the mechanical condition of equipment
- Design data on instruments and control valves
- Inspection records for the equipment
- Plot plans and electrical area diagrams
- Electrical one-line diagrams
- Meteorology data for the site
- Information on the fire protection systems at the site
- Design duties and controlling contingency for all relief devices
- Information on the safety relief and blow-down systems on the site
- Control and instrumentation philosophy
- History of safety and environmental incidents at the site
- Knowledge of the communication systems at the site
- An assessment of the external risks to which the site might be exposed such as earthquakes, falling aircraft, and so on

The above list is not all-inclusive for specific sites. There may well be other information that should be gathered before beginning to attempt to identify the remaining risks and the potential ways to reduce them. In some cases, a significant portion of the information described here will not be available formally.

In such cases, it is necessary to distill this information from the memories of plant people and verify the details. It would be wise to document the verified data for future use.

Several methods of identifying risks will be discussed in the following sections of this chapter. No single approach is universally better than any other. Indeed, several approaches are normally used at any one facility depending on the nature of the processes being examined and their stages of development. In all plants, learning from history is a valuable source of lessons that can be used to prevent future difficulties. Therefore, this is usually an ongoing process used in addition to any other specific risk identification program for either new or existing facilities.

Consistent with our definitions from Chapter 1, *hazards* are inherent properties of the process materials, equipment, or utilities *that remain* after the effort to make a process system inherently safer. Risks are the combination of the likelihood of an incident resulting from any remaining hazard coupled with the consequence of the incident. In this context, this chapter will explore ways to identify *risks* even though some of the techniques discussed are conventionally referred to as ways to identify *hazards* (e.g., HAZOPS). Risks are controlled by a combination of facilities and procedures in use in the plant. Hazards are either eliminated or remain to cause risks.

Incident Reporting and Analysis

The system for incident reporting and analysis should be designed to capture all incidents and to understand their causes to prevent recurrence. The objective is to obtain as much data as possible to understand and measure trends and concerns. Through this understanding, preventive measures can be developed or improved to minimize the likelihood of future incidents.

The scope of the incident reporting system should include all sorts of incidents, such as accident investigations, personnel injury reports, and unsafe conditions reported during safety surveys or casual observation. Some plants call the latter "near misses." All personnel in the enterprise should feel empowered and obligated to report their observations and the management must feel obligated to follow up.

Clear specifications for the data of interest are necessary. These are normally provided by *incident reporting forms.* The data are entered into the incident data base for analysis after appropriate review by management. Some incidents stand alone in indicating remedial action from a single incident. Others require integration with those of a similar nature before a root cause becomes clear and effective remedial action can be taken.

Several pitfalls need to be overcome if the benefits of incident analysis are to be had. The first is the temptation to curtail incident reporting out of embarrassment. Management can impact this by taking a positive view of reporting as a sign of awareness and an opportunity to learn how to run its business better. It can never, repeat never, yield to the temptation to criticize a group for reporting too many "near misses" or "potentially unsafe conditions or acts." Such an approach will kill potential learning before it starts.

The second pitfall is too much paperwork and too much management review. Getting the facts down needs to be a simple "fill in the blanks" one-page form. Management's review needs to be focused entirely on getting to the root cause in a timely way. The syntax of the words, any potential discipline or embarrassment, and the temptation to handle the report on a low-priority basis all must be put aside. The analysis and learning cannot begin until the report is completed and issued.

A third pitfall is concern for the legal ramifications of any observation. Lawyers tend to want to limit the distribution of any information that may be potentially incriminating even if it is not related to an actual injury, but rather only to a potential one. Flagellation is inappropriate in any enterprise. Therefore, I believe we should find a way to learn from any incident without inviting lawsuits or notices of violation or condemning ourselves to lose any litigation by our own paperwork. The standard approach to getting some protection in noninjury cases is to carry out the investigation under attorney–client privilege, limiting the distribution of the complete report as required by attorneys, but preparing a "safe," "lessons-learned" summary for wider distribution. While this approach is far from bulletproof, and everyone must listen to their own attorneys, experience says that a relatively small percentage of incident reports require even this level of security. Of course, serious accidents involving injuries and major community impact require the full treatment, but sanitized versions of both the Phillips and ARCO explosions in Texas were available not too long after the events.

The scope of the report will vary with the nature of the event. For all events, there must be a classification of the type of incident and the root and contributing causes to help with the analysis. For unsafe conditions or actions, where there were no injuries or breach of containment, a one-page report is usually sufficient. The events can be recorded on the front and the classification on the back. For more complex events, a short investigation may be appropriate, and for major events, a full, independent, multidisciplinary team will be required. In any case, the data-gathering and analysis must be timely lest the lessons get lost in the fog of fading memories.

Illustration 5.1 provides a sanitized example of a simple incident reporting form drawn from those actually in use at an Exxon plant. It is titled "Potentially Unsafe Occurrence Report" partially in deference to the lawyers, but also to broaden the scope of its use. It is intended to encompass preventive actions as well as those where an actual incident occurred, but *luckily* nothing bad resulted (i.e., the traditional "near miss").

The essential elements of the incident narrative and analysis are included, as are the names of the observer and perpetrator in case further follow-up is required. The cause of the potential incident (i.e., the lesson learned from it) was either ignorance or disregard by the contractor of plant safety regulations. The "recommendations" section was purposely left blank here to give the reader

Illustration 5.1 Example of Exxon Chemical Americas "Near Miss" report.

Potentially Unsafe Occurrence Report Report No. 01-91-XX
Descriptive Title: Nonconforming Work Practice on Tank Construction
Data Entered By: G. C. Doe
Date of Accident: 01/08/91 (Tuesday) *Time of Accident:* 09:40 A
Incident Potential: Injury
Business Group: Paramins *Unit:* Conservation
Employee Group: Contractor (Jones TK Fab) *Area On Unit: Separators*
Job Title: Other *Job Post:* Foreman
Operating Mode: Special Maintenance *Agency Involved:* Other
Permits Issued: Entry
Primary Cause Of Incident: Procedures Not Followed By Contractor
Root Cause(s) Of Incident:
Lack of knowledge of safety regulations on site (observation 1)
Inattention to obvious and generally accepted construction practice (observations 2,3)
Employee Narrative:

1) Contractors seen entering the bottom section of tank #6 (under construction) with no safety watch.

2) Several employees were seen without required personal protective equipment.

Observations/Conclusions:

1) No permits had been issued for confined space, cold work or hot work.
2) Scaffold brackets were being lifted over the head of a welder at grade and he was without hardhat.
3) Several contractors were not equipped with goggles, side shields, glasses, fall protection, and hardhats.
4) This contractor is clearly operating in violation of our safety procedures and sound construction practice.

Recommendations (w/respons. & timing):

Recommend review by: No further review necessary
Reviewed by: Sam Spade Date: 1/12/91
Manager

---

a chance to write down his or her own view of what to do. We shall return to the question of remedial action later.

Illustration 2 gives a typical accident investigation report drawn from a real event at the same plant. Again, it has been sanitized a little and the management summary section has been omitted to avoid duplication.

Illustration 5.2 Typical Accident Report

Narrative

On March 13, 1989, a West Side machinist working in the machine shop, following the completion of work on the 10TK mixer, attempted to move the mixer via the shop's hoist. This top-entry agitator was resting on a work table after having the motor recoupled. The mixer was rigged with slings and cables. The choke sling was placed around the mixer bearing farthest away from the motor and attached to a shackle. A cable and shackle were attached to the mixer motor's eyebolt. With this rigging arrangement, the machinist began to raise the mixer off the bench using the hoist. Moments after the support of the mixer had been transferred to cable, the eyebolt on the mixer motor (through which a support cable was connected) broke. The mixer and motor crashed to the floor striking a tool box and narrowly missing the machinist.

Failure Analysis

As part of the accident investigation, a failure analysis of the fractured motor eyebolt was performed. This analysis revealed that the eyebolt experienced a low-ductility failure as a result of an overload. The microstructure suggested that the eyebolt had been heat-treated for strength. This heat treatment, while increasing the mechanical strength, left the material extremely brittle. In addition, the loading on the eyebolt during the lift was in bending rather than straight tension. Essentially all of an eyebolt's load-bearing capabilities are in vertical loading. Eyebolts possess little strength in bending. For example, if the load-bearing capability in vertical tension is 2,500 lb., in bending the load-bearing capability may be as low as 25 lb.

Based on the analysis, we concluded that the combination of the eyebolt's brittleness and the bending load resulted in the failure. Because of the bending and the overloaded condition, the bolt may well have failed regardless of the toughness of the material. However, the brittleness of the bolt left no room for error. The full report documenting the failure analysis is attached.

Evaluation of Rigging Operation

Discussions with West Side riggers and a former rigging supervisor indicated that the mixer was improperly rigged. As described in the failure analysis section, eyebolts are designed for vertical loads. Loads imposed in other directions greatly reduce the load-bearing ability of the eyebolt. The arrangement recommended for safely rigging the mixer/motor assembly for a short-distance "shop" lift instead of relying on the eyebolt was defined. For longer distances, even more secure rigging procedures were recommended.

Conclusions/Recommendations

Our investigation revealed that this incident resulted from the use of an improper rigging arrangement. It is clear from discussion with the machinist involved in the incident that his use of this arrangement did *not* stem from negligence on his part, but rather from a lack of familiarity with rigging heavy, awkward equipment configurations.

As a by-product of the investigation, we found that the eyebolt attached to the motor possessed poor mechanical toughness. This eyebolt was supplied by the motor manufacturer and appears to be inferior to those routinely used by the plant.

Based on our investigation, we recommend that the following actions be taken to prevent another such incident:

1. Development of rigging procedures for awkward equipment configurations such as top-entry mixers. (F. G. Doe, mechanical section supervisor)
2. Reexamination of the rigging portion of mechanical training, including identification of the level of training required for the different mechanical disciplines. (F. G. Doe)
3. Refresher discussion in mechanical-gang safety meetings of the particular hazards associated with eyebolt lifting. (All mechanical first-line supervisors
4. Evaluation of eyebolts supplied with equipment. Motors are an example of such equipment. If eyebolts appear inferior to those purchased for routine use in the plant, they should be replaced. (All mechanics and mechanical FLS)

A more extensive investigation was undertaken on this incident because something serious had actually occurred and more extensive technical analysis was required to establish the root and contributing causes of the event so as to avoid similar problems in the future.

The lessons learned from this incident were:

- Machinists were likely to lack more than very rudimentary rigging skills.
- They may also not understand the technical aspects of using eyebolts properly.
- Not all eyebolts are created equal.

The recommendations made to help prevent future problems of this type included procedure development, training, and inspection. The requirements for procedures and training are fairly clear. Since the plant had a limited number of types of "awkward equipment," practical solutions to these problems could be developed in a timely fashion, and a mechanism existed (safety meetings) to quickly bring the procedures to the affected workforce.

The inspection recommendation left something to be desired, however. If it took a microstructure analysis to identify the brittleness of the eyebolt in question, how is a mechanic going to determine if a vendor-supplied eyebolt "appears inferior" to those supplied by the plant. Clearly, more is required in this area of prevention. A lesson was learned, but the remedial action proposed did not completely address the problem. An appropriate recommendation might have been to order all equipment without eyebolts so that only those brought to plant specifications would be used. That way, perhaps, the cost of the vendor eyebolt might be saved, and the hazard of potentially inadequate eyebolts would be eliminated.

These illustrations were chosen for two reasons. The first was to show that risks are associated with the equipment, utilities, procedures, and people we use to implement a process, not just with the process itself. The second reason is that the quality of the analysis in any incident investigation is critical to preventing future repititions. The latter applies even more strongly to analysis of process incidents since causes may be even more subtle.

Illustration 5.3 comes from an article on troubleshooting published in *Chemical Engineering Progress* (April 1991) by Gans, Kohn, and Palmea. It demonstrates that thorough process data and analysis are essential if the real cause of some difficulties is to be learned. Accident investigation specialists say that it takes five "why?"s to unearth the root cause of complex issues. They refer to "peeling the onion" to remove apparent causes until the real cause is exposed. Once this is done, often the solution is simple and inexpensive.

This illustration embodies a series of incidents. The first involved plugging in the tower and repeated safety valve releases. The cause of the safety valve releases was the plugging in the tower by polymer (one layer of the onion). The cause of the polymer was thought to be catalyst carryover (two layers). When this lesson was seemingly confirmed by laboratory tests, the theory that reducing the severity of process conditions would reduce polymerization was tried on the commercial scale. This turned out to be a costly mistake, both to the plant and to the environment (more safety valve releases).

Peeling another layer of the onion would have involved some laboratory exploration of whether the presence of catalyst was the *only* condition required

Illustration 5.3 A more complex example.

A first-of-its-kind process, one using a newly developed catalyst in a novel reaction system (Fig. 5.1), was commissioned to produce a well-known product. The product has a tendency to polymerize. Shortly after start-up, the vacuum refining column for the purification of the product showed a dangerous tendency to plug because of the polymerization of the product. The column pressure drop increased suddenly, the relief valves on the kettle reboiler blew repeatedly, and inspection showed that the lower trays were full of solids and buckled because of an excessive pressure drop. Laboratory analyses of the solids from the trays showed trace quantities of catalyst from the novel reaction system. Rough laboratory tests showed that the product would polymerize in the presence of this catalyst.

Removing the trace quantities of the catalyst prior to distillation threatened to be excessively expensive and it seemed obvious that a major process problem had developed. Attempts were made to run the critical refining column at a lower throughput, at a lower boil-up, and at a still-lower absolute pressure in order to reduce the column temperatures and to avoid polymerization. Attempts to lower the pressure, however, were not successful. A careful check of the performance characteristics of the vacuum jet system showed that it should have been capable of attaining a substantially lower absolute pressure.

The possibility of a leak into the tower was explored on two fronts: a pressure test of the tower system, and laboratory tests to determine if the presence of air contributed to the tendency to polymerize the product. The leak test revealed a substantial leak on the mechanical seal of the bottom circulation pump. The pump seals were changed and provided with nitrogen padding so that only nitrogen would be sucked into the column if the seal failed. The laboratory tests soon showed that air, as well as the trace catalyst, was required to cause polymerization of the product. The column never plugged again.

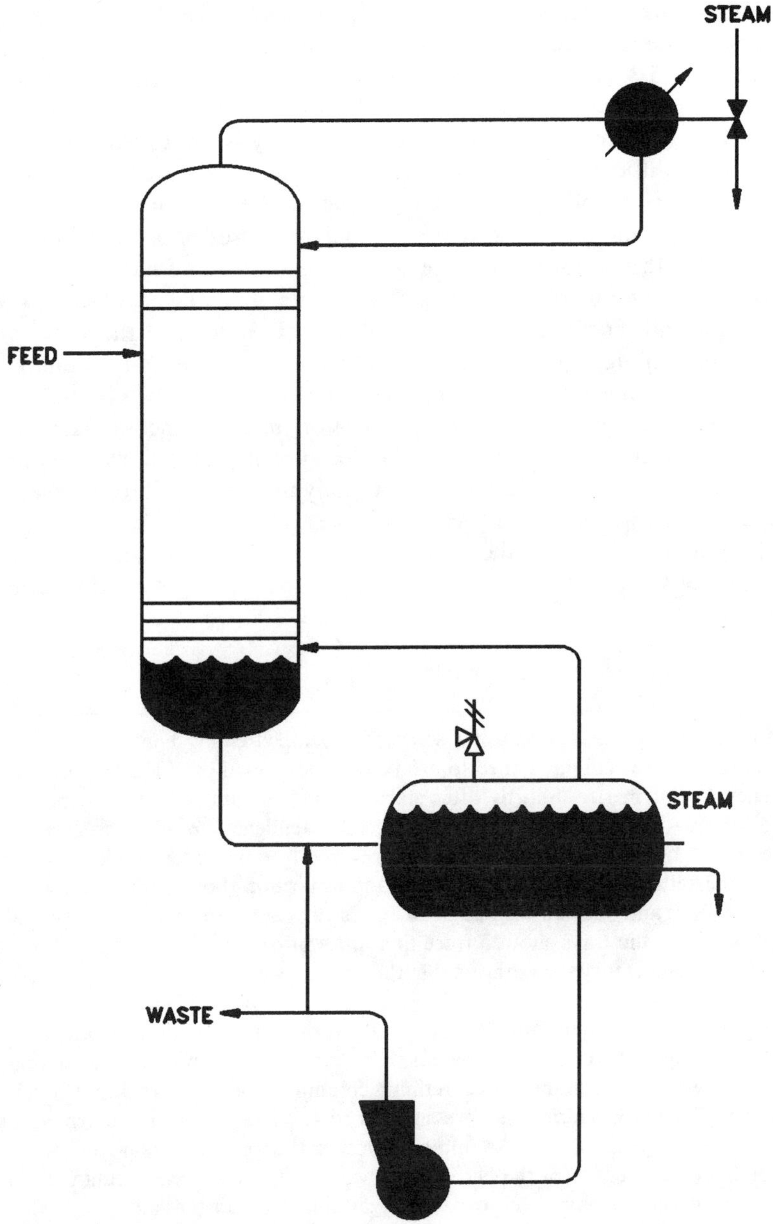

Figure 5.1. Novel reaction system of Illustration 5.3.

to result in polymerization *at lower temperatures.* Presumably, by peeling back another layer of the onion in the laboratory, it would have been found that air was also necessary for polymerization. The field investigation could then have gone directly to the leak test phase (four layers) with considerable savings. Once the leak was discovered, the solution was simple.

In a more general sense, there are other lessons to be learned from this example. Obviously, neither the plant nor the development folks understood the process well enough before start-up. Some contingency planning was needed to cover unknown areas. Also, the interaction of the mechanical and process systems was not appreciated. Finally, experience says that major problems are often solved by simple solutions once all the necessary information is available. Rushing to start up without all the process information is a risk. It needs to be assessed and, *if judged tolerable,* contingency plans need to be made beforehand.

Illustration 5.1 Revisited

Let us return to Illustration 5.1 to see what recommendations might be made to prevent reoccurence of this type of incident. It seems obvious that the contractor involved needs his or her attention to safety reinforced. If this company was a habitual offender, they might be thrown off the job and barred from future ones. If it was a first offense, enhanced training is clearly indicated for the group. So much for the obvious.

Why was no permit issued for the job? There are no data in the report to explain this, so some follow-up was needed. As it turned out, one group at the site told the contractor that the work site was in an "open" or "permit-free" zone because of its location within the site. What this person did not know (or at least tell the contractor) was that plant safety procedures required the desig- nation of an "open" area to be in writing and signed by management *and* that entry into a confined space without a permit required a waiver signed by the plant manager. Thus, one root cause of this incident was the ignorance by a plant person of plant safety procedures. Therefore, to deal with all aspects of this incident, refresher training of plant personnel about permitting rules is also required.

The overall lesson is that one cannot expect all incident reports—especially those for minor events—to be complete and to "peel the onion" thoroughly. As a result, those who would learn from incidents must be prepared to follow up to flesh out the details and get to the root cause(s). To get good coverage of incidents, all mechanics and operators need to be empowered to write up their observations. These folks are rarely experienced writers, so some reading between the lines and questioning should be expected if the full story is to be understood.

Another issue requiring some attention is obtaining the truth. Questions must be simple and easily understood. Answers need to be verified, if possible, from other sources for critical facts. I am not advocating training investigators as police detectives, but my observation is that the naive are often easily misled (deliberately or accidentally) by those with a personal position to protect.

Facilities and Procedures Analysis

While continuing analysis of current experience is an essential element in continuously improving the understanding of remaining risks, periodic formal analysis of facilities, control systems, procedures, transportation, and disposal methods is also required. By "required," I mean more than dictated by company policy. At this writing, three states—New Jersey, California, and Delaware—require it as a matter of law for certain chemicals. New regulations by OSHA will make such risk studies mandatory nationwide for yet another list of chemicals. In addition, the CMA Responsible Care Program declares, in less specific terms, that hazard analyses should be performed periodically.

These legal requirements are often quite specific about which techniques are to be used. In New Jersey, for example, "what if" analysis, hazard and operability studies (HAZOP), and failure mode and effects analysis (FMEA), are specifically identified as acceptable techniques. If the potential for a large release (five times the registration quantity for the chemical in question in one hour) is identified, then a quantitative risk assessment (e.g., fault tree) is required by the New Jersey law.

The first three techniques are qualitative in nature. They are based on questions asked by a multidisciplinary team and on an evaluation of the answers. The questions asked are somewhat different and documentation methods differ somewhat. "What if" questions generally take a less-structured approach, while HAZOP questions focus on certain key words, and the FMEA approach concentrates on all the possible failures for each component in the system.

The Definition of a "Single Contingency"

All of the techniques are generally based on a single failure, not multiple, unrelated failure scenarios. This is sometimes a little tricky to rationalize. For example, a power failure would be a single contingency of concern. However, if all cooling water pumps are driven by electric motors, then cooling water flow would also cease in a power failure, and both events would have to be considered a single failure.

Sometimes, time is a subtle variable in the definition of a single contingency. Consider the explosion that occurred in Russia when two trains passed through a gas pipeline leak simultaneously and set off a terrible explosion. My first reaction to that story was that it was a very unusual multiple contingency. I still believe that two trains passing the point of the leak was a double contingency. However, once there was a gas leak, it was *inevitable* that one train would pass by sooner or later *because the gas line ran alongside the tracks.* In my view, the possibility of a gas leak *and* a train passing had to be considered a single contingency connected by an uncertain time. To break the connection, it would have been necessary to stop all trains until the leak was found and fixed. In a plant situation, certain inevitable chains of events also exist. The most obvious is the limited availability of battery backup power for instruments, computers, and so on, in a power failure. The time when control is available must be spent putting the process into a safe condition, not hoping that the power will come back soon enough. Similarly, if a refrigeration system fails, it is often only a matter of time before safety valves in the system that had been cooled begin to release as it warms up. Thus, inevitable, but not necessarily coincident, events must be considered as part of a single failure.

With this background, let us consider the three types of qualitative risk identification techniques. Note that those can all be applied to either existing plant units or new designs.

"What if" analysis

The concept of this approach is perhaps the most intuitive of the three. It consists of a knowledgeable group asking, "What if this goes wrong?" based on their understanding of the process and equipment in place and how they interact with the plant complex as a whole. This examination can include buildings, power systems, raw materials, products, storage, material handling, in-plant environment, operating procedures, work practices, management practices, plant security, and so on. This is a powerful procedure if the staff is experienced. Otherwise, the results are likely to be incomplete.

The formulation of the exact questions is left up to those conducting the examination and thus depends very much on their skill. The questioning usually starts at the input to the process and follows the flow of the process. Alternatively, the questioning can center on a particular consequence category—for example, personnel safety or public safety. Usually a small group of two or three experts conducts the examination and reports the findings. The findings are usually accident event sequences that result from the "what if" questions. The questions essentially suggest an initiating event, and perhaps a failure from which an undesirable event sequence could occur. For example, a question might be, "What if the raw material is the wrong concentration?" The team would then attempt to determine how the process would respond. For example: "If the concentration of acid were doubled, the reaction could not be controlled and a rapid exotherm would result."

Thus, a potentially serious risk has been identified and some remedial action is needed. The tendency is to leap immediately to problem-solving, but this is often wasteful until all problems have been identified. Sometimes a single solution will remove several problems. Consequently, it is recommended that the group proceed through the risk identification procedure until all risks are identified before any evaluation or problem-solving begins.

Before beginning the study, the kinds of risks to be identified must be defined. Broadly, there are three types: employee risk, production or economic risk, and community impact. These broad categories usually need some further breakdown, particularly in the community area. Clearly, the team will be concerned with major explosions or fires that would cause physical damage to the neighbors. They will also be concerned with releases of chemicals that might impact the health of the community as a whole or some segment of it (e.g., a nursing-home population). They may also be concerned about odor or noise due to local circumstances or history. Similar breakdowns can be developed for the other broad categories of risk. These boundaries will serve to focus the team, but hopefully not narrow their vision so that an unusual specific risk is overlooked.

Other boundaries must also be considered. These involve interactions with other parts of the plant. These can cover a very wide range of possibilities and can test the team's expertise. Many of these interactions are propagated through the utility system. For example, a rapid dump of a large low-pressure steam load in one unit may upset a steam turbine in the boiler house or another process unit that supplies it, causing a "domino" effect. Or, a failure of a product cooler might cause hot product to be introduced into an off-site storage vessel or a feed drum to another unit that is not designed for the temperature, causing a safety valve release or worse. The team must evaluate all potential interactions as well.

Another set of boundaries involves the extent to which procedures and people are to be probed. In many cases, safeguards against risk involve maintenance or operating procedures and the knowledge possessed by the mechanic or operator. A frequent cause of accidents is shortcuts practiced in the field that are unknown to management. These risks can usually only be assessed by interviewing people in the field to see if they understand the critical procedures. The quesiton, "What if the safety relief device does not function when it is needed?" only defines a double contingency *if* the preventive maintenance program for safety relief devices is carried out as scheduled by capable people. Similarly, if responses by operators are critical to prevent a minor upset from becoming a major emergency, these will only occur in timely fashion if the operators know what to do *and* periodically practice the response. In many cases, this dimension must also be explored by the team to cover all of the risks remaining in a plant.

The final step in the risk identification procedure is to record the results. This amounts to defining a series of problems that should be solved. In some cases, the solutions might be obvious and be suggested by the team. In others, there may be several possible solutions that need to be studied. In any event, responsibilities and target dates need to be assigned for the development and implementation of the required remedial action plan.

Examples of "what if" analysis do not abound in the literature. The Center For Process Safety (AIChE) in its monograph *Guidelines for Hazard Evaluation Procedures* presents a simple illustration that focuses on a few process upsets and recommends solutions that involve both controls and procedures.

HAZOPs

Like "what if" analysis, HAZOPs are aimed at identifying risks in either an existing plant or a new design (or addition). The goals of a Hazop are:

- Identify the nature and scale of risks faced
  - Foresee the expected
    - Chemistry
    - Operating envelop vs. mechanical design
  - Foresee the unexpected
    - Upsets, contaminants, and so on
    - Unintended operation
- Identify the critical features of design and risk control systems needed
  - Instruments and procedures
- Provide definitive documentation for future use
  - By plant staff
  - For authorities
- Create a broader understanding of the risks of the plant

Clearly, these focus on risk identification/problem definition. Again, most solutions need to be developed and implemented in subsequent phases of activity.

HAZOPs operate on certain assumptions. The first is that risks are detectable by careful review. The second is that these risks are controllable by a combination of equipment and procedures that are then defined as "safety critical." Another is that the HAZOP exercise is conducted with openness and good faith by all parties. The last is that a plant that is designed, built, operated, and maintained to appropriate standards will not suffer catastrophic loss of containment if operated within design parameters. This last assumption recognizes that even the best designs can, over time, be compromised by poor maintenance and changes to the operation. Much more will be said about controlling these risks later in this book.

There are two types of HAZOPs described in the literature: The Lawley method and the Exxon Chemical approach. Both have the same objectives and are based on systematic analysis of the process and instrument diagrams (P&ID) for the unit being studied. Both suffer from the same general weaknesses; to wit, they:

- Create a tendency to overreact
- Create a tendency to add hardware
- Have validity only to the extent that construction, operation, and maintenance match design intent
- Can be invalidated by subsequent changes to facilities or procedures

The Lawley method uses specific guide words to help formulate questions about each element of the P&ID. These are listed in Table 5.1. All questions are asked at each point on the diagram and the likely consequences are recorded. The advantage of the Lawley method is its systematic formal approach to avoid overlooking anything. It is zero-based so that no assumptions are needed, but that means the capabilities of the team are critical to its success. The approach is very versatile. It can be applied to any process and even be performed early in the development before good P&IDs are available. However, it is very time

Table 5.1. Lawley Guide Words

Guide Word	Deviations
None	No forward flow when there should be—no flow or reverse flow
More of	More of any relevant physical property than there should be— higher flow (rate or total quantity), higher temperature, higher pressure, higher viscosity
Less of	Less of any relevant physical property than there should be— lower flow (rate or total quantity), lower temperature, lower pressure
Part of	Composition of system different from what it should be—change in ratio of components, component missing
More than	More components present in the system than there should be— extra phase present (vapor, solid), impurities (air, water, acids, corrosion products)
Other than	What else can happen apart from normal operation—start-up, shutdown, uprating, low rate running, alternative operation mode, failure of plant services, maintenance, catalyst change

consuming and requires large blocks of time from senior people, such as the plant manager, operations superintendent, plant engineer, process designers, instrument engineers, safety coordinator, research chemists, applications engineers, project engineer, and an independent chairperson.

The Exxon Chemical approach is not zero-based. It assumes that Exxon's design practices and mechanical standards represent accepted risks for the corporation. Thus, the HAZOP comes in two parts: (1) a check that any deviations from corporate standards have been thoroughly analyzed (and are appropriate for the local site) and (2) an imaginative look for unforeseen problems. The mechanism is similar to the Lawley method—that is, a systematic critical walk through of the P&ID to identify risks and the provision of appropriate documentation.

The advantages of the Exxon Chemical approach, which is only applicable if comprehensive process design and mechanical standards are available, come from faster, lower cost results. Since the team is not starting from zero, less time and manpower are required. Since corporate standards are well defined, less-experienced team members are required and a consistent basis for judgment is used.

The disadvantages are that good P&IDs are required to accomplish the task. The procedure cannot be used early in project development. In addition, batch processes are sometimes hard to relate to standards based on large continuous processes.

Presumably, any company with well-defined corporate standards could use the Exxon Chemical approach and its own standards. However, the use of general standards (e.g., American Society of Mechanical Engineers (ASME) code) may not be appropriate unless the company has evaluated the risk associated with the standard as being correct for the site being studied.

Exceptions are frequently granted from any company's standards in a specific case. The key element is that each exception has been thoroughly evaluated by competent, independent people, and the decision documented.

HAZOP Methodology

The methodology used in a HAZOP is simply a line-by-line progression through the flow plan, examining the consequences of all foreseeable independent failures. The interaction of the team members is vital. No one's input can be suppressed and no one can be allowed to dominate. A thorough understanding of the operating procedures is essential. In the Lawley method, all guide words are applied at each potential failure point in the process. In the Exxon approach, a meticulous comparison to standards and a selective use of the guide words is made. Special emphasis is placed on consequences which might result in loss of containment. A typical sequence is listed in Table 5.2.

The scope of the system to be studied will depend on its interaction with other systems at the site. Those that cause little impact on or are little impacted by a given unit maybe excluded. For example, if a unit uses only relatively nonvolatile materials far from their flash points, the firefighting system will probably be largely excluded from the study. On the other hand, if large volumes of LPG are present, the opposite will be true.

The ground rules and guidelines of the study may be identical, or may not, due to local conditions. The rules of baseball are the same for each league, but sometimes local conditions require additional definitions. Similarly, the corporate guidelines for carrying out a HAZOP may need some elaboration to meet local conditions. For example, local conditions may make it important to consider odors as an undesirable consequence, while this may not be important at other company sites.

The consequences of all forms of mal-operation need to be examined. This might include a single operator error, an instrument failure, utility failure, pump failure, valve failure, and so on. It is normally assumed that a physical failure of a containment component will not occur unless design conditions are exceeded.

As the HAZOP continues, concerns and questions are recorded on a worksheet or directly into a computer data base. Some investigations record all results, both of concern and not, others chose to record only those causing concern or question. Some include estimates of severity and probability at this point. A key element is an indication of acceptability of the risk and identifi-

Table 5.2. Implementation of a Hazard and Operability Study (HAZOPS)

1. Develop guidelines
2. Determine scope of study
3. Obtain necessary information
4. Prepare necessary ground rules
5. Organize schedule and participants
6. Perform study
7. Documentation of results
8. Action items/follow-up

cation of those items needing remediation or further study. A typical worksheet is shown in Table 5.3.

Documentation

Final documentation is prepared to summarize and prioritize the worksheet data. It usually starts with a narrative of the principal hazards of the process and then lists the risks of concern discovered in the HAZOP. These are often prioritized by the team. More will be said about doing this in the next chapter. A list of safety critical equipment and procedures is added to provide focus for the maintenance and operating organizations. In cases where the worksheets are computerized, the final report might be prepared directly on the machine and its issue greatly expedited.

A key element in the final report is an assessment of the probability and severity of any consequence. These are generally assessed qualitatively as high (H), medium (M), or low (L), and the resulting residual risk is characterized as acceptable (A), obviously unacceptable (U), or needing further study (C) (for concern). A somewhat more quantitative approach will be discussed in the next chapter as well.

The judgments assigned could well vary from location to location. Normally, one would characterize an event that occurred once a month as very high frequency and one that happened once in ten years as very low frequency. However, the scale may have to be shifted to include earthquakes, floods, or airplanes falling from the sky, so a "very low" category may be needed.

Similarly, the consequence scale may be affected by history or local regulation. Odors or noises may be of high consequence at one site and low at another. The release of even tiny amounts of certain substances may require government reporting, and, therefore, escalate in severity, even though no health effects are foreseen.

All such items should be covered in the guidelines prepared for the study.

The notes taken during the team discussion are a key element in prioritizing the risks identified developing remedial action plans, and for future reassessments of the unit. Often these are not captured (except cryptically) on the worksheet, so the scribe must make a diligent effort to capture the meat of each discussion in decipherable form.

The large amount of notes and data that accumulate in a HAZOP leads to the obvious application of computers to record keeping and report writing. Two papers presented at the AIChE meeting in San Diego, California, in August 1990 discussed available software for this purpose while others merely use less-specific commercial software such as Lotus 123. Gordon, Hessian, Greenberg, and Early of Stone and Webster *("Utilization of PC-Based Hazard and Operability Study Data")* discussed some of the values of their proprietary PC HAZOP program, and Maker, Morris, and Sloane of Westinghouse *(Tools to Enhance the Flexibility of Risk Assessment and Hazards Identification)* considered the characteristics of HAZTEK, HAZOP-PC, HAZSEC, and HAZOP-

Table 5.3. Hazop Worksheet

UNIT __________ SECTION __________ DWG # __________ DATE __________

ITEM	EQUIPMENT	CONCERN	F/Q	RISK	S	P	INDICATION/PROTECTION	A	COMMENTS/ACTION	RESP

F/Q - FACT OR QUERY
S - SEVERITY: LOW, MEDIUM, OR HIGH
P - PROBABILITY: LOW, MEDIUM, OR HIGH

A - ACCEPTABILITY:
A - Acceptable risk, no further action required.
U - Unacceptable for long term, must take corrective action
C - Remains a concern, requires action to establish if acceptable or not.

Do - Simple Items

TIMIZER. Properly applied, all of these programs have value in reducing HAZOP manpower and enhancing the quality and timeliness of the result. Some examples of specific benefits for a computer approach mentioned in these papers are:

1. Use of the program in pre-planning efforts by the facilitator saves time during the HAZOP study. As plant drawings are being segmented for the HAZOP discussion, records with the identifying information can be entered in advance, thus saving time during the HAZOP.
2. A customized input form displayed on the PC screen provides a template to record and organize the information during the HAZOP review.
3. After each HAZOP session, team members can have a printed sort of their assigned action items so that they can prepare responses for the next session.
4. After completion of the HAZOP study, the data base can be sorted variously to assist the HAZOP facilitator in development of the HAZOP Study Report.
5. Plant management can be kept informed of the HAZOP progress during the study in a report format suitable to their needs.
6. By linking the HAZOP data base, the reports required by internal and external groups can be quickly prepared.
7. After the HAZOP, the data base is a permanent document that can be easily updated as necessary.
8. Built-in examples can assist with training team members.
9. More complete and legible record keeping is possible.
10. It is possible to project the computer display to a larger screen to allow simultaneous team review, correction, and updating.

In general, the cost of software (specific or general) and the necessary familiarization investment by the team leader or scribe is an investment with a very attractive return if more than a single, simple HAZOP is planned.

Leadership Is Critical

The leader of a HAZOP, or other risk identification study, has a great deal of preparation work to do. He or she must assemble all the necessary information and verify its accuracy insofar as possible. The meetings must then be scheduled at an acceptable place and schedule conflicts must be eliminated. In some cases, training for team members may have to precede the actual study.

The first job for the leader is to select a team from those made available by the management. In most cases, this will consist of several full-time members and some part-time specialists. A typical list of candidates is given in Table 5.4, drawn from a paper entitled "Hazops as a First Step," presented by H. L. Phillippi of Westinghouse at the AIChE meeting in August 1990 (San Diego, CA). Team selection may be the first of many challenges for the leader, as we shall see. It is impossible to do a quality job without a quality team. If one is not made available, refuse to do the job until one is.

Table 5.4. Personnel

• Recorder (scribe)
• Technical representatives
• Materials/mechanical engineer
• Process/chemical engineer
• Chemist
• Instrumentation engineer
• Electrical engineer
• Operations
• Maintenance

The radical statement just made was for a purpose: HAZOP leaders need much more than technical skill to do the job well. McKelvey and Zerafa of Technica discussed HAZOP leadership at length at the AIChE meeting mentioned earlier in a paper entitled "Vital Hazop Leadership Skills and Techniques." Their definition of the desired characteristics of a HAZOP leader include:

Technical Skills and Knowledge

- Broad and in-depth, hands-on process experience and knowledge, including an appreciation of the reality and practicality of day-to-day plant operations accumulated through years of field experience with many types of processes
- Process hazards knowledge and experience, with a special awareness acquired through learning from bad experience of others
- HAZOP experience, a first-hand understanding of the capabilities and limitations of the HAZOP tools

Nontechnical Skills and Knowledge

- Leadership skills—an example of competence, concern, responsibility, and professionalism that will provide motivation for the team to follow and to keep the proceedings orderly
- Management skills—knowledge and ability in applying proven management techniques in planning, organizing, implementing, and controlling the HAZOP
- Interpersonal skills—an understanding of group dynamics and the skill to elicit a spirit of cooperation and teamwork
- Commitment to high ethical standards—seeking optimum results rather than merely adequate solutions that are in the end more costly

Assuming an appropriate leader and team have been assembled, the leader still has much to do. The first important step is to plan ahead for all of the details from obtaining and evaluating the accuracy of the necessary information to the details of the meeting schedule and amenities for each team member.

The second critical element is to ensure the HAZOP is a genuine team effort. *All* team members are important and must participate. In many cases, this needs to be reinforced periodically. In some cases, some training in HAZOP techniques or other team-building activity may be necessary at the first meeting to get started properly. Other specifics to consider are:

1. In the event of conflict between team members, consider:
   a. Leaving the controversial area until later in the study
   b. Changing the composition of the team
   c. Changing the team leader.
2. In many cases, team members will be senior to the team leader. This situation must not deter a team leader from carrying out the full procedure or adopting a modified procedure with which he does not agree. The leader must be strong, yet diplomatic.
3. Speak clearly, and never be afraid to say, "I do not understand the technical point you are making" or "That point does not relate to the area under discussion," and so on. This is sometimes difficult if you are junior to the team members.
4. Before leading a team for the first time, it is advisable to sit in as a team member on some other project. This will give an element of confidence and some guidance on how to phase questions.

Another critical element is to provide balance to the judgments being reached. Overkill is a great temptation, as is the tendency to rush through because there are "more important" things to do. The leader may have to play various roles to draw out the real importance of any item, such as taking the blame for not making the question clear or playing the "devil's advocate." Sometimes, it may be necessary to cut off long discussions by deciding that a particular issue will be resolved outside the meeting. Similarly, bringing side discussions to the attention of the full group will be essential from time to time.

Another challenge is to stop the team from redesigning the process. Team activities should be confined to identifying problems and defining what is needed for a successful solution. Focus on risk *identification.* Let solutions be developed as part of the follow-up activity.

The leader will frequently be under pressure not to study certain minor drawings on the basis that they are safe systems. The decision not to analyze what is said to be "minor" must be taken carefully. A minor system, although safe in itself, could cause risks in other systems.

This list only covers a few points of guidance that leaders may find useful. Some other helpful hints on the process of the HAZOP are included in Table 5.5.

The final challenge for the leader is to evaluate the effectiveness of the study. This is largely a matter for experienced judgment. If the team seems to be addressing the safety concerns noted by the leader without prompting, he or she can be reasonably sure that other, more subtle, risks are also being uncovered. In doubt, perhaps a very experienced HAZOP leader can be induced into sitting in on a team session to help the leader in this judgment.

Table 5.5. Helpful Hints for HAZOP Process

- Process overview for team members
- Review of reactive chemistry
  - + Exotherms
  - + Corrosion
  - + Contaminants
- From selected starting point on flow plan
  - + Track coverage with "hi-liter"
  - + Identify question/comment by number on flow plan
  - + Useful to segregate safety from operability questions
- Look carefully for sources of pressure, temperature, and contaminants
  - + How will deviation be detected?
  - + How will deviation be controlled?
  - + Is the team satisfied?
  - + Use checklist to assist
- Take time necessary to capture all concerns
  - + Resolve on spot if practical, or
  - + Defer for later resolution by team, or others
- Keep sessions short enough to avoid "burnout"

HAZOP Illustration

After such a long discussion, a short illustration is essential to clarify the process and some results. Figure 5.2A B shows a P&ID for a railcar unloading system. The product is sulfur dichloride. It is toxic, volatile, corrosive in air, and fed directly into a process reactor through Drum 6D-15, which provides a small inventory in the piping to signal when a railcar is empty. The motive force for the transfer is N_2 pressure. An excess flow valve (2) is provided in the product line to stop flow immediately at a set value higher than process requirements since this might be the result of a broken feed line. The HAZOP worksheet is shown in Table 5.6. The prioritization of any remaining concerns identified in this table will be discussed further in the next chapter.

Failure Mode and Effects Analysis (FMEA)

The Failure Mode and Effect Analysis (FMEA) is similar to a HAZOP and requires the same strong team input and leadership but uses a somewhat different approach. This methodology has been endorsed by both the Chemical Manufacturers Association (*Process Safety Management, Control of Acute Hazards,* May 1985) and the American Institute of Chemical Engineers (*Guidelines for Hazard Evaluation Procedures,* 1985). It is a thorough and systematic approach for evaluating the impact of a single equipment failure or human error on the entire system. A distinct advantage of the FMEA method is that no assumptions are made as to how the failure occurs, be it due to human error, an equipment mechanical malfunction, or a site-wide utility failure. The method assumes that if a failure mode can occur, it should be evaluated to ascertain the adequacy of the facilities to handle safely the consequences of the

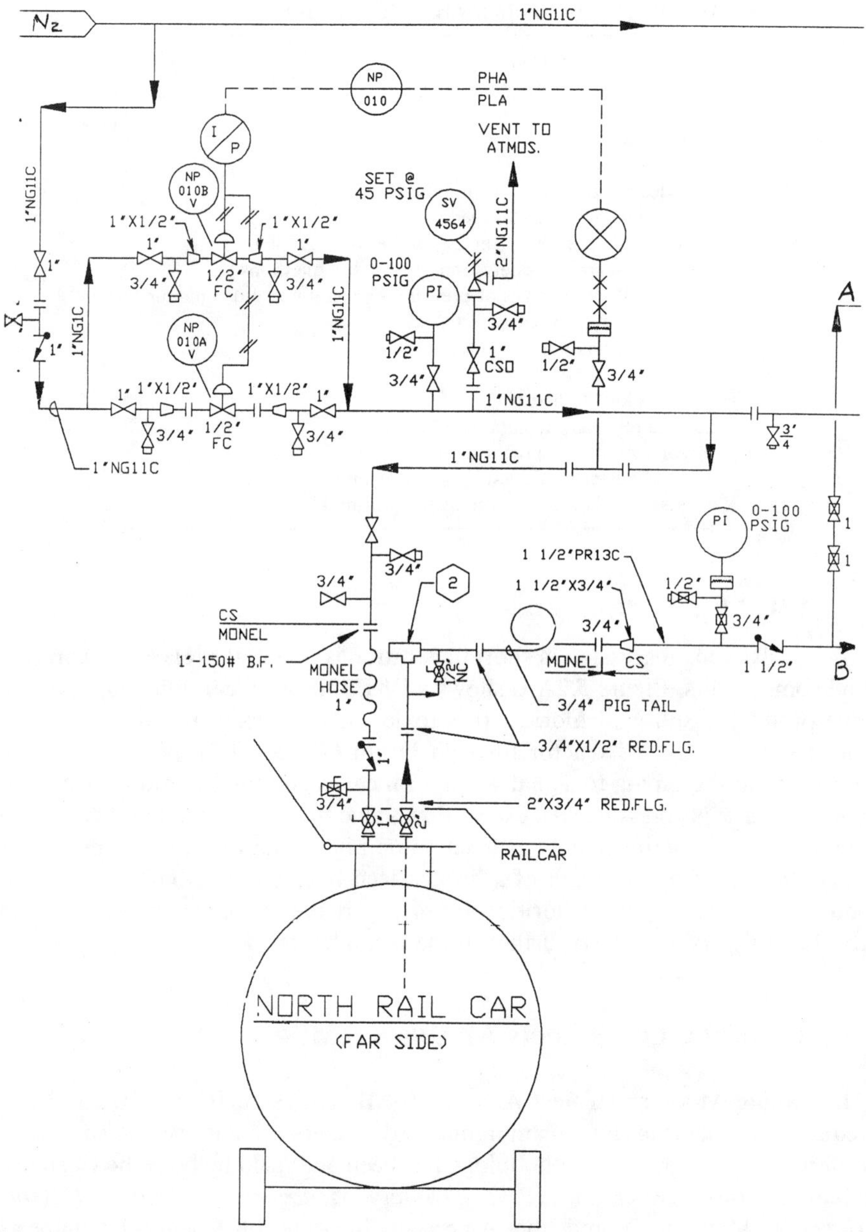

Figure 5.2. A railcar unloading piping and instrument diagram.

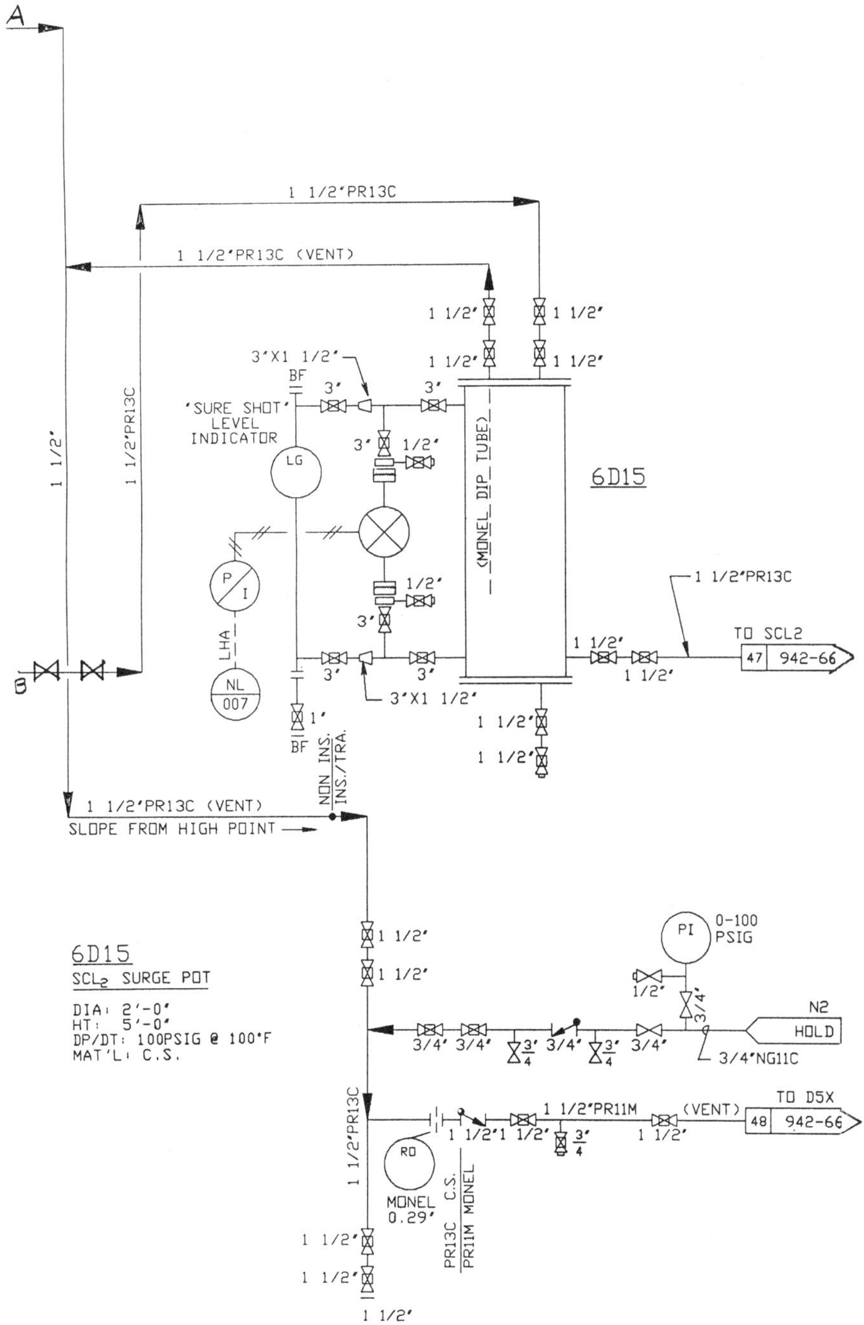

Figure 5.2 *(continued)*

Table 5.6. Hazop Worksheet

UNIT: NPS SECTION SCl2 R/C DWG #: DATE:

ITEM	EQUIPMENT	CONCERN	F/Q	HAZARD	S	P	INDICATION/PROTECTION
S1	SCl2 RAILCARS	BACKFLOW OF SCl2 INTO THE N2 SYSTEM	F	Cl2/SCl2 IN N2 SYSTEM	M	M	CHECK VALVE ON R/C'S CONNECTION HIGH PRESSURE ALARM
S2	N2 PIPING TO R/C	SUPERFLUOUS BLEEDER	F	LEAK			VALVES NORMALLY CLOSED
S3	SMALL PIPING CONNECTIONS	BREAKING OFF OF SMALL PIPE	F	LEAK	H	L	APPLICABLE SMALL PIPE STANDARDS FOLLOWED
S4	N2 HOSE TO R/C	RUPTURE OF HOSE	F	LEAK	H	L	HOSE IS DESIGNED FOR 100 PSIG @ 150 DEGREES F MONEL USED FOR CORROSION PROTECTION
S5	SCl2 RAILCAR	WATER/CONTAMINANTS INTO CAR	Q	REACTION, RELEASE TO ATMOSPHERE	H	L	PROCEDURE - KEEP HOSES DRY AND CLEAN
S6	N2 TO RAILCARS	OVERPRESSURE	F	LIFT SV RELEASE TO ATMOSPHERE	M	L	PRESSURE CONTROLLER ALARM & SV ON N2 SYSTEM
S7	EMERGENCY VENT	VENT TO SCRUBBER WITH INSUFFICIENT CAUSTIC	F	RELEASE TO ATM	M	L	SCRUBBER SOLUTION SAMPLED TWO TIMES/DAY. NaOH MONITORED. ALL RELEASES TO TOWER ARE SAFETY RELATED AND DETECTABLE.
S8	SCl2 LIQUID CONNECTION	LEAK ON LIQUID SIDE	F	LEAK	M	L	PRESSURE TEST ALL FITTINGS W/N2 PRIOR TO OPENING VALVE CHECK GASKETS ON ALL FLANGES
S9	SCl2 PIGTAIL	RUPTURE	F	LEAK	H	L	MONEL SELECTED. FULL STRENGTH TUBING USED EXCESS FLOW CHECK VALVE W/LIMIT RELEASE PROCEDURES

F/Q - FACT OR QUERY
S - SEVERITY: LOW, MEDIUM, OR HIGH
P - PROBABILITY: LOW, MEDIUM, OR HIGH

A - ACCEPTABILITY
DO - SIMPLE ITEMS
U - UNACCEPTABLE FOR LONG TERM, MUST TAKE CORRECTIVE ACTION
C - REMAINS A CONCERN, REQUIRES ACTION TO ESTABLISH IF ACCEPTABLE OR NOT
A - ACCEPTABLE RISK, NO FURTHER ACTIONS REQUIRED. MEETS TCPA CRITERIA.

(Continued)

failure. The results of the evaluation are reported in a standardized manner to facilitate both a complete review and documentation. The reporting format to be used for this study is the Typical FMEA Summary Sheet (Table 5.7).

The FMEA is carried out by conducting a thorough review of the process flow diagrams, including P&IDs (Piping and Instrument Diagrams) of the existing equipment. For an individual piece of equipment, each pertinent failure mode will be considered. An evaluation will be made of how the failure is to

Table 5.6. *(Continued)*

UNIT: NPS SECTION SCl2 R/C DWG #: DATE:

ITEM	EQUIPMENT	CONCERN	F/Q	HAZARD	S	P	INDICATION/PROTECTION
S10	SCl2 R/C'S	FIRE UNDER CAR	F	RELEASE	H	L	VIDEO MONITORING OF CARS LEAK DETECTORS OPERATOR ROUNDS
S11	SCl2 R/C'S	LEAK	F	LEAK	H	L	EXCESS FLOW CHECK VALVE ON LIQUID CHECK VALVE ON VAPOR SPACE CONCRETE PAD TO CATCH DRIPS SPILL-X FOR CLEAN-UP
S12	SCl, R/C	OPERATOR TRAPPED ON PLATFORM DURING EMERGENCY	Q	EXPOSURE	H	L	SECONDARY ESCAPE VIA R/C LADDERS ESCAPE PACK ON PLATFORM
S13	SCl2 LIQUID LINE	PLUGGED LINE	F	POSSIBLE LEAK WHILE CLEARING	M	L	6D-15 AVAILABLE AS BLOWDOWN VESSEL TO CLEAR LINE
S14	SCl2 6D-15 LEVEL IND.	SINGLE BLOCKS TO ATMOSPHERE ON CALIBRATION VALVES AND DRAIN	F	LEAK	M	M	NORMALLY CLOSED
S15	SCl, R/C	RECYCLE OF SCl2 FROM CAR TO EMPTY ONE	F	LEAK, UNEXPECTED PRESENCE OF PRODUCT	L	L	PROCEDURES DOUBLE BLOCK VALVES
S16	SCl2 VENT LINE	BOTH VENT & FEED VALVES OPEN	F	D5X OPERABILITY	L	L	PROCEDURES P5-X INSTRUMENTS EXCESS FLOW VALVE
S17	SCl VENT	LIQUID TRAPPED IN LINE	F	UNEXPECTED PRESENCE	M	M	PROCEDURES
S18	SCl2 LIQUID LINE	TRAPPED LIQUID THERMAL EXPANSION	F	LEAKS	H	M	HEAD SPACE IN 6D-15 FOR EXPANSION
S19	SCl2 VENT AT PDX	THERMAL EXPANSION OF NPS IN PIPE	F	LEAKS	M	M	PROCEDURES

F/Q - FACT OR QUERY
S - SEVERITY: LOW, MEDIUM, OR HIGH
P - PROBABILITY: LOW, MEDIUM, OR HIGH

A - ACCEPTABILITY
DO - SIMPLE ITEMS
U - UNACCEPTABLE FOR LONG TERM, MUST TAKE CORRECTIVE ACTION
C - REMAINS A CONCERN, REQUIRES ACTION TO ESTABLISH IF ACCEPTABLE OR NOT
A - ACCEPTABLE RISK, NO FURTHER ACTIONS REQUIRED. MEETS TCPA CRITERIA.

be detected (alarms, instrument signals, operator observation of the unit, etc.), the effect on other equipment, and the response of the pertinent safety systems to the failure. Each failure will be evaluated for its potential to cause a release with potential threat of injury to workers and to public health or welfare. Failures that could result in such a release are subject to further study for possible remedial action.

For those failures that affect other equipment, an additional assessment for

Table 5.7. Typical FMEA Summary Sheet

SYSTEM:	PID NO.:	BY:		DATE:		PAGE OF	
				EQUIPMENT AFFECTED			
NO.	COMPONENT IDENTIFICATION	FAILURE MODE	DETECTION METHOD	EQUIPMENT IDENTIFICATION	EFFECTS	SAFETY SYSTEMS RESPONSE	COMMENTS

the affected equipment is made of the detection method, safety system response, and potential for an environmental release.

Table 5.8 is a representative listing of equipment and its failure modes for the expected types of equipment and failure modes in many process plants. Those failure modes that are pertinent to the actual equipment will be evaluated.

Table 5.8. A Failure Mode Checklist for FMEA Analysis

I. *Exchangers*
   A. Tube leak from high- to low-pressure side—E
   B. Tube side-flow restriction—E, V
   C. Shell side-flow restriction—E, V

II. *Pumps/Compressors*
   A. Stops/stopped while running—E, H
   B. Fails/failure to start—E, H
   C. Fails/failure to stop—E, H
   D. Low flow—E, H, V
   E. High pressure—E, H, V
   F. Low pressure—E, H, V

III. *Manual Valves*
   A. Valve left open—H, E
   B. Valve left closed—H, E

IV. *Reactor/Agitators*
   A. Agitator stops while operating—E, H
   B. Agitator fails to start—E, H
   C. Agitator fails to stop—E, H
   D. Addition of components in wrong order—H
   E. Leak of jack or coil—E
   F. Insufficient heating—E, V
   G. Insufficient cooling—E, V

V. *Instruments/Control Valves*
   A. Insufficiently opened—E, H
   B. Insufficiently closed—E, H
   C. Fails/failure to open—E, H
   D. Fails/failure to close—E, H
   E. Gives wrong reading, high or low—E

VI. *Drums*
   A. Failed separator internals—E

VII. *Towers*
   A. Internal damage—E

VIII. *Tanks*
   A. Wrong material unloaded/pumped to tank—H
   B. Overfilling of tank—E, H, V
   C. Bottom leak—E
   D. Insufficient heating/cooling
   E. Coil leak—E

IX. *Human Factors*
   A. Bypassing of critical instruments/alarms
   B. Improper temporary hose connections

E = Equipment failure. H = Human error.
V = Failure can also be due to faulty instrument valve or wrong manual valve position. To be evaluated when considering specific instrument or valve.

Other equipment types might have the same or additional possible failure modes.

By virtue of the use of FMEA, it will be possible to evaluate thoroughly the consequence of an equipment failure on:

- The safe operation of upstream and downstream equipment
- The capability of the equipment to handle excursions
- The adequacy of safety facilities
- The adequacy of the equipment to contain upsets and prevent potential releases

Documentation for the FMEA is similar to that for a HAZOP. Normally, all reactions to all failures are documented for an FMEA. This has the advantage of fully characterizing system response, but makes for a very bulky report that needs a separate interpretation summary.

Some teams will try to add a more quantitative assessment of any risks discovered to the report. If this is to be attempted, a qualitative cut must be made first because of the bulk of the data.

FMEA Illustration

An example of the use of the FMEA method follows. Figure 5.3 is a diagram of a hypothetical operation and Table 5.9 gives the results of the analysis. The operation to be analyzed consists of a continuous stirred reactor being fed components *A* and *B* via two flow recorder controllers (FRC-1 and FRC-2) and producing product *C*. Product *C* normally contains a low level of residual raw material *A* and zero concentration of *B*. The level in the reactor is controlled by a level recorder controller (LRC-1) on the discharge of the pump (P-1). The pump also provides the circulation required to pass the reaction liquid through the heat exchanger (E-1) in order to remove the heat of reaction and maintain the reaction temperature. Also provided are a safety valve (SV-1) to vent the reactor to the flare system, level alarms (LHA-1, LLA-1), an agitator (MA-1), a dial thermometer (TI(d)-1), and a temperature high alarm (THA-1). Also present are four manually operated valves (V1, V2, V3, V4).

For the purpose of this particular example, we will consider that a high concentration of chemical *B* in the equipment downstream of LRCV-1 would be a concern.

The failure modes evaluated for each piece of equipment are tabulated in Table 5.9. This example shows that item Nos. 4, 5, 6, 8, 10, and 12 could potentially result in a significant environmental concern. These items are subject to further study for possible remedial action, based on the qualitative judgment of the team.

Note that the events leading the potential releases identified involve both operating procedures and equipment malfunctions. The temptation could be to add additional hardware safeguards as the immediate solution to any problems, just as in a HAZOP analysis. However, the addition of hardware introduces another risk in that too many safeguards may dilute operator attentiveness and

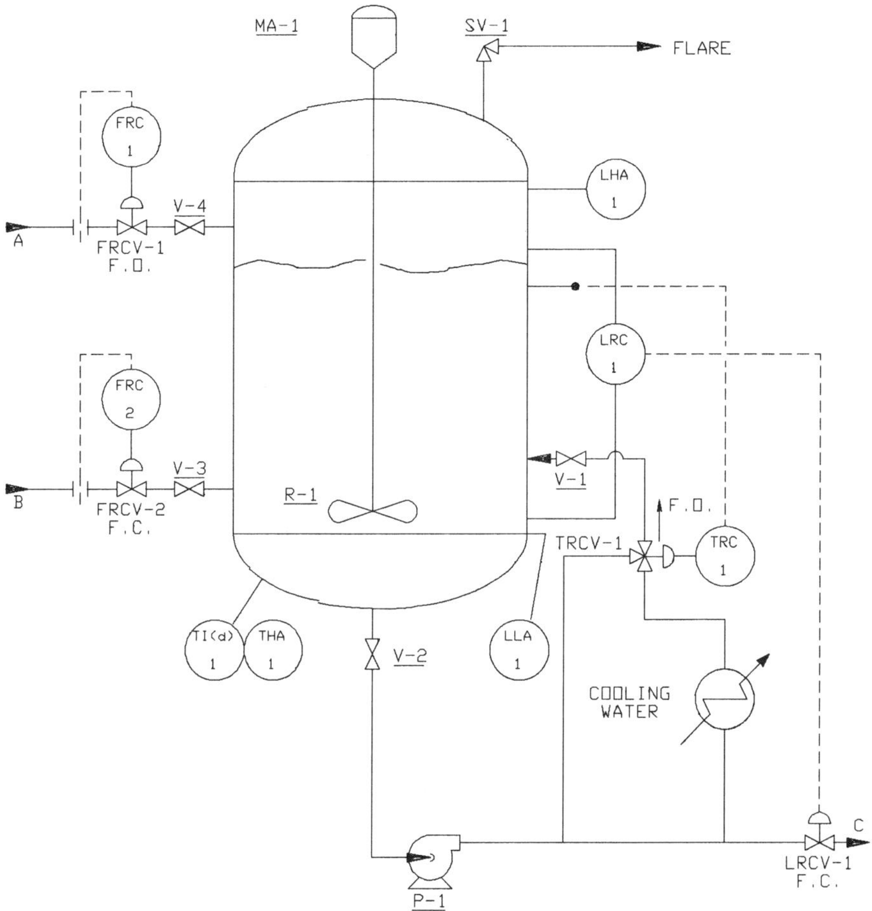

Figure 5.3. FMEA example.

also exceed the capacity of the preventive maintenance staff, which must ensure that it performs its task when needed. Therefore, the table presents these risks as items for further study only, not with first-pass suggestions for solving the problem. This is doubly important if the study is to be presented or discussed with a government agency. Experience has shown that these first-pass solutions become a standard and any improvements to the solutions have to be *justified* to the agency.

Summary

In this chapter, several ways of identifying the risk remaining in a process plant were discussed. Regardless of the names attached to these methods, they are all aimed at uncovering the risks associated with operating a given facility. As

Table 5.9. FMEA Analysis

Unit: XYZ Chemical: B

NO.	EQUIPMENT ID AND DESCRIPTION	FAILURE/ERROR MODE	OTHER EQUIPMENT POTENTIALLY AFFECTED: EQUIPMENT ID	OTHER EQUIPMENT POTENTIALLY AFFECTED: EFFECT	DETECTION METHOD	SAFETY SYSTEM RESPONSE	REMARKS
1	R-1 REACTOR	AGITATOR STOPS	NONE	--	MOTOR AMPERAGE INDICATION	--	AGITATOR NOT CRITICAL FOR CONTROLLING REACTION. ADEQUATE AGITATION PROVIDED BY PUMPAROUND LOOP.
2	"	AGITATOR FAILS TO START	"		"	--	SEE #1.
3	"	AGITATOR FAILS TO STOP	"		"	--	NO IMPACT.
4	"	ADDITION OF COMPONENTS IN WRONG PROPORTION	"	--	FRC 1/2 LABORATORY ANALYSIS OF PRODUCT C	--	SEE #5.
5	FRC-1 FLOW SIGNAL FOR A	INSUFFICIENTLY OPENED OR CLOSED	DOWNSTREAM OF LRCV-1	CONTAINS HIGHER THAN NORMAL CONCENTRATION OF B	FRCV-1, LAB ANALYSES OF C	--	TOO LOW A FLOW OF A FOR A SIGNIFICANT PERIOD OF TIME COULD RESULT IN A HIGH CONCENTRATION OF B IN DOWNSTREAM EQUIP-MENT AND COULD RESULT IN A SIGNIFICANT ENVIRONMENTAL RELEASE.
6	FRCV-1 FLOW CONTROL VALVE FOR A	FAILS TO OPEN	"	CONTAINS HIGHER THAN NORMAL CONCENTRATION OF B	TRC-1 FRC-1	--	STARTUP OPERATION REQUIRES FLOW OF A BEFORE INTRODUCING B, BUT TOO LOW FLOW OF A COULD RESULT IN SAME EFFECT AS #5.
7	"	FAILS TO CLOSE	"	CONTAINS HIGHER THAN NORMAL CONCENTRATION OF A	FRCV-1 LRC-1	--	OPERATOR WOULD MANUALLY SHUT V-4 AS PART OF S/D PROCEDURE. HIGH CONCENTRATION OF NON-TOXIC MATERIAL A PRESENTS NO HAZARD.
8	FRC-2 FLOW CONTROLLER FOR B	INSUFFICIENTLY OPENED OR CLOSED	DOWNSTREAM OF LRCV-1	CONTAINS HIGHER THAN NORMAL CONCENTRATION OF A OR B	FRCV-2 LABORATORY ANALYSIS OF PRODUCT C	--	TOO HIGH A FLOW OF B FOR A SIGNIFICANT PERIOD OF TIME COULD RESULT IN A HIGH CONCENTRATION OF B IN DOWNSTREAM EQUIP-MENT AND COULD RESULT IN A SIGNIFICANT ENVIRONMENTAL RELEASE.

(Continued)

Table 5.9. *(Continued)*

Unit: XYZ Chemical: B

NO.	EQUIPMENT ID AND DESCRIPTION	FAILURE/ERROR MODE	OTHER EQUIPMENT POTENTIALLY AFFECTED: EQUIPMENT ID	OTHER EQUIPMENT POTENTIALLY AFFECTED: EFFECT	DETECTION METHOD	SAFETY SYSTEM RESPONSE	REMARKS
9	FRCV-2 FLOW CONTROL VALVE FOR B	FAILS TO OPEN	"	CONTAINS HIGHER THAN NORMAL CONCENTRATION OF A	TRC-L LABORATORY ANALYSES OF C	--	DOWNSTREAM EQUIPMENT NOT IMPACTED BY HIGH CONCENTRATION OF NON-TOXIC MATERIAL A.
10	"	FAILS TO CLOSE	"	CONTAINS HIGHER THAN NORMAL CONCENTRATION OF B	FRCV-2	--	SHUTDOWN PROCEDURE REQUIRES STOPPING FLOW OF B FIRST. OPERATOR CAN CLOSE V-3. A SIGNIFICANT ENVIRONMENTAL RELEASE COULD RESULT IF THIS IS NOT DONE.
11	V-4 BLOCK VALVE ON FEED A	VALVE OPEN	NONE	--	OPERATOR INSPECTION FRC-1	--	FRCV-1 WOULD BE CLOSED.
12	"	VALVE CLOSED	DOWNSTREAM OF LRCV-1	CONTAINS HIGHER THAN NORMAL AMOUNT OF B	FRC-1, LOW REACTION TEMPERATURE	--	OPERATOR OPENS V-4 OR SHUTS DOWN OPERATION. HIGH CONCENTRATION OF B IN DOWNSTREAM EQUIPMENT COULD RESULT IN A SIGNIFICANT ENVIRONMENTAL RELEASE.
13	V-3 BLOCK VALVE ON FEED B	VALVE OPEN	NONE	--	OPERATOR INSPECTION, FRC-2	--	FRCV-2 WOULD BE CLOSED.
14	V-3 BLOCK VALVE ON FEED B	VALVE CLOSED	NONE	--	FRC-2, LOW REACTION TEMPERATURE	--	IMPACTS ON PRODUCT QUALITY ONLY.
15	V-2 R-1 OUTLET	VALVE CLOSED	P-1 E-1	NO FLOW	LRC-1	LHA-1	--
16	"	VALVE OPEN	NONE	--	OPERATOR INSPECTION	LLA-1	LRCV-1 CLOSED ISOLATING R-1 FROM EQUIPMENT DOWNSTREAM OF LRCV-1.
17	P-1	STOPS	R-1	HIGH TEMPERATURE	TRC-1	THA	STOPPING FEEDS A & B IS ADEQUATE TO CONTROL THE REACTION TEMPERATURE.
18	"	"	HIGH LEVEL	LRC-1	LRC-1	LHA-1	STOPPING FEEDS WILL CONTROL THE LEVEL.
19	P-1	FAILURE TO START	R-1	HIGH TEMPERATURE	TRC-1	THA	STOP S/U OPERATION.

(Continued)

Table 5.9. *(Continued)*

Unit: XYZ Chemical: B

NO.	EQUIPMENT ID AND DESCRIPTION	FAILURE/ERROR MODE	OTHER EQUIPMENT POTENTIALLY AFFECTED: EQUIPMENT ID	OTHER EQUIPMENT POTENTIALLY AFFECTED: EFFECT	DETECTION METHOD	SAFETY SYSTEM RESPONSE	REMARKS
20	"	"	R-1	HIGH LEVEL	LRC-1	LHA-1	"
21	"	FAILURE TO STOP	NONE	--	OPERATOR INSPECTION	--	NO IMPACT.
22	"	LOW FLOW	R-1	HIGH LEVEL	LRC-1	LHA-1	S/D FEEDS IF NECESSARY.
23	"	"	"	HIGH TEMPERATURE	TRC-1	THA	"
24	E-1	TUBE LEAK	R-1	PRODUCT QUALITY	SAMPLE APPEARANCES LABORATORY ANALYSES	--	LEAK WOULD BE FROM WATER (HIGH PRESSURE) TO PROCESS (LOW PRESSURE SIDE).
25	E-1	LOSS OF COOLING WATER UTILITY FAILURE	R-1	HIGH TEMPERATURE	TRC-1	THA	S/D FEEDS, STOP REACTION.
26	TRC-1 - R-1 TEMPERATURE CONTROLLER	INSUFFICIENTLY OPENED/ CLOSED	R-1	HIGH/LOW TEMPERATURE	TRC-1. LABORATORY ANALYSES TI(D)-1	THA	WILL AFFECT PRODUCT QUALITY ONLY. OPERATE IN MANUAL USING TI(D)-1.
27	TRCV-1 - R-1 TEMPERATURE CONTROL VALVE	FAILS TO OPEN (BYPASS OPEN)	R-1	HIGH TEMPERATURE	TRC-1. TI(D)-1	THA	WILL AFFECT PRODUCT QUALITY ONLY.
28	TRCV-1 - R-1 TEMPERATURE CONTROL VALVE	FAILS TO CLOSE (BYPASS CLOSED)	R-1	LOW TEMPERATURE	RTC-1. TI(D)-1	--	WILL AFFECT PRODUCT QUALITY ONLY.
29	LRC-1 R-1 LEVEL CONTROLLER	INSUFFICIENTLY CLOSED DOWNSTREAM OF LRCV-1	R-1	LOW LEVEL	LRC-1	LLA	OPERATE ON MANUAL OR S/D.
30	"	INSUFFICIENTLY OPEN	"	HIGH LEVEL	FRCV-1	LHA	S/D.
31	LRC	ERRONEOUS READING	"	HIGH OR LOW LEVEL	LRC-1	LLA, LHA	S/D.
32	LRCV-1 - R-1 LEVEL CONTROL VALVE	FAILS TO OPEN	R-1	HIGH LEVEL	LRC-1	LHA	STOPS S/U.

(Continued)

Table 5.9. *(Continued)*

Unit: XYZ Chemical: B

NO.	EQUIPMENT ID AND DESCRIPTION	FAILURE/ERROR MODE	OTHER EQUIPMENT POTENTIALLY AFFECTED: EQUIPMENT ID	OTHER EQUIPMENT POTENTIALLY AFFECTED: EFFECT	DETECTION METHOD	SAFETY SYSTEM RESPONSE	REMARKS
33	"	FAILS TO CLOSE	NONE	--	LRC-1	LLA	FOR S/D CLOSE V-1. FEEDS HAVE ALREADY BEEN STOPPED.
34	V-1	VALVE OPEN	--	--	OPERATOR INSPECTION	--	DURING S/D LRCV-1 WOULD ISOLATE THE UNIT FROM DOWNSTREAM FACILITIES.
35	"	VALVE CLOSED	R-1	HIGH TEMPERATURE	TRC-1	THA	STOP FEEDS.
36	INSTRUMENT AIR	LOSS OF SUPPLY PRESSURE	ALL CONTROL VALVES	CONTROL VALVES TO FAILURE POSITION	FRC-1, 2 LRC-1	LHA-1	STOP THE FLOW OF A AND SHUT DOWN.

a reminder, risks are present because hazards in the process, in the materials and utilities used, and in the limitations of equipment design have not been eliminated.

The first identification process discussed was fundamentally an attempt to learn from the history of the site by analyzing what has previously gone wrong, or almost gone wrong. This is the classic analysis of any incident or near-incident, and should be an ongoing process at any site. Indeed, under the requirements of the New Jersey Toxic Catastrophe Prevention Act, other laws, and recent OSHA regulations, "accident" investigation is required.

More formal studies of existing or new facilities and any proposed changes to existing facilities are also required. Some form of "hazard analysis" (we are calling these risk identifications) is required under the CMA Responsible Care Program and under the OSHA regulations about process safety. The three types of formal analyses discussed in this chapter are specifically mentioned in most of these legal requirements. They all depend on a knowledgeable team of people asking intelligent questions about how any process is designed and intended to be operated. The composition of the team and the ability of the leader are critical in the effectiveness of any of these studies. Their intent is to find risks remaining in the operation of the process and not to solve all of those risks during the study. The three methods are labeled "what if" analysis, HAZOP, and FMEA (Failure Modes and Effects Analysis).

All of these methods are primarily qualitative in nature. However, risk assessment is usually undertaken once all of the risks have been identified. This usually takes the form of a simple priority list, or perhaps a qualitative assessment of whether the severity and probability of an incident are high, medium, or low. Attempts to provide more quantitative risk assessment will be discussed in the next chapter.

A follow-up system is essential if the risk identification process is to bear fruit. The follow-up system must include an assessment or prioritization of the risks identified, a remedial action plan, and a mechanism for implementation of the remedial actions decided on. Remedial action may include changes to materials, facilities, procedures, or personnel.

The several facets of the follow-up program will be discussed separately in following chapters.

CHAPTER

6

Risk Assessment

Introduction

Assessing the risks identified by any approach is a complicated task. This is because risk (the product of severity times probability or frequency), while sometimes represented by a number, has at least as many social aspects as scientific. While scientific sources have most of what data exist, there is still much missing, especially about the health impacts of chemicals. This necessitates making informed judgments in the risk assessment process. Since the impact of the same data on various communities can vary, the uncertainty of any risk assessment multiplies.

The CMA and the American Chemical Society (ACS, 1155 16th St., NW, Washington, D.C.) have prepared excellent reports addressing the "perception" aspects of risk assessments. In the following five paragraphs, I have drawn heavily from the ACS report, *Chemical Risk: A Primer,* in an effort to address the public perception issues succinctly.

Why is it that certain people will accept the risk of developing cancer by smoking cigaretes but will protest construction of a nuclear power plant near their home? How can we explain the fact that individuals will avoid airplanes because of the risk of a crash, but will drive an automobile, despite overwhelming evidence that motor vehicle accidents are a more frequent (and kill far more people) than plane crashes? Are these decisions irrational? Obviously, statistical measures of death or injury are not necessarily the crucial factor. Instead, it is the "perceived" risk that influences, not only the public's risk decisions, but also the public's acceptance of risk decisions made by the government (and industry). Nine criteria have been identified as influencing public perceptions of risk. These are listed in Table 6.1. The characteristics listed in the center column have been found to contribute to perceptions of lower risk, while those on the right are factors contributing to perceptions of higher risk.

Several general observations about perceptions of risk have been made. People tend to judge exposure to involuntary activities and/or technologies as riskier than voluntary ones. The obvious reason for this perception is that a voluntary risk can be avoided whereas an involuntary one cannot. The amount of pesticide residues in food or the concentration of contaminants in drinking water is an involuntary decision for the public. Therefore, because the decision is "out of its hands," the public must turn to the government to regulate these activities and technologies and is less tolerant.

In general, the man-made factor is viewed as riskier than the natural, even though the risks may be identical. For example, aflatoxin, one of the most potent carcinogens known to man, is produced naturally by molds on peanuts and other foods. Yet studies have shown that many people would accept a cancer risk of one in 10,000 to one in 100,000 from using peanut butter which may contain aflatoxin, but would reject using a synthetic material with a cancer risk factor of one in 1,000,000. This phenomenon explains, in part, the general aversion the public has toward chemicals—viewing them all as man-made and hazardous.

Catastrophic events are perceived as riskier than ordinary events. This explains the perception that flying in a plane is riskier than driving in an automobile. In addition, delayed effects, such as developing cancer, are dreaded more than immediate effects such as poisoning.

The public has developed its own system of evaluating the risks associated with activities and technologies. Regardless of the "true" risks, it is the public's perceptions that largely determine those risks that are acceptable and those that are not.

Rather obviously, it is the public's perception of risk that influences the legislation with which all producers and users of hazardous materials must live.

Table 6.1. Criteria Influencing Public Perception of Risk

Criteria	Characteristics Perceived as Lower Risk	Characteristics Perceived as Higher Risk
Origin	Natural	Man-made
Volition	Voluntary	Involuntary
Effect manifestation	Immediate	Delayed
Severity (number of people affected per incident)	Ordinary	Catastrophic
Controllability	Controllable	Uncontrollable
Benefit	Clear	Unclear
Familiarity of risk	Familiar	Unfamiliar
Exposure	Continuous	Occasional
Necessity	Necessary	Luxury

While no absolute answers may be available to the question, "What is safe enough?" intelligent estimates of risk and the effective communication of those estimates to those who need to know these are essential. That is not to say that quantitative risk assessment is essential.

Quite the contrary is true. Some risks and some situations are more susceptible to quantitative measures; some represent impossible challenges. Explosion risks are subject to more quantitative analysis because the chemistry of the reactions is well known and a number of tests have been run on the energy release in explosions and detonations, the conditions at which they can occur, and the effect of blasts on buildings and people. Still, assumptions are required about blast efficiency, the impact of obstacles, partial containment, and so on. In addition, explosions are both an instantaneous phenomenon and one with which the public is somewhat familiar. Given all of this, uncertainty still exists in the probability of any explosion and in predicting the severity of the result.

Estimating the risks associated with exposure to chemicals is very much more complex. Explosions cause what are called acute—that is, immediate—effects. Chemical exposures can be both acute and/or chronic—resulting from prolonged exposure to lower doses.

Estimating the impact of chronic exposures introduces a number of additional uncertainties into any risk assessment. Some of these were mentioned in Chapter 3, but bear repeating. They include:

- The uncertainty of measuring low doses and their duration
- The extrapolation of high-dose animal tests to chronic impact on humans
- Sensitizing reactions with other chemicals
- Genetic proclivities or sensitization
- Interaction with lifestyles—smoking or diet
- The presence or absence of threshold exposure for certain effects
- The effectiveness of the body's immune reaction to a given situation
- The impact of other diseases/exposures

It seems reasonable to conclude that even when appreciable data are available, estimating the risk of any given chemical exposure is fraught with uncertainty. There are some cases where quantitative approaches are worthless. Indeed, some are worse than worthless because they provide misleading information.

Another aspect of trying to evaluate chemical risks is the statistical nature of the research. An LD50 is the exposure at which fifty percent of the subjects die in either an acute or chronic study. Obviously, most of these studies are done on animals. Researchers still cannot predict which fifty percent of the population studied will die. In human studies, exposures are often back-calculated after an effect is suspected. Most companies these days do epidemiological studies on their employees and annuitants. They track the health of these people to see if any statistical excess of any disease or early death occurs. If one is suspected, possible exposures are reconstructed looking for a possible cause. This is very frustrating because the cause is discovered only after a series of irreversible effects is already in progress. Yet, this was the route that identified a number of risks that are the basis for current exposure standards.

In spite of all of these caveats, dose-exposure curves are calculated for many exposures: chemical, biological, and physical. Given these data, the question still remains: How safe is safe enough? 10^{-4}, 10^{-5}, 10^{-6}, what? The nuclear industry tried to convince society of its efficacy on a scientific basis and failed. Each enterprise must assess its situation for itself.

Qualitative Approaches to Risk Assessment

Qualitative assessments are often the first step in any program and may be all that is required in a given situation. Often, the qualitative judgments are prioritized by adding an assessment of the probability and severity of any impact to the HAZOP or other risk identification process. These are usually high (H), medium (M), or low (L), or a simple numerical scale—for example, 1 through 5.

Consider the HAZOP illustrated in the previous chapter. Two approaches might be used to assess whether the risks identified are in need of further analysis or remediation. The first is to assign a value (high, medium, or low) to both the probability and severity of any given event. Then a matrix of probability vs. severity is constructed to depict graphically the events with highest risk. Such a diagram is shown in Figure 6.1.

Events that fall in the upper left sector (black) clearly merit immediate attention. Those that fall lower in the matrix (the gray area) may be subject to remedial action or not depending on the plant situation and other relative risks in the community. Those in the lower right (white) boxes are probably acceptable. Indeed, in a qualitative assessment, it is hard to say that any item with potentially severe consequences should not be addressed in some way because, even at low probability or frequency, the *investigator's perception* of the risk is that it is unacceptable.

The nature of remedial action can vary from modification of the hardware design or operating/maintenance procedures to simply verifying a point in either the design, the procedures, or the management system.

Another approach (described by Verdiano and Sloan in "Implementation of Hazard and Operability Reviews in the Sun Toledo Refinery," which was presented at the AIChE meeting in San Diego, CA, 8/21/90) would be to assign a numerical value to each severity and probability as follows:

Probability 1 (very frequent) - - - - - - - - - - 5 (very rare)
Severity 1 (very severe) - - - - - - - - - - - - - - 5 (very mild)

These assessments are then multiplied together to give a single number for the risk. The investigators are still faced with the need to define which number merits remedial action. In addition, the severity rankings should be maintained separately, as, for the reason given above, any item with a severity of, say 2, may be addressed regardless of probability.

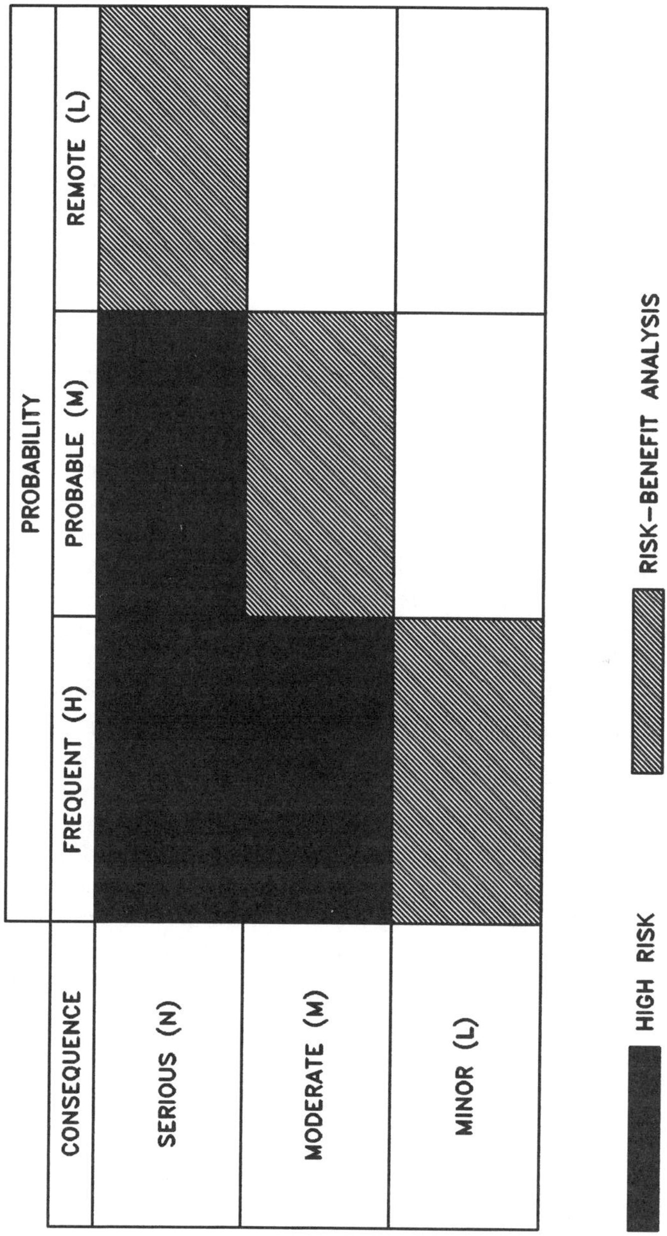

Figure 6.1. Qualitative risk assessment matrix.

Clearly, these ranking techniques can be applied to any risk identification approach to help in prioritizing remedial action. How such techniques can contribute to justifying the investigators' judgments to a community or government agency will depend on a plant's relationship to these groups and on local laws. In New Jersey, for example, if the magnitude of a release of certain substances (Extraordinarily Hazardous Substances—EHS) were to exceed five times the threshold quantity for that substance in an hour, a quantitative risk assessment would be required by the Toxic Catastrophe Prevention Act.

Increasingly, company lawyers are anxious about risk assessments and the lack of responsible follow-up documentation. In the United States in particular, corporate lawyers have nightmares about a prosecutor learning, once an incident has already occurred, that the potential problem had been identified in HAZOP (or the like), but apparently went unaddressed. If the plant manager has been shielded from understanding the risks by subordinates who felt empowered to decide the fate of recommendations for remedial action, the situation becomes even worse.

Common sense (and the new OSHA process safety regulations) require that each recommendation be evaluated and responded to. The response must give cogent reasons for the action decided upon and the projected schedule for completion. Periodic follow-up reports also serve to prove what the lawyers call "due diligence." While it is possible on occasion to protect some evaluations and conclusions from discovery in a litigation, counting on this protection to avoid major problems is usually considered imprudent. Decisions on remedial action to reduce risks must be endorsed by top management and properly documented. Thereafter, an effective follow-up program is needed to provide the best possible defense should an incident result in spite of prevention efforts.

Table 6.2 adds assessments of acceptability to the HAZOP summary table for SCl_2 unloading (Table 5.6) presented in Chapter 5. Many items have been judged acceptable (A) or were so simple, obvious, and inexpensive that they were simply marked "Do." The former judgments were based on the design meeting accepted standards of the company and/or the industry. Alternatively, a specific configuration could be judged "state of the art"—that is, the best you can do with reasonable economics. In New Jersey, "state-of-the-art" technology is required for the facilities that handle the EHS covered under NJTCPA. In effect, the HAZOP verified that the design met these standards and assessed whether the associated procedures recognized every credible accident.

In many cases, this assessment resulted in "Do" items related to procedures, maintenance, and training. These tend to cost little or nothing, and round out a complete prevention system for dealing with the risk. Note that a number of potentially high-severity items have procedures as a component of the total risk management system. (We shall treat verification that procedures are followed in a later chapter.)

Aside from meeting design standards or being easy to do, two other factors influenced the judgments listed in the table. These were vulnerability (i.e., the potential consequences of an accident) and time.

Table 6.2. Hazop Worksheet

UNIT NPS SECTION SCl2 R/C DWG # ____ DATE ____

ITEM	EQUIPMENT	CONCERN	F/Q	HAZARD	S	P	INDICATION/PROTECTION	A*	COMMENTS/ACTION
S1	SCl2 RAILCARS	BACKFLOW OF SCl2 INTO THE N2 SYSTEM	F	Cl2/SCl2 IN N2 SYSTEM	M	M	CHECK VALVE ON CONNECTION HIGH PRESSURE ALARM	A/DO	CONFIRM THAT THE BEST TYPE CHECK VALVE IS USED
S2	N2 PIPING TO R/C	SUPERFLUOUS BLEEDER	F	LEAK			VALVES NORMALLY CLOSED	DO	REMOVE OR PLUG UNNEEDED VALVES
S3	SMALL PIPING CONNECTIONS	BREAKING OFF OF SMALL PIPE	F	LEAK	H	L	APPLICABLE SMALL PIPE STANDARDS FOLLOWED	A	NONE, MECHANICAL STD'S ADEQUATE
S4	N2 HOSE TO R/C	RUPTURE OF HOSE	F	LEAK	H	L	HOSE IS DESIGNED FOR 100 PSIG @ 150 DEGREES F	A	NONE, AS ABOVE
							MONEL USED FOR CORROSION PROTECTION	A	
								DO	ASSURE THAT HOSE INSTALLATION MINIMIZES STRESS ON FITTINGS
								DO	ESTABLISH AN INSPECTION PROGRAM
S5	SCl2 RAILCAR	WATER/CONTAMINANTS INTO CAR	Q	REACTION, RELEASE TO ATMOSPHERE	H	L	PROCEDURE - KEEP HOSES DRY AND CLEAN	A DO	ENSURE PROCEDURES ARE ADEQUATE CHECK W/SUPPLIER ABOUT REACTIONS
S6	N2 TO RAILCARS	OVERPRESSURE	F	LIFT SV RELEASE TO ATMOSPHERE	M	L	PRESSURE CONTROLLER ALARM & SV ON N2 SYSTEM	A	SV SET POINT HIGH

F/Q - FACT OR QUERY
S - SEVERITY: LOW, MEDIUM, OR HIGH
P - PROBABILITY: LOW, MEDIUM, OR HIGH

A - ACCEPTABILITY
DO - SIMPLE ITEMS
U - UNACCEPTABLE FOR LONG TERM, MUST TAKE CORRECTIVE ACTION
C - REMAINS A CONCERN, REQUIRES ACTION TO ESTABLISH IF ACCEPTABLE OR NOT
A - ACCEPTABLE RISK, NO FURTHER ACTIONS REQUIRED. MEETS TCPA CRITERIA.

** MEETS STATE OF THE ART*

(Continued)

Table 6.2. *(Continued)*

UNIT NPS SECTION SCl2 R/C DWG # DATE

ITEM	EQUIPMENT	CONCERN	F/Q	HAZARD	S	P	INDICATION/PROTECTION	A*	COMMENTS/ACTION
S7	EMERGENCY VENT	VENT TO SCRUBBER WITH INSUFFICIENT CAUSTIC	F	RELEASE TO ATM	M	L	SCRUBBER SOLUTION SAMPLED TWO TIMES/DAY. NaOH MONITORED. ALL RELEASES TO TOWER ARE SAFETY RELATED AND DETECTABLE.	A DO DO	NEED TO ESTABLISH COMMUNICATIONS WITH OTHER UNITS HOSES TO BE MADE "SAFETY EQUIPMENT" AND TREATED AS SUCH
S8	SCl2 LIQUID CONNECTION	LEAK ON LIQUID SIDE	F	LEAK	M	L	PRESSURE TEST ALL FITTINGS W/N2 PRIOR TO OPENING VALVE CHECK GASKETS ON ALL FLANGES	A DO DO	STORE PROPER GASKETS ON UNIT DISCONNECT N2 PRESSURE TEST HOSE WHEN NOT IN USE (PROCEDURES)
S9	SCl2 PIGTAIL	RUPTURE	F	LEAK	H	L	MONEL SELECTED FULL STRENGTH TUBING USED	A DO	EXPERIENCE WITH Cl2 SET UP INSPECTION PROGRAM SIMILAR TO Cl2 PIGTAILS
							EXCESS FLOW CHECK VALVE WILL LIMIT RELEASE, LEAK DETECTORS	A	TO BE CONSIDERED A CRITICAL INSTRUMENT & INSPECTED ACCORDINGLY
							PROCEDURES	A DO	MONEL PIGTAILS TO BE COVERED BETWEEN USES PER PROCEDURES ASSOCIATED FITTINGS MUST BE STORED ADEQUATELY
S10	SCl2 R/C'S	FIRE UNDER CAR	F	RELEASE	H	L	VIDEO MONITORING OF CARS LEAK DETECTORS, CLEAR AREA OPERATOR ROUNDS	A	PROVIDE GOOD INSTRUCTIONS FOR NIGHT SUPERVISORS MAINTAIN VIDEO & DETECTORS

F/Q - FACT OR QUERY
S - SEVERITY: LOW, MEDIUM, OR HIGH
P - PROBABILITY: LOW, MEDIUM, OR HIGH

A - ACCEPTABILITY
DO - SIMPLE ITEMS
U - UNACCEPTABLE FOR LONG TERM, MUST TAKE CORRECTIVE ACTION
C - REMAINS A CONCERN, REQUIRES ACTION TO ESTABLISH IF ACCEPTABLE OR NOT
A - ACCEPTABLE RISK, NO FURTHER ACTIONS REQUIRED. MEETS TCPA CRITERIA.

** MEETS STATE OF THE ART*

(Continued)

UNIT NPS SECTION SCl2 R/C DWG # ______ DATE ______

ITEM	EQUIPMENT	CONCERN	F/Q	HAZARD	S	P	INDICATION/PROTECTION	A*	COMMENTS/ACTION
S11	SCl2 R/C'S	LEAK	F	LEAK	H	L	EXCESS FLOW CHECK VALVE ON LIQUID CHECK VALVE ON VAPOR SPACE CONCRETE PAD TO CATCH DRIPS SPILL-X FOR CLEAN-UP	A	SMALL DRIPS CONTAINED
S12	SCl, R/C	OPERATOR TRAPPED ON PLATFORM DURING EMERGENCY	Q	EXPOSURE	H	L	SECONDARY ESCAPE VIA R/C LADDERS ESCAPE PACK ON PLATFORM	C	STUDY W/SAFETY COORDINATOR
S13	SCl2 LIQUID LINE	PLUGGED LINE	F	POSSIBLE LEAK WHILE CLEARING	M	L	6D-15 AVAILABLE AS BLOWDOWN VESSEL TO CLEAR LINE	DO	ENSURE PROCEDURES ARE ADEQUATE ENSURE ADEQUATE CHEM CLEANING PROCEDURES ARE AVAILABLE
S14	SCl2 6D-15 LEVEL IND.	SINGLE BLOCKS TO ATMOSPHERE ON CALIBRATION VALVES AND DRAIN	F	LEAK	M	M	NORMALLY CLOSED	C	CONSIDER DOUBLE BLOCK OR BLIND FLANGE, SMALL DRIPS
S15	SCl, R/C	RECYCLE OF SCl2 FROM CAR TO EMPTY ONE	F	LEAK, UNEXPECTED PRESENCE OF PRODUCT	L	L	PROCEDURES DOUBLE BLOCK VALVES	A DO DO	 EXPAND PROCEDURES TO COVER THE CASE TO DISCOVERING POSSIBLE BACKFLOW PROCEDURES PLACARD ON PLATFORM
S16	SCl2 VENT LINE	BOTH VENT & FEED VALVES OPEN	F	D5X OPERABILITY	L	L	PROCEDURES P5-X INSTRUMENTS EXCESS FLOW VALVE	A DO DO	 CONFIRM THAT EXCESS FLOW VALVE WILL GET PREVENTIVE MAINTENANCE

F/Q - FACT OR QUERY
S - SEVERITY: LOW, MEDIUM, OR HIGH
P - PROBABILITY: LOW, MEDIUM, OR HIGH

A - ACCEPTABILITY
DO - SIMPLE ITEMS
U - UNACCEPTABLE FOR LONG TERM, MUST TAKE CORRECTIVE ACTION
C - REMAINS A CONCERN, REQUIRES ACTION TO ESTABLISH IF ACCEPTABLE OR NOT
A - ACCEPTABLE RISK, NO FURTHER ACTIONS REQUIRED. MEETS TCPA CRITERIA.

** MEETS STATE OF THE ART*

(Continued)

Table 6.2. *(Continued)*

UNIT NPS SECTION SCl2 R/C DWG # DATE

ITEM	EQUIPMENT	CONCERN	F/Q	HAZARD	S	P	INDICATION/PROTECTION	A*	COMMENTS/ACTION
S17	SCl VENT	LIQUID TRAPPED IN LINE	F	UNEXPECTED PRESENCE	M	M	PROCEDURES	DO	EXPAND PROCEDURES TO COVER WAYS TO DEAL W/LIQUID IN VENT SYSTEM
S18	SCl2 LIQUID LINE	TRAPPED LIQUID THERMAL EXPANSION	F	LEAKS	H	M	HEAD SPACE IN 6D-15 FOR EXPANSION	A C	 ENSURE VAPOR SPACE IN 6D-15
S19	SCl2 VENT AT D5X	THERMAL EXPANSION OF NPS IN PIPE	F	LEAKS	M	M	PROCEDURES	DO C	CHECK PROCEDURES FOR ADEQUACY CONSIDER LEAVING VALVE OPEN AT D5-X

F/Q - FACT OR QUERY
S - SEVERITY: LOW, MEDIUM, OR HIGH
P - PROBABILITY: LOW, MEDIUM, OR HIGH

A - ACCEPTABILITY
DO - SIMPLE ITEMS
U - UNACCEPTABLE FOR LONG TERM, MUST TAKE CORRECTIVE ACTION
C - REMAINS A CONCERN, REQUIRES ACTION TO ESTABLISH IF ACCEPTABLE OR NOT
A - ACCEPTABLE RISK, NO FURTHER ACTIONS REQUIRED. MEETS TCPA CRITERIA.

** MEETS STATE OF THE ART*

Vulnerability really has to do with the plant's location and history. If the plant is located in the middle of nowhere or next to any army camp where soldiers are trained to survive in gas warfare, the plant location is much less vulnerable than if it is located near a nursing home or grade school. If part of the plant encompasses a labor-intensive operation, such as canning or fiber spinning, workers are more vulnerable to a release or explosion than in the modern petrochemical plant, where many of the operators function in an explosion-resistant control center. If the plant has a history of releases that have set off a community reaction—even if there was no damage—simply an odor or noise, the plant is more vulnerable due to its past "sins." Thus, judgments about what constitutes a significant risk will be influenced by the vulnerability of the site, as well as the scientifically calculable consequences of any event.

The second factor has to do with the potential for taking remedial action to mitigate the consequences of an accident. This constitutes two parts: detection and reaction. Detection may be the operators, control instrumentation, or specific detectors positioned as a safeguard. The available reaction time will depend on the nature of the accident. There is no time for an explosion, very little for a gas release, much more for a nonvolatile liquid spill, and very long for nonvolatile solids.

In our example, item S9 has potentially high severity, but no additional facilities are required and proven procedures can be adapted to the new system. This is true because while a pigtail rupture would be a major event, early detection and immediate reaction (excess flow check valves) are provided. What remains then is to ensure that the procedures used for similar C12 unloading equipment are extended to the SC12 railcars. If the plant in question had no previous successful experience with unloading tankcars of volatile toxics or had a bad history of leaks, it might have been moved to provide secondary containment (e.g., a building and a scrubber) for the unloading operation.

The example clearly illustrates the subjective nature of qualitative risk assessments. More might have been done to try to ensure that no toxic gas cloud would endanger either plant workers or the community. Would more have been better? Would the reduction of risk be worth the cost? How might this judgment that the proposed procedures and facilities are adequate be justified to the community and relevant agencies? In the investigator's view, vulnerability is not high (there is no nursing home across the fence), the proposed approach has worked successfully on a more toxic material, immediate detection is provided, and operators are trained in emergency response. The state agency accepted this logic; it has never been tried on the community at large.

Let us now consider those items marked "C"—that is, in need of further study. Four items are so marked, two are of potentially high severity, the rest are in the medium/medium catagory. One of the high-severity items involves a potential leak (S18), the other potential worker injury in case of a leak (S12).

Relative to the other items, these two merit high priority. For item S12, the safety coordinator believed that an escape air pack on the loading platform was adequate protection. The basis for this judgment was that, in addition to the low probability of a leak at the time an employee was on the platform, all employees were subject to a "no beard" policy, annual refresher training on air pack use, and periodic simulated emergency drills. In addition, a systematic program was in place to test and maintain all air packs.

Items S18 and S19 were dealt with together by ensuring a vapor space in D5 and an open path to that drum. This was a more complex solution involving both procedure and equipment changes. As it turned out, the original solution was only partly correct. During a shutdown, backflow of reactor liquid occurred followed by a period of time during which the system was not drained. When the SC12 feed was restarted and entered the point where the reactor liquid accumulated, a reaction took place that resulted in a leak. Additional facilities were provided to prevent this apparent coincidence from reoccurring.

The purpose of the above was to emphasize the point that a single HAZOP may not identify all of the real risks. Both events were considered in the HAZOP but not the combination of the two. Perhaps this was judged a double contingency. Perhaps it was not considered. If you remember my earlier comments about what constitutes a real double contingency as opposed to two single contingencies separated by an insufficient amount of time to be safe, you will see a parallel here. The key point is that periodic HAZOPS that incorporate all available experience are essential. Finding no real possibilities in the first pass needs to be confirmed by periodic reappraisals to minimize risk.

To summarize the action plan resulting from this HAZOP, it consists of all items marked "Do" plus the studies and resulting actions discussed above, all properly documented to make the rationale clear.

The Concentration Approach—a More Quantitative View of Consequence

Earlier, I referred to dose-response curves as a means of defining acceptable risks. Sometimes government agencies or corporations use this type of datum to set concentrations in the air or water that must be met as a matter of policy. This has the effect of forcing the assessment team to calculate concentrations in the controlling medium one way or another.

Specific calculations are not always required. For some chemicals, the impact of certain types and sizes of releases have already been published in the literature. Chlorine Institute pamphlet 92 plots the air concentrations for several typical leaks at several atmospheric conditions as a function of distance from the release to help chlorine users assess risks on a consistent basis. All the user has to do is compare his or her potential releases to those calculated and decide if the resulting concentration is acceptable. In some cases, governments will set the acceptable limits. Government tests aimed at verifying dispersion-model calculations also present data of this sort. Air and water permit cases often present public information that might be used to assess similar situations. Clearly, with the development of this information comes a commitment to act on it, but I am not espousing deliberate ignorance.

In cases where specific calculations are required, using the correct correlation is important. Government agencies may require that a specific air-dispersion model be used or specify certain biological oxidation plant efficiencies and so on. In any case, where these calculations might be made public, it is wise to use accepted correlations for the type of release.

Making specific calculations for a case has the advantage of providing insights on how to modify design or response procedures to mitigate the severity. A restriction orifice or excess flow valve might be strategically placed, for example. Similarly, detection devices and rapid response procedures, equipment, and training might be installed to minimize the impact. The challenge is to find a basis for selecting concentrations for each chemical at which action is required.

F. M. Renshaw (Rohm & Haas Co., Bristol, PA) outlined the company's approach to risk assessment and responses in the July 1990 issue of *Plant/Operations Progress,* an AICHE publication. He defines three action levels related to the Emergency Response Planning Guidelines (ERPG) of the American Industrial Hygiene Association (Akron, OH). These are:

Level 1 (ERPG-1)	Threshold of irritation
Level 2 (ERPG-2)	Threshold of irreversible effects
Level 3 (ERPG-3)	Threshold of life-threatening effects

The ERPG thresholds deal with toxicity and are available for only thirty to forty substances at this writing (1Q92). (Note that NJDEP has defined "acute toxic concentrations" for the 104 chemicals covered by NJTCPA. These roughly correspond to ERPG-2 and are based on severe impact on twenty percent of the exposed population.) Renshaw defined equivalent fire and explosion consequences for each level. These are:

	Level 1	Level 2	Level 3
• Fireball, heat release	1,270 BTU/hr Ft^2	2,850	3,800
• Pool fire, heat release	320 BTU/hr Ft^2	950	950
• Unconfined vapor cloud explosion, overpressure	0.3 psia	2	3

These equivalences were derived from damage potential estimates obtained by the use of a personal computer program called PHAST, a product of Technica, Ltd.

For scenarios resulting in level 1 damage projections the company ensures that appropriate training and procedures are in place and evaluate further damage reduction possibilities. For levels 2 and 3 a team is assembled to develop strategies and recommendations to reduce the frequency and severity of credible incidents. For level 3 possibilities quantitative risk assessment is considered to provide more insight on preventive steps. Once the remedial measures are in place, risks are reassessed to ensure they are acceptable.

To summarize, level 3 risks are unacceptable. Any remaining level 2 risks are reevaluated periodically with the goal of improving damage potential to level 1. On a qualitative basis, I believe this philosophy is followed by a number of companies—that is, they wish to tolerate potential consequences outside the fence no more severe than temporary discomfort. This kind of policy clearly represents "due diligence" on the part of management if follow-through is good.

In all of this, the identification and definition of the event to be assessed are critical. There is a temptation to insert "worst case scenarios" into the analysis even if no one believes the probability of such a case is significantly above zero. Somewhere a judgment must be made, and be bought into, about the realism of the events to be considered. Difficulties on this point may lead to the use of more quantitative assessment techniques.

Quantitative Risk Assessment

Quantitative risk assessment (QRA) goes by many names in the literature. These include HAZAN, probabilistic risk assessment, fault trees, event trees, hazard quantification, and so on.

In a quantitative risk assessment, the severity/consequence and probability/frequency factors are evaluated numerically and combined into a single quantitative assessment of potential damage, injury, or death. This can be done on either an absolute or a relative basis. Some examples of the kinds of results derived from the two approaches are given in Table 6.3.

Absolute vs. Relative Values

At the end of this chapter is a brief reading list dealing with risk assessment and quantitative risk assessment. All the entries are fairly general in nature and many give examples of studies found useful. The consensus of this modern literature is that relative studies are of more practical value than absolute numbers.

Part of this consensus is based on the prevalence of misconceptions about

Table 6.3. Results

Factor	Absolute	Relative
Frequency	With the base design, an explosion would be expected 5×10^{-4} times per year	With the improved design an explosion would be expected 10^{-4} times per year less often
Consequences	2,000 people will be exposed to an ammonia concentration > 400 ppm if overpressure occurs	The exposed population can be reduced by a factor of 10 if a scrubber is added
Overall risk	The risk to employees from the base process is 5×10^{-6} fatalities per year	The modified process has a risk of fatality 15 times lower than the base case OR This process has a risk of fatality which is 10 times greater than the average of our other processes using toxic materials

QRA. Some of these are listed in Table 6.4, taken from the CMA report cited in the reading list. Because of these misconceptions, *communicating* the interpretation of absolute risk assessments effectively to either management or communities is very difficult. This is not to say that communication is the only or even major difficulty in getting value from QRA effort. There are very real technical and management problems. Some of these, such as data availability and cost control, can be dealt with by developing (or hiring) expertise and carefully defining the problem to be worked.

There are some very real problems with QRA in the chemical business. As Guymer et al. (see reading list at end of chapter for reference cited and other pertinent background) point out, process-industry QRA is much more complex than that for the nuclear industry, and we know what has happened to many nuclear plants. French et al. point out that while the calculations can be carried out with great *precision,* the human inability to conceive every possible event sequence significantly detracts from the absolute value of any result. Even though sometimes required by governments, these studies are approached with skepticism even by those skilled in the art.

Relative studies, on the other hand, can pinpoint high-risk areas in a process and evaluate potential improvements on a consistent basis. The same discipline as for an absolute study is imposed and engineering talent can be focused in areas where the greatest overall risk is present.

Perhaps the clincher in the debate is that Air Products, a company that has trained hundreds of its professionals in QRA, still uses relative assessments to make decisions.

Some of the other problems associated with QRA are also formidable. These include:

- Not reproducible—because of differences in assumptions, different investigators will get different absolute results with the same process.
- Inscrutable—QRA is so complex that those without expertise have difficulty understanding and therefore do not "own" the results.
- Uncertain—the range of possibilities may lead to so much uncertainty that firm decisions cannot be made.
- Incomplete—there is a limit in any study to the level of detail that can be included. Almost anyone can think of another case.

Table 6.4. Misconceptions About QRA Technology and Risk

- A QRA can prove that the plant is safe or unsafe
- If we do a QRA, we can reduce our risk to zero
- QRA is expensive
- QRA is cheap
- We can usually predict risk to an accuracy of a factor of 2 or better
- If we could measure risk accurately, our decisions would be easy
- We analyzed all possible accidents
- We do not have enough data so we do not need to do QRA
- QRA is a totally objective way to understand risk
- QRA is pure science

A decision to attempt QRA is a serious one and should not be taken without proper consideration.

Making the Decision

The first criterion in deciding to use QRA is a belief that it will provide a better understanding of the process under consideration and how to operate it safely. Qualitative approaches should have been tried and found inadequate for decision-making needs and/or the needs to convince the public or government agency of the efficacy of plans.

A second reason to do QRA may be a legal requirement. As previously stated, New Jersey requires a quantitative risk assessment if the potential exists for a release of an EHS that is five times its registration quantity in one hour. Thus, an enterprise has two choices: Reduce the size of the potential release or do QRA. Some companies have chosen to limit size of the potential or eliminate the release—probably the better choice for the community. Others have chosen to do QRAs.

As an aside, one company submitted comprehensive QRAs on its many hazardous processes. This formidable document caused such delight at NJDEP that its QRA expert recycled it with requests for more detail and more cases before it was accepted. My conclusion was that the DEP went to school on the company's submission. Such are some additional risks of undertaking QRA for government reasons.

Given a need for greater understanding and a desire to address internal concerns about employee or community safety, further motivation may come from several sources. These might be classified as internal or external, but these days, customer concerns about security of supply may be quite important.

Internal motivation might come from a corporate audit, union pressure, an accident at your (or a similar) plant, or as a result of plant initiative to upgrade its situation.

We have already discussed legal requirements as an external force. Other external forces include the corporate insurance provider, permit requirements, local emergency planning committee requests, the need to respond to public outrage, and the desire to reach out to the community on a more quantitative basis. The latter pressures almost invariably result from a well-publicized incident in the region.

Given a positive response to enough of the above considerations, two further factors need to be considered before a decision is reached. The first is an assessment of the available data on which the study would be based. Is there enough pertinent information? Is it likely that the study can be carried through to provide the answers desired? The second factor is the source of expertise with which to carry out the study. Is it available in-house? If not, is there a suitable outside source? Finally, does the company wish to divulge to an outside firm the information about its operations that it will need to do a good job?

If enough positive answers are there, the next step is to establish a system to manage the study.

Managing the Study

A broad discussion of this process is developed in the CMA report cited in the reading list. What follows are highlights of some of the critical control needs in a QRA study.

Once QRA is chosen to satisfy a particular need, three key tasks require attention, the responsibilities for which are shared between the group requesting the study and the experts who carry it out. These are:

- Defining the problem
- Selecting appropriate techniques
- Understanding the assumptions and limitations

The ultimate user must carefully define written objectives for the QRA project team. The selection of specific analytical techniques involves decisions that the user will influence, but that should be left to the team's discretion. Understanding the limitations of the study will require careful interaction and discussion between the QRA team and management to ensure that the final product meets the defined needs.

These tasks are interrelated and decisions about one affect the others. Also, periodic reviews of each as intermediate results are developed are a good idea to ensure that the QRA remains on track.

If a QRA is to satisfy management's requirements, it must have a specifically defined charter for the QRA project team. Defining the problem requires an understanding of the reason for the study, a description of the manager's needs, and an outline of the type of information required from the study. Sufficient flexibility must be built into the analysis scope, technical approach, schedule, and resources to accommodate later refinement of any undefined charter element(s) based on knowledge gained during the study. The QRA team must understand and support the analysis charter; otherwise, a useless product may result.

The key elements that must be defined in a charter for the team include the study objectives, the scope of facilities and hazards to be assessed, the general technical approach, the resources available, the schedule, and the review process.

An important and difficult task is concisely translating management's requirements into study objectives. For example, if the objective is to decide between two methods of storing a hazardous chemical in a plant, the charter should precisely define that what is needed is the relative difference between the methods. Asking the QRA team for more than is necessary to satisfy that particular need is counterproductive. For any QRA to produce the necessary types of results efficiently, the user must clearly communicate his requirements through well-written objectives.

Some examples of realistic and practical objectives are the following:

- Determine whether a catastrophic failure of the NH_3 storage tank is likely to cause irreversible health effects on a nearby neighborhood.
- Compare three process designs and rank them according to their risk to the community.

- Investigate the potential for vapor-cloud explosions resulting from LPG storage operation.

Defining the scope of the study boundaries may also require some interaction with the QRA team as the study develops. Early study data (generally from approximate methods) will help define the areas of the process that are the largest contributors to risk and their possible interactions with other parts of the plant. At this point, more detailed analysis can be directed toward the areas most likely to provide improved understanding of the remaining risks.

While the detailed technical approach for the study must be left to the QRA team, management's ability to use the results of the study will depend on understanding the important assumptions in the study and the limitations they impose on its usefulness and validity. Many techniques will provide more information than is needed to meet the basic objective. This needs to be avoided. External quality assurance reviews may be useful in developing the required technical understanding.

Just as the technical scope of the study can expand if not carefully controlled, so can the resources devoted to it. QRA can be like a new toy to an inexperienced team or management. As new insights surface, new facets of the study and/or new demands for other expertise are generated. Scientific curiosity is a wonderful thing but can expand a modest effort into a major one if left unfettered. On the other hand, the QRA must be adequately staffed if it is to be successful. In addition to the QRA expertise, the study team will need process scientists and engineers and operating and maintenance input, and may require input from specialists in such fields as such as machinery, instrumentation, and inspection at least part-time.

Presenting/Using the Results

Perhaps the most critical decisions and controls involve the way in which the results will be presented. It is entirely possible to emerge with gobbledygook understandable only to the experts who produced it. To achieve the best perspective, it is generally desirable to present the results with one or more graphical methods. One such approach is an *F-N* curve. It plots the cumulative frequency (probability) of events that cause harm to a specified number of subjects (e.g., people or damage). With an *F-N* curve (see Fig. 6.2), you can see the expected frequency of events that would hurt more than a fixed number of people. For example, in Figure 6.2, the frequency of accidents that would injure ten people is calculated to be approximately one in 100 years and that of an accident where 1,000 people would be hurt less than one in 10,000 years.

Another approach is a risk profile. This graph shows the individual frequency predictions of accidents that would have a particular level of consequence. In the example in Figure 6.3, the frequency of events that would hurt one to ten people (or cost one to ten cost units to repair) is about one in ten years. Other frequencies decrease with severity except for the bar between 100 and 300

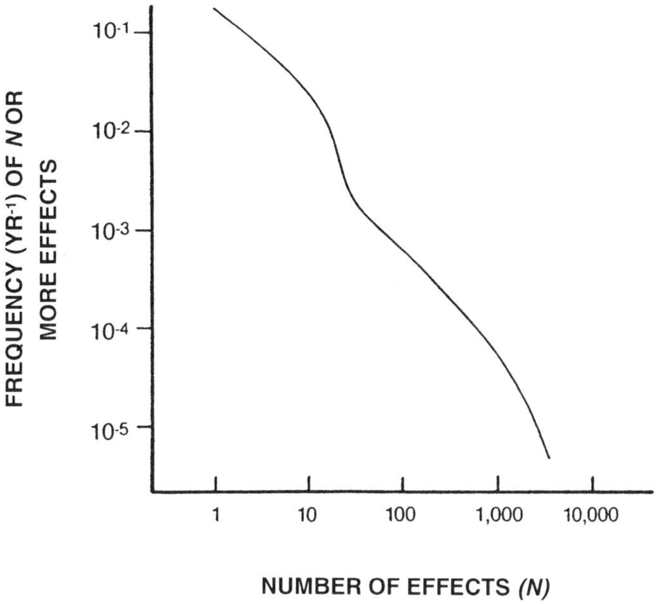

Figure 6.2. Example of an *F-N* curve.

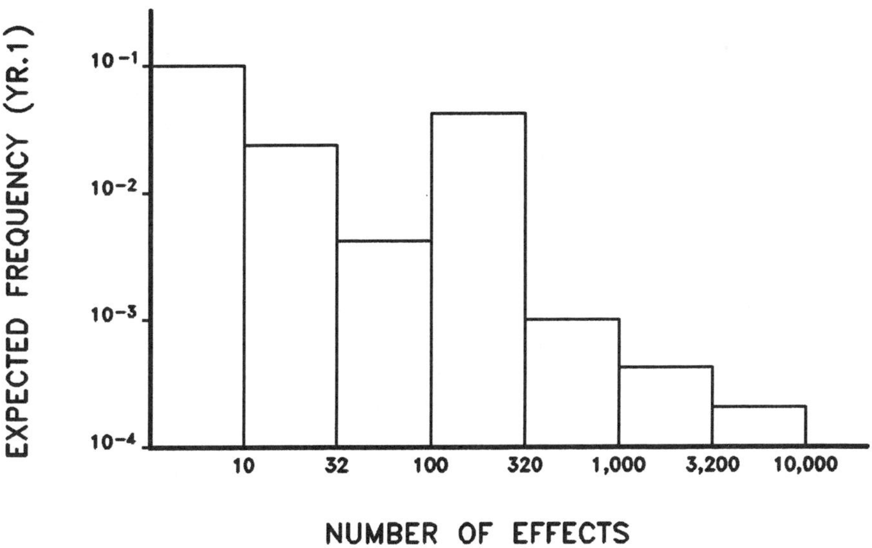

Figure 6.3. Example of a risk profile.

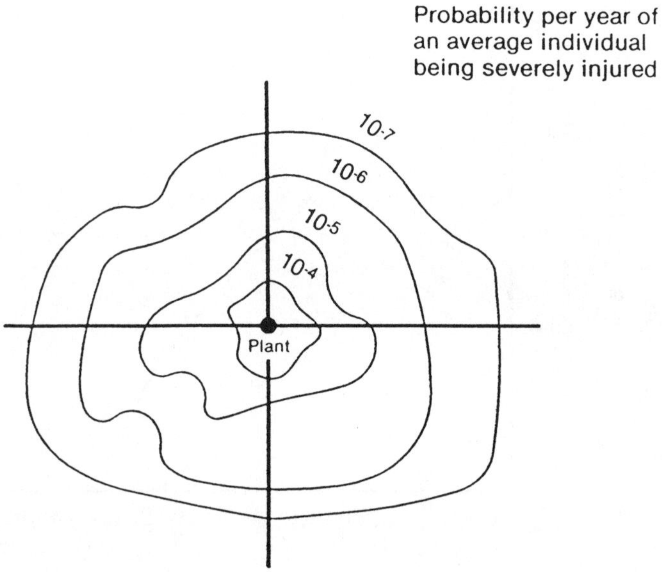

Figure 6.4. Example of a risk contour.

severity. This has a frequency of about one in fifty years—much more than those on either side.

The purpose of this graph is to identify areas of greater risk. As the consequences of an event go up, its frequency should go down if the overall risk (frequency times consequences) is to be constant. Clearly, the 100–320 bar represents an area that violates this principle. The causes of this increase need examination, explanation, and/or correction.

A third approach is a risk contour. In this approach, probability lines of an accident of a given severity are plotted on a map with the plant studied in its center. The severity measure could be illness, deaths, odor threshold, damage factors, and so on. Figure 6.4 shows a typical contour plot. Of itself, it represents an absolute risk estimate. If applied to alternate process or loss-prevention configurations or alternate plant sites, it can define the lower risk configurations. An interesting approach is to use it to define the necessary exclusion area around a plant that is consistent with that at other plant sites.

Beyond these tools that capture the overall situation in a form that can be more easily assimilated by the nonexpert, more detailed analysis of the components of risk generated by the computer can contribute more guidance to those who would reduce risk efficiently. An important option available to managers for evaluating risk estimates calculating the importance of various components, human errors, and accident scenarios to the total risk. For example, two accident scenarios may contribute ninety percent of the total risk. With this data, it is obvious that your loss-prevention resources should be focused on reducing the potential for those accidents. In other cases, all of the accident scenarios may have comparable risks, but failure of a process control computer is required for every scenario. The process computer will show up as the most risk-important component, and loss-prevention resources might best be spent in providing a backup computer. To use QRA to assist in decision-making, risk-importance results can be analyzed to understand the basis for the major risk contributors.

Another way to evaluate the risks is to calculate the sensitivity of the total risk estimates to changes in assumptions, frequencies, or consequences. Risk analysts tend to be conservative in their assumptions and calculations. The cumulative effect of this conservatism may be a substantial overestimation of risk. For example, always assuming that short-term exposure to chemical concentrations above some threshold limit value will cause serious injury may severely skew the calculated risks of health effects. Consequently, management needs to understand the sensitivity of the risk results to this and other assumptions or it may misallocate loss-prevention resources or misinform the company or the public about the actual risk.

Risk sensitivity results are also very useful in identifying key elements in existing loss-prevention program. For example, suppose the fire protection system was assumed to have a very low probability of failure because it is tested weekly. Fire protection failures may not show up as an important contributor to total risk (because failure is so unlikely), but the total risk estimate may be extremely sensitive to any change in the probability of fire protection failures. Hence, management should not divert resources away from testing the fire protection system unless the alternate use of funds will decrease risk more than the reduced testing will increase risk.

The lack of reproducibility of QRA studies is another sensitivity that needs to be understood in a given case. In a recent study, ten analysts were given identical systems to study and freedom to choose methods and data. The teams calculated "probability of failure" results that ranged from 10^{-7} years to 10^{-3}. Analysis showed that the differences were due to differences in base assumptions. When coached to use similar assumptions, the ten groups produced estimates that differed only by a factor of five. In a given company, or in dealing with a single consultant, teams can be guided to "standard" or realistic assumptions. Using different teams/consultants without guidance would seem to be difficult.

The work required to evaluate risk results will be a function of the objectives of the study. For relative risk studies, this evaluation is usually not very time-consuming. For absolute-risk studies, in which many uncertainty and sensitivity cases may have been produced, the risk evaluation step may account for ten percent to thirty-five percent of the total effort of a large-scale QRA.

Communicating Results

As indicated earlier in this book, much has been documented on effective risk communication. None of that will be duplicated or explored here. Instead, I will try to focus on a single point that I believe to be the critical parameter in the process.

Having thus set the stage for a magnificent pronouncement, I shall underwhelm you with this single truth: *Know your audience.* I mean *know,* almost in the biblical sense.

The results of a risk assessment study must be communicated to a number of audiences for a number of purposes. Anyone who thinks the same approach will be effective in all venues is sure to be disappointed. The presentation appropriate to convince management to act on the most important contributors to potential accidents will be totally unsuitable to give to the town fathers interested in the total risk of a toxic material release in the community, and would signal Armageddon to a group of activists concerned about the possibilities of such an event.

There are at least five audiences to consider:

- Your management and that of neighbor companies
- Your employees
- Local and state agencies and emergency preparedness groups
- Neutal community groups/neighbors
- Local activists and their broader support groups

For all of these audiences, the following four principles apply:

- Presenting the results in perspective
- Recognizing the factors that influence perceptions of the meaning of the results
- Credibly communicating risk information
- Avoiding common pitfalls in using the results to facilitate the "right" decisions

It is often helpful to talk to the QRA team members to determine their personal impressions and conclusions about the study. Many times a great benefit of a QRA is the insight the analysts gain from having gone through this exercise. The more you can absorb these insights and develop a "feel" for the real situation, the better able you will be to interpret confidently the results and communicate them effectively.

Although the facts will not change, the most effective communication approach will be influenced by the changes in perspective and perceptions of the various audiences and their relative ability to understand the facts objectively.

Even *knowing* your own management can be more difficult than you might anticipate. Normally, one would want to inspire them to support remedial action for the risks assessed to require it. However, overkill in this effort may result in negative reactions due to profitability pressures and/or feelings that somehow the ethics of the company have been impugned by finding that it had been taking all these risks for so many years. Doing some research on management's real attitudes toward reducing risk will help to determine the most effective level of intensity to employ.

Knowing your employees may not be as easy as it sounds either. Some companies have effective mechanisms for communicating openly with their employees; some have these processes encumbered by paternalism or union/management strife. One of the action programs under CMA's Responsible Care codes promotes the dialogue between company and its employees, especially in the area of risk. In reality, such dialogue is a win/win situation for both parties. While both sides need to recognize the inherent commitment to do something with what is learned, everybody is usually ahead when understanding is increased.

The mechanism for accomplishing an effective dialogue will vary with plant culture. At our plant, we used three mechanisms. The first was regular safety meetings. The effectiveness of these depends very much on the skill of the supervisor who conducts the meeting. This skill can obviously be enhanced with training. Material for these meetings is provided by the safety group regularly, but the main benefits occur when the operators and mechanics show their views. The second mechanism was regular monthly meetings between plant and union managements devoted to safety and the environment. These are more formal but provide an opportunity to share views and actions on these issues. The third mechanism was to assign a union operator to work in the safety/environmental group. This has brought a more realistic component to field safety surveys and an informal mechanism with which to exchange views with at least some of the operating folks. We also had union representation at the close-out meeting of our first NJTCPA audit to allow unfiltered input.

The net result was an improved—but not complete—understanding by employees of what the plant is doing to prevent accidents and to comply with the applicable laws. More importantly, it is helping to build trust on both sides, but the discussion of quantitative risk analysis may still be a bit much for the current state of mutual understanding. Since little of this sort of analysis has been done at this site, it will be some time yet before this test is made.

Knowing the safety/environmental agencies may be easier than you think. I have recently come to believe that the only real hope we have for rational safety/environmental policy is through viable agency activity. These are the only people charged by law to balance the extremes of the environmentalists and "green" legislators against the views of profit-minded industry and lobbyists. The right support and input to agency thinking can do much to create a rational basis for public policy. More than that—it is essential that companies take the time to make such input.

We all know that government agencies are run by political appointees, some of whom know little about the technical content of their responsibilities. We know these people will seek headlines whenever possible. We know that if you have an accident, the agencies will be in to cite violations after the fact, and we know that agency lawyers are often inexperienced zealots who will push for the most far-reaching interpretation of any regulation regardless of whether the agency can administer it.

What we should not forget is that the senior technical people at most agencies are usually capable and experienced. These folks have continuity and a great deal of influence on the ultimate resolution of many issues. These are the people everyone needs to get to know. Many agencies have formal advisory panels at which input is sought from all groups early in the process of formulating regulations. Companies need to invest the time to participate regularly with knowledgeable technical people. These same agency people are usually willing to discuss problems and interpretations before the fact. They often participate in negotiations on contentious issues and they are anxious to learn about current industry practice. In my view, these people are a critical audience for all of us to know well and with whom to communicate effectively. In reality, they may be the most knowledgeable and receptive audience for risk studies.

Knowing the community and the activist groups is a formidable task. In effect, this amounts to knowing the perceptions of the neighborhood and its potential spokespeople. In "company towns," this might be accomplished with in-house staff quite effectively. Plant neighborhood nights, school projects, newsletters, and personal visits to the affected neighborhoods after any plant incident go a long way to help a plant understand the perceptions of the community. In many cases, *citizen advisory panels* provide a more formal approach to exchanging views and probably provide the best mechanism within which to discuss plant risks and response procedures.

In more populous and diverse areas, I believe professional help is required to get a reasonable insight into both the community's viewpoint and the most effective way to communicate with the people at large and with their leaders. Many groups are available to test community attitudes and recommend communication approaches. This can be done without divulging any specific risk information to the consultant. While these studies can uncover attitudes and avenues of communication, they are not cheap. Most of these companies will also offer to help structure a plant's actual message(s) to the community, but at some point, the management of the plant must take charge of this step and be visible.

I recommend professional help with the community because experience has shown LEPCs and municipal governments to not be effective. These groups are a separate audience with their own agenda. The LEPCs tend to think like industry and be task-oriented. The local politicians tend to be interested in their own solutions to their perception of the city's overall problems, whether those solutions be more trees, more taxes, or a public alarm system. The most effective way to work with these groups is to participate on the LEPC and in community emergency drills. During the latter, the major plant risks can be played out and effective countermeasures can be developed.

Avoid Classic Mistakes in Communication

Much has been written about effective risk communication. The CMA, in particular, has reduced much of the research of Covello, Sandman, and others into some rational guidelines for getting a message across. More importantly, they point out the many pitfalls into which the inexpert may fall. Yet, even in CMA publications, all of the potential pitfalls are not observed, which emphasizes that some risk remains in any risk communication.

We know that the public gets more upset about involuntary man-made risks than voluntary ones. Therefore, *do not* compare the risk to the health of the community of a release from your plant of dimethyl death (DMD) to that resulting from a neighbor's choice to smoke or drive a car. Nor should the release risk be compared to that of being hit by lightning or other naturally occurring phenomenon. The community might be interested in knowing that the risk has been reduced by a factor of ten by the addition of a scrubber and that you have never had a leak of DMD.

We know that the public will be more outraged by a catastrophic event, such as a plane crash, than by a series of small ones, such as auto accidents. Thus, one might compare the risk of an explosion at the plant to the chance of a plane falling from the sky, since both are potentially catastrophic, both are perpetrated by a profit-making enterprise, and people realize that planes do not crash very often.

We know that the public fears unperceived chronic damage, such as long-term toxic or cancer-producing effects, more than it does obvious risks such as a fire or explosion. Since health effects data have a large component of uncertainty in them because of their very nature, discussing such risks is doubly difficult. Even the projection of a single extra cancer case as a result of plant activity may cause serious outrage. Being sure that concentrations at your fence line are well below accepted (or suggested) third-party standards is an approach that provides some hard data. Using this approach may be the best anyone can do. If standards are unknown, working with a credible university or medical research group to develop a suggested standard may go a long way toward setting a basis for communication.

These highlights represent only a few of the instinctive errors we may be tempted to commit in risk communication—qualitative or quantitative. This is a fast-growing, multivariable art, and all efforts at it should be carefully planned and frequently evaluated with the help of experts.

QRA Technology—Very Briefly

Once hazards have been identified, QRA involves quantification of the frequency and consequences of an accident. Several of the references in the reading list describe specific methods and many computer programs are available to facilitate the process. I have neither the expertise nor the space to evaluate these or discuss them in detail. However, a brief outline of the overall processes used in QRA may be useful.

Frequency or Probability

Frequency assessment involves estimating the probability or frequency of each of the undesired situations defined in the hazard identification step. Most accidents analyzed by QRA are so rare that the frequences must be synthesized using frequency estimation methods and models.

Synthesizing the frequencies of rare events involves (1) determining the important combinations of failures and circumstances that can cause the accidents of interest, (2) developing basic failure data from available industry or plant data, and (3) using appropriate probabilistic mathematics to determine the frequency estimates. The most frequently used models are event trees and fault trees. An event tree is often used to define all of the possible accident scenarios that could result from a particular initiating event. Fault trees can be used to estimate the frequency or probability of individual events in an event tree. Though limited, a few industry data bases are available from which to obtain generic data on component failure, and AIChE recently sponsored a project to develop a data base specifically for the chemical industry.

The AIChE guidebook *Guidelines For Hazard Evaluation Procedures,* which is included in the reading list, provides descriptions of the two approaches, as do other references in the list, such as Ozog and Bendixen.

Fault Tree Analysis

FTA breaks down an accident into its contributing equipment failures and human errors. The method, therefore, is a "reverse-thinking" technique; that is, the analyst begins with an accident or undesirable event that is to be avoided and identifies the immediate causes of that event. Each of the immediate causes is examined in turn until the analyst has identified the basic causes of each event. This type of development requires that the analyst have a complete understanding of the system/plant operation and the various possible failures. The fault tree is a diagram that displays the logical interrelationships between these basic causes and the accident.

The result of the FTA is a list of combinations of equipment and human failures that are sufficient to result in the accident called *minimal cut sets.* Each minimal cut set is the smallest set of equipment and human failures that is sufficient to cause the accident if all the failures in that minimal cut set exist simultaneously.

Logic Symbols for Fault Tree Analysis

The fault tree is a graphical representation of the interrelationships between equipment failures and a specific accident. The following symbols are used in fault tree construction to display these relationships.

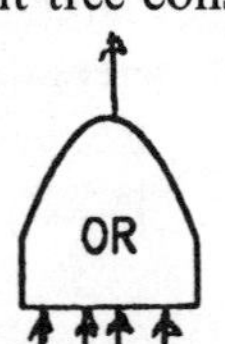

OR Gate: The *OR* gate indicates that the output event occurs if any of the input events occur.

AND Gate: The *AND* gate indicates that the output event occurs only when all the input events occur.

DELAY Gate: The *DELAY* gate indicates that the output event occurs when the input event has occurred and the specified delay time has expired.

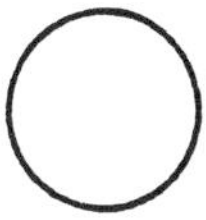

BASIC Event: The *BASIC* event represents a basic equipment fault or failure that requires no further development into more basic faults or failures.

INTERMEDIATE Event: The *INTERMEDIATE* event represents a fault event that results from the interactions of other fault events that are developed through logic gates such as those defined above.

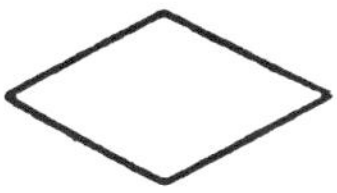

UNDEVELOPED Event: The *UNDEVELOPED* event represents a fault event that is not examined further because information is unavailable or because its consequence is insignificant.

EXTERNAL Event: The *EXTERNAL* event represents a condition or an event that is assumed to exist as a boundary condition for the fault tree.

Other symbols are used by the expert in complex fault trees. There will be not attempt to cover more than a simple introduction in this text.

A very simplified example of a fault tree is shown in Figure 6.5. It shows that a reactor incident (the top event) can occur only if three events occur simultaneously: (1) loss of reactor cooling, (2) failure of the feed shut-off, and (3) failure of reactor dump mechanism (*AND* gates). Each of the intermediate events leading to the failure of the dump mechanism can only be caused by the simultaneous occurrence of two events (*AND* gates).

To complete the picture, numerical probabilities/frequencies have to be assigned to each event and then be combined to develop the overall frequency of a reactor accident. There are some caveats in the literature about how to combine the individual data to reach the whole. Shorthouse recommends simply multiplying the probabilities for the inputs to an *AND* gate. Ozog and Bendixen point out that if the data are listed as frequencies, not probabilities, life is more complex.

At an *OR* gate, Shorthouse recommends adding the probabilities of the input events and then subtracting their product. Ozog points out that all data at an *OR* gate must be in the same units. Available software generally takes care of these problems, and for complex trees, a computer is essential.

Event Tree Analysis (ETA)

ETA evaluates potential accident outcomes that might result following an equipment failure or process upset known as an initiating event. Unlike fault tree analysis, the analyst begins with an initiating event and develops the following sequences of events that describe potential accidents, accounting for

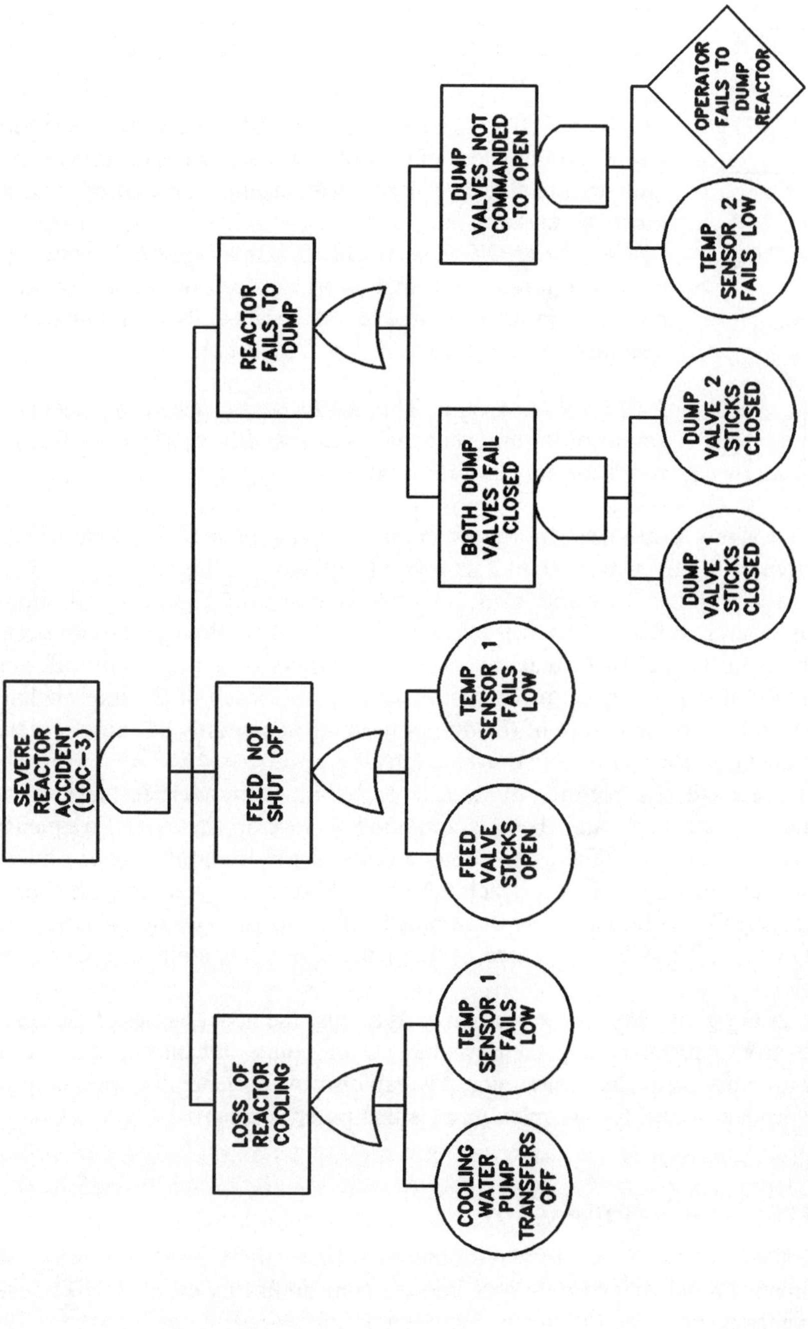

Figure 6.5. Simplified example of fault tree model.

both the successes and the failures of the safety functions as the accident progresses.

Event trees provide a precise way of recording the accident sequences and defining the relationships between the initiating events and the subsequent events that combine to result in an accident.

Event trees are well suited for analyzing initiating events that could result in a variety of effects. Each branch of the event tree represents a separate effect (event sequence) that is a clearly defined set of functional relationships.

Selecting the initiating event is an important part of event tree analysis. The event should be a system or equipment failure, human error, or a process upset that can result in a variety of effects, depending on how well the system or operators respond to the event. If the selected event results directly in a specific accident, it is better suited for a fault tree analysis that will determine its causes. In most applications of event tree analysis, the initiating event is "anticipated"; that is, the plant design includes systems, barriers, or procedures that are intended to respond to and mitigate the effects of the initiating event. For example, consider "loss of cooling water to reactor," as we did in the FTA example.

The safety functions (safety systems, procedures, operation actions, etc.) that respond to the initiating event can be thought of as the plant's defense against the occurrence of the initiating event. These safety functions usually include:

- Safety systems that automaically respond to the initiating event (including automatic shutdown systems)
- Alarms that alert the operator when the initiating event occurs
- Operator actions designed to be performed in response to alarms or required by procedures
- Barriers or containment methods that are intended to limit the effects of the initiating events

For our example, we consider the following safety functions in response to the initiating event "loss of cooling water to oxidation reactor":

- The reactor high-temperature alarm alerts operator.
- Atomatic-feed shutdown system stops reaction.
- Operator dumps the reactor.

The first step in constructing the event tree is to enter the initiating event and safety functions that apply to the analysis. The initiating event is listed on the left-hand side of the page and the safety functions are listed, in chronological order, across the top of the page. Figure 6.6 shows this completed step for the example. The line underneath the initiating event description represents the progression of the accident path from the occurrence of the initiating event up to the first safety function.

The next step is to evaluate the safety function. Only two possibilities are considered: success of the safety function and failure of the safety function. This evaluation assumes that the initiating event has occurred. The analyst must

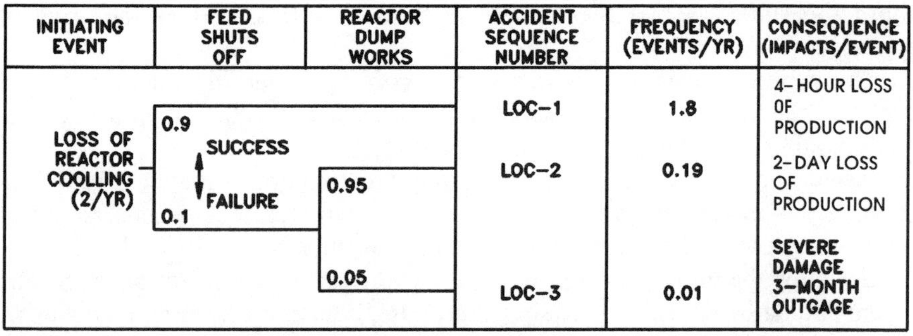

Figure 6.6. Simplified examples of event tree model.

decide how the success or failure of the safety function affects the course of the accident. Different consequences are obviously possible. In the example, the probability is shown as ninety percent that the feed is shut off and only a four-hour production loss is sustained. On the other branch of the event tree, the chances are ninety-five percent that the reactor dump works but five percent that nothing works and much more severe consequences result. The probability of each path can be calculated independently.

The problems with event trees include the potential to be incomplete if all initiating events cannot be identified, the difficulty of handling more than a yes/no action of a safety action (i.e., delayed response), and the complexity of developing the probabilities of some of the branches. Indeed, it is sometimes necessary to do FTA on some of the event tree branches to establish these numbers. Some analysts feel event trees are best for exploring the consequences of a failure and identifying possible intervention opportunities.

The frequency assessment step results in an estimate of an accident's statistically expected occurrence frequency. The estimates often take the form of very small numbers (e.g., 2×10^{-5} per year).

Interpreting small numbers such as these is often a difficult task when evaluating risk-related results.

If there is a lack of specific, appropriate data for a process facility, there can be considerable uncertainty in a frequency estimate. QRA teams often want to express their results as a range (e.g., ninety percent confidence limits 1×10^{-4} per year to 8×10^{-6} per year) rather than as a single point estimate (e.g., 2×10^{-5} per year). For this reason alone, it may be necessary that an uncertainty analysis be performed.

For most decisions, managers will have to rely on best estimates, compensating for any uncertainty with good judgment and intuition. Frequency assessment can typically require twenty-five percent to fifty percent of the total effort in a large-scale QRA study. If an uncertainty analysis is performed, the effort required for the frequency assessment can be much greater.

Consequence Assessments

The consequence assessment step involves four activities:

- Characterizing the source of the release of material or energy associated with the hazard being analyzed.
- Measuring (through costly experiments) or estimating (using models and correlations) the transport of the material and/or the propagation of the energy in the environment to a target of interest.
- Identifying the effects of the propagation of the energy or material on the target of interest.
- Quantifying the health, safety, environmental, or economic impacts on the target of interest.

Many sophisticated models and correlations have been developed for consequence analysis. Millions of dollars have been spent researching the effects of exposure to toxic materials on the health of animals; the effects are extrapolated to predict effects on human health. A considerable empirical data base exists on the effects of fires and explosions on structures and equipment. And large, sophisticated experiments are sometimes performed to validate calculated dispersion results and explosion impacts.

Like frequency estimates, consequence estimates can have very large uncertainties. Estimates that vary by orders of magnitude can result from (1) basic uncertainties in chemical/physical properties, (2) differences in average vs. time-dependent meteorological conditions, and/or (3) modeling uncertainties. In any case, like frequency assessment, examining the uncertainties and sensitivities of the results to changes in boundary conditions and assumptions provides greater perspective.

A common risk evaluation and presentation method is simply to multiply the frequency of each event by the consequence of each event. The results of an uncertainty analysis, if performed, can be presented as a range defined by upper and lower confidence bounds that contain the best estimates.

One danger in only using risk estimates presented as the product of frequency and consequence is losing perspective on the types of accidents contributing to the risk. Are they low-frequency/high-consequence accidents that would be catastrophic? Potentially severe accidents usually generate greater concern than smaller accidents, even though the risk (product) may be the same. We return then to the challenge of communication of these results, which has been discussed earlier.

Summary

Having identified the risks in a plant or operation, a responsible management must decide what to do about them. To make the study involves a commitment to do something with the results. Therefore, risk assessment follows risk identification and some form of remedial action must follow for the risks judged in need of additional control. All of these decisions must be documented and followed up to demonstrate "due diligence."

Qualitative approaches have generally been able to satisfy management's needs for information on which to base decisions. Verifying compliance with accepted standards, ensuring that risks are comparable to those at units that have been operated successfully, and adjusting the scope of risks considered in need of action to correspond with special situations at a given plant are all useful approaches to responsible risk management. Simplified calculations of airborne concentrations of toxics and energy release rates for fires and explosions can provide further insight if realistic thresholds can be established. This is particularly true if a company has an active program for risk reduction and for communicating its results for the public. In many cases, even risk management regulations do not require any more detail than that provided by a thorough qualitative assessment.

In the future, there is likely to be an increased need to use quantitative risk assessment (QRA). In special situations, QRA could be required by law, by the need to explain risks better to a given community, or simply by management's demands for a more detailed understanding of a situation before making a decision. Whenever possible, QRA studies should be designed to produce relative, rather than absolute, results to avoid many of the "numbers" problems associated with trying to predict absolute estimates of risk. If absolute numbers are required, the results must be carefully analyzed and interpreted rather than accepting the numbers at face value. Proper interpretation of the uncertainty inherent in the study must also be made and communicated. Companies that use QRA intelligently have a better base of information on which to base decisions and can demonstrate enhanced concern for the safety of workers, neighbors, and the environment.

Typical pitfalls associated with QRA are:

- Inadequately defining analysis scope and objectives
- Using QRA in situations where qualitative approaches would suffice
- Overworking the problem: analyzing more cases and using more complicated models than required to produce the necessary information for a decision
- Dictating that inappropriate analysis techniques be used.
- Using inexperienced or incompetent practitioners
- Choosing absolute results when relative results would suffice
- Selecting an incorrect risk characteristic as a factor of merit
- Not providing sufficient resources
- Having unrealistic expectations
- Being overly conservative
- Failing to acknowledge the importance of the analysis assumptions and limitations

Effective internal and external communication of QRA results is fraught with challenges even for a well-designed and executed study. This phase of the study needs to be as thoroughly researched as any other, and the perspective of each audience to whom results are to be presented must be well understood. A single "canned" presentation is likely to be a failure.

The techniques, data bases, and software systems for QRA are developing rapidly. While the use of fault trees and/or event trees is fairly standard, data on failure rates and for consequence analysis continue to be developed. Thus, last year's conclusion regarding a specific QRA may well not be correct this year. Further, the growth of public expectations and legal requirements indicates that developing a source of QRA expertise (either inside or outside) might be a prudent decision for a forward-looking company handling hazardous materials. It is the author's hope that such an effort would not be undertaken without also developing the necessary understanding to interpret and communicate the results of any QRA study.

Risk Assessment Reading List

1. A. Amendola, "Uncertainties in Systems Reliability Modeling: Insight Gained Through European Benchmark Exercises," *Nuclear Engineering and Design,* Vol. 93, Elsevier Science Publishers, Amsterdam, Holland, 1986, pp. 215–225.
2. P. K. Andow, F. P. Lees, "Process Plant Alarm Systems: General Considerations," *Loss Prevention and Safety Promotion,* Vol. 1, 1974, p. 229.
3. Bernard L. Cohen, I-Sing Lee, "A Category of Risks," *Health Physics,* Vol. 36, Pergamon Press Ltd., London, England, June 1979, pp. 707–722.
4. Vincent T. Covello, et al., editors, *The Analysis of Actual Versus Perceived Risks,* Plenum Press, New York, NY, 1983.
5. Edmund A. C. Crouch, Richard Wilson, *Risk/Benefit Analysis,* Ballinger Publishing Company, Cambridge, MA, 1982.
6. George Eads, Peter Reuter, *Designing Safer Products: Corporate Responses to Product Liability Law and Regulation,* The Rand Corporation Institute for Civil Justice, Santa Monica, CA, 1983.
7. F. R. Farmer, "Reactor Safety and Siting: A Proposed Risk Criterion," *Nuclear Safety,* Vol. 8, No. 6, Oak Ridge, TN, November–December 1967, pp. 539–548.
8. R. W. French, G. L. Peloquin, R. E. Olsen, "Quantified Risk as a Decision Aid," (internal Exxon report, Baytown, TX).
9. J. B. Fussell, J. S. Arendt, "System Reliability Engineering Methodology: A Discussion of the State-of-the-Art," *Nuclear Safety,* Vol. 20, No. 5, Oak Ridge, TN, September–October 1979.
10. Clayton P. Gillette, Thomas D. Hopkins, *Federal Agency Valuations of Human Life—A Report to the Administrative Conference of the United States* (Draft), Boston, MA, April 1988.
11. Vernon L. Grose, *Managing Risk—Systematic Loss Prevention for Executives,* Prentice Hall, Englewood Cliffs, NJ, 1987.
12. P. Guymer, G. D. Kaiser, T. C. McKelvey, G. W. Hannaman, "Probabilistic Risk Assessment in the CPI," *Chemical Engineering Progress,* January 1987.
13. D. F. Haasl et al., *Fault Tree Handbook,* USNRC, NUREG-0492, Washington, DC, January 1981.
14. Steven R. Hanna, Peter J. Privos, *Guidelines for Use of Vapor Cloud Dispersion Models,* American Institute of Chemical Engineers, New York, NY, 1987.
15. George L. Head, *The Risk Management Process,* Risk and Insurance Management Society, Inc., New York, NY, 1978.
16. R. E. Knowlton, *Hazard and Operability Studies, The Guide Word Approach,* Chemetics International Company, Vancouver, BC, 1981.

17. F. P. Lees, *Loss Prevention in the Process Industries,* Butterworths, London, 1980.
18. D. R. T. Lowe, "Major Incident Criteria—What Risk Should Society Accept?" *I. Chem. E. Proceedings of Eurochem Symposium,* 1980.
19. William W. Lowrance, *Of Acceptable Risk,* William Kaufmann, Inc., Los Altos, CA, 1976.
20. H. Ozog, L. M. Bendixen, A. D. Little, Inc., "Hazard Identification and Quantification", *Chemical Engineering Progress,* April 1987.
21. C. G. Ramsey, R. Sylvester-Evans, and M. A. English, "Siting and Layout of Major Hazardous Installations", *I. Chem. E. Symposium Series No. 71,* 1982.
22. Joseph V. Rodricks, Susan Brett, Grover Wrenn, *Significant Risk Decisions in Federal Regulatory Agencies,* ENVIRON Corporation, Washington, DC, 1987.
23. William D. Rowe, *An Anatomy of Risk,* John Wiley and Sons, New York, NY, 1977.
24. Barry O. Shorthouse, "Using Risk Analysis to Set Priorities for Pollution Prevention," *Pollution Prevention Review,* Winter 1990–91.
25. Chauncey Starr, "Social Benefit versus Technological Risk," *Science,* Vol. 165, September 19, 1969, pp. 1232–1238.
26. C. H. Vervalin, "Hazard Evaluation," *Hydrogen Processing,* December 1986.
27. Ray A. Waller, Vincent T. Covello, editors, *Low Probability/High Consequence Risk Analysis,* Plenum Press, New York, NY, 1984.
28. Chris Whipple, "Redistributing Risk", *Regulation: Journal on Government and Society,* American Enterprise Institute, Washington, DC, May/June 1985, pp. 37–44.
29. William C. Wood, *Nuclear Safety: Risks and Regulation,* American Enterprise Institute for Public Policy, Washington, DC, 1983.
30. John L. Woodward, editor, *Text Proceedings of the International Symposium on Preventing Major Chemical Accidents,* American Institute of Chemical Engineers, New York, NY, 1987.
31. John L. Woodward, editor, *International Conference on Vapor Cloud Modeling,* American Institute of Chemical Engineers, New York, NY, 1987.
32. *Canvey: An Investigation,* Health and Safety Executive, HMSO, 1978. *Canvey: A Second Report,* Health and Safety Executive, HMSO, 1981.
33. "CEFIC Views on the Quantitative Assessment of Risks from Installations in the Chemical Industry," European Council of Chemical Manufacturers' Federations, Brussels, Belgium, April 1986.
34. "Chemical Risk: A Primer," Department of Government Relations and Science Policy, *American Chemical Society.*
35. Department of Labor, Bureau of Labor Statistics, Washington, DC, 1976.
36. *Guidelines for Qualitative Hazards Evaluation Procedures,* American Institute of Chemical Engineers, New York, NY, 1985.
37. *Guidelines for Chemical Process Quantitative Risk Analysis,* American Institute of Chemical Engineers, New York, NY, 1989.
38. *Guidelines for Process Equipment Reliability Data,* American Institute of Chemical Engineers, New York, NY, 1989.
39. *Manual of Industrial Hazard Assessment Techniques,* The World Bank, October 1985.
40. Nuclear Regulatory Commission, *PRA Procedures Guide—A Guide to the Performance of Probabilistic Risk Assessments for Nuclear Power Plants,* NUREG/CR-2300, January 1983.
41. *Process Safety Management,* Chemical Manufacturers Association, Washington, DC, May 1985.
42. *Risk Analysis in the Process Industries,* European Federation of Chemical Engineering Publication No. 45, The Institution of Chemical Engineers, Rugby, England, 1985.
43. *Risk Analysis of Six Potentially Hazardous Industrial Objects in the Rijnond Area: A Pilot Study,* D. Reidel Publishing Company, Dordrecht, Holland, 1982.

CHAPTER

7

Management of Change

Introduction

Having identified a plant's risks, assessed their importance, and reduced those in need of it, the last thing anybody wants is the erosion of this status by small changes. There are numerous examples in the literature (e.g., Flixborough) in which ill-conceived changes have led to disaster. These changes could come in facilities, procedures, processes, instrument set points, computer applications, maintenance practices, operator training and/or manuals, or even in the names of those responsible for certain activities. Conversely, failure to change some things—such as process and instrument diagrams and operator training manuals—to correspond to what actually exists in the field could also contribute to erosion of risk margins and to inadequate emergency response procedures. Thus, a management-of-change system must (1) appropriately review the proposed change, (2) update the appropriate documentation, and (3) train operators and mechanics to use the new systems safely and effectively.

The need to manage change is recognized by management, by CCPS, by NJ TCPA, and by other regulations/government agencies. The tendency of regulators has been to require specific procedures and team compositions for reviewing changes that are suitable mostly for large changes. In my view, any management-of-change system has to be practical for *all* changes.

Thus a system is needed that includes the definition of modifications that mandate a *Change Review* plus a definition of the appropriate extent of the Change Review. The AIChE has given an outline of a change management system as part of their *Guidelines For Technical Management of Chemical*

Process Safety. The system discussed in this chapter includes more specifics than those included in the *Guidelines.* Its underlying principles are described below.

A Change Review should always be carried out for the following changes:

- Addition, deletion, or modification of any equipment, including piping, valves, vessels, instrumentation, except for routine maintenance replacement of equipment that does not change the function or operating limits of the equipment
- Change of operating limits—temperature, pressure, composition, chemical specifications—outside the documented proven operating range
- Bypass of critical safety alarms and shutdown systems
- Change of sequence of operating steps
- Reduction of operations staffing

It is possible that minor changes in operating conditions and instructions, which do not increase the probability or the magnitude of a release, would not require a Change Review, but this approach provides a loophole that might lead to abuse.

Examples of changes that might not need to be reviewed are limited, but include the following:

- Changes of operating conditions or compositions within established safe ranges for the process
- Periodic updating or reorganization of operating instructions to clarify them or to add additional hazard precautions

The Change Review would apply to both permanent and temporary changes. Its depth and the composition of the review team would vary, depending on the nature of the change proposed. As a minimum, all changes proposed would be reviewed and approved by the responsible operations supervisor and a central safety/risk management officer. The latter would ensure that both an independent view of the proposed change is undertaken and that the appropriate depth of review is being carried out.

For simple changes, approval and documentation by these managers would be adequate. Examples of simple changes might include a temporary leak repair, a simple control valve bypass, the dismantling of unused piping or equipment, the addition of an instrument, a short-duration bypass of a critical safety system, the car sealing of a valve, and so on.

As the complexity of a proposed change increases, a more extensive review procedure would be required by the safety/risk management officer. For example, if metallurgy is to be changed, a review by a metallurgist would be required.

Similarly, if changes to operating conditions or compositions outside the proven range are proposed, a process technologist would be added to the team. The form of review at this stage might be a simple safety checklist or a "what if" analysis done individually or as a group.

As the level of complexity increases, it will ultimately reach the point where the type of safety analysis and hazard assessment now defined in the NJ TCPA regulations is reached—a HAZOP, FMEA, or QRA. This type of analysis would be needed for such modifications as the introduction of a new toxic substances, significant expansions of existing systems, the introduction of new products, and all new capital projects.

A mechanism to update the documentation affected by the change must be included as an inherent component of the Change Review system. This includes the P&IDs, operating instructions, process operating limits, and batch composition sheets.

For temporary changes, the Change Review procedure must include a definition of the duration of the change and a system for verifying that the system has been returned to its original condition by the date set.

While it is normally intended that Change Reviews be completed and documented prior to implementation, in some emergency situations, it may be necessary to process a change prior to formal documentation. The Change Review procedures recognize that possibility by providing for on-the-spot verbal analysis but requiring an immediate formal analysis at the appropriate level, complete with documentation at the earliest practical time. The scope of the review must not be compromised even in emergency circumstances, but the documentation may follow.

Research has shown that different companies approach the implementation of these principles in different ways but that the fundamental criteria are still met. We submit that maximizing the use of existing systems and procedures is the best way to ensure that change reviews are applied rigorously and that no changes are made without an appropriate review.

Philosophy and Scope

A system for management of change is necessary to ensure that adequate consideration has been given to the impact of the change on the facility and its operating personnel, and to ensure that the change can be accomplished without compromising safety. Changes occur constantly. Each plant must understand the nature of changes, their magnitude, their frequency, and how they affect the safety of operations. To implement this philosophy, procedures must be established to document and review changes and to request their approval. A review is always carried out for the addition, deletion, or modification of equipment, piping, instrumentation, and so on, except for routine maintenance or the in-kind replacement of the equipment, which does not change the function or operating limits of the equipment, and/or the sequence of the operating steps. A review is also required to the bypass critical safety instruments and shutdown systems and to deviate from established work practices or from the company standards, ASME, and other standards. Also covered must be changes in critical instruments, instrument ranges, alarm set points, control-system software, temporary repairs, and operating and maintenance procedures and the introduction of new chemicals and/or products. A noninclusive list of changes, which should be reviewed and documented, is given in Table 7.1. Depending on the nature of a given plant, there may be more classes of items to add to this basic list.

A Change Review System

It is appropriate to make Change Review procedures as simple as possible. In addition, it is important to base them on procedures already in use and recognized in the plant.

Most plants have a process design specification and project engineering system to cover items listed first in Table 7.1—that is, significant changes to equipment, new projects, and major maintenance jobs. Safety reviews and HAZOPS are generally carried out, and the scope of project hardware is carefully monitored, albeit for financial and schedule reasons primarily. In any event, these procedures/documents normally provide a sound foundation on which to base a Change Review system for larger changes.

What of the rest of the list, then? It is possible to deal with all of these in a single system, but I recommend using two. I would cover all other items in Table 7.1 except #8 in one "field change" system and would initiate another, separately controlled approach for the bypass of critical safety systems, alarms, and equipment. These two systems have many things in common, but the timing aspects of safety systems bypasses merit special consideration.

Table 7.1. Partial List of Changes That Require Review and Documentation

1. Significant changes to process, safety or utility equipment, whether required by a capital project or maintenance activity
   — This includes changes to the facilities specified in a process design resulting from mechanical engineering developments
2. Minor changes to equipment, piping, instrumentation
3. Changes to procedures
4. Changes to process sequence, operating conditions, alarm set points, computer control programs, materials of construction, instrument ranges, electrical systems, the list of critical safety equipment and instruments
5. Special cleaning or operating procedures—a plant test run
6. Temporary repairs, jumpovers, bypasses
7. Introduction of new chemicals/products into existing equipment
8. Temporary bypass of a critical safety system if not covered by a standard operating procedure or permit (e.g., removal of a safety valve for maintenance)
9. Changes to equipment already in operation that might affect its safety or environmental performance (e.g., changing type of pump seal)
10. Changes in the scope of preventive maintenance (PM) practices on critical equipment; note frequency of PM should be guided by the analysis of historical performance
11. Changes in specific personnel protective equipment for a given operation
12. Changes in the mechanical or design standards to be applied to a given operation—granting an exception to a company or industry standard

The characteristics of any field change system include:

- Simple but effective documentation
- Visibility
- Timeliness—that is, responsiveness to operations needs
- Knowledgeable, appropriate, and independent review/authorization
- Follow-up

For most changes, these can be accomplished with the system described below, called a *field change system.* Some companies call these process changes; others, simply changes, or configuration management. M. L. Griffin described Procter and Gamble's Cincinatti, Ohio, "Maintaining Plant Safety Through Effective Process Charge Control," system at the AIChE meeting in Houston, Texas, in April 1989. He describes a number of possible pitfalls that can beset even those who are vigilant in monitoring changes. For example, the use of unstabilized SO_3 vs. stabilized material can provide some serious problems if the SO_3 freezes and must be melted.

The system described below uses different terminology than that described in Griffin's paper but describes similar requirements.

An Example of a Change Management System

The following six-part system covers all of the items in Table 7.1 except item 8, as mentioned earlier.

1. *Field change authorization (FCA) approval and timing requirements*

   All FCAs are to be approved, as a minimum, by the operations-section supervisors (shift supervisors during off-hours) and by the FCA System Manager. FCAs can, if necessary, be implemented at the discretion of the operations-section supervisor prior to FCA system manager approval, but the implementation is subject to the subsequent conditions stipulated by those reviewing the FCA. If an FCA is implemented prior to the review, the review process must be initiated no later than the first regular business day after the field change.

   The approval of the FCA by the operations-section supervisor should represent his/her thorough review of the proposed changes from an operational and safety point of view (to the best of his/her knowledge) and should not be merely an indication that he/she wishes to initiate an FCA.

   In general, it is desirable that all reviews be conducted prior to the review by the FCA system manager. However, in many instances, the FCA originator will contact the FCA system manager to determine if the FCA is the appropriate document for the proposed change, and if so, to determine the appropriate approval schedule. For instrument changes, an instrument review is required. For materials of construction, piping layout or repairs, operating conditions, or service changes, an inspection and materials (IMS) review is required. Any changes with TCPA implications require a risk management review. New raw materials or a change of chemicals in the specific process need to be approved by the Environmental and the Industrial Hygiene departments. The SOC Design Subcommittee chairperson and other reviewers (electrical, rotating equipment, etc.) will be involved in the review process as necessary.

The FCA originator is responsible for circulating the FCA to the required approvers. Special colored envelopes are useful to speed this process.

Sometimes approvals will be conditional on an improvement or modification of the FCA as first proposed. For example, special technical coverage around the clock may be required for a test run of a new product or process. The FCA system or risk manager will stipulate this condition and it would have to be acknowledged by the operations-section supervisor before the FCA is approved.

2. *Temporary changes*

   A leaky pipe patch, a hot (or cold)-weather bypass, and the insertion of an equipment component in short-lived materials are all examples of temporary changes that must be approved before installation. (Remember the temporary flexible joint installed at Flixborough that caused that disaster?) The time limit for such temporary repairs should be based on the probability and consequences of a failure in the temporary change and compliance with legal requirements. At our plant, we developed a matrix of failure severity vs. permitted time that we use as a guide. In most cases, permitted times are less than three months. Temporary FCAs are tracked monthly and the supervisor is reminded the month before any temporary FCAs of his are scheduled to run out.

   For temporary FCAs that will expire and still need to be in use, an extension must be requested using a new FCA form. For this purpose, only the following need to be provided: unit, date, originator, original FCA number, and revised expiration date. As a minimum, an extension requires the operations-section supervisor and FCA system manager's signatures. In some cases, an extension will require that the FCA be reviewed by other individuals to ensure that it can be extended without any safety or environmental implications.

3. *FCA contents*

   FCAs are documented on a one-page form available on the computer network or in hard copy. An example is shown in Figure 7.1. The originator should provide the FCA category, the location, a brief description, and a proposed title (on the upper line) for the change; the FCA system manager will verify that the proper category has been used and will provide a new title (on the lower line), if necessary. Additional pages and documents, including P&IDs and instrument specification sheets, should be attached as necessary. The quality of the backup material should be equivalent to that required for a process design specification. The FCA system is intended to

UNIT	FCA CATEGORY (See FCA Procedures for Code No.)	DATE	ORIGINATOR (Print Name)

FCA NO ______ (By FCA System Secretary)	TITLE ______ (By FCA System Mgr.)

CHANGE TYPE

☐ PERMANENT ☐ TCPA APPLIES ☐ AIR PERMIT

☐ TEMPORARY EXPIRATION: ______ ☐ MAINTENANCE RECORDS UPDATE

☐ EXTENSION OF TEMPORARY FCA NO: ______ ☐ REVISED EXPIRATION ______

LOCATION of CHANGE (Equipment Number, Block Number, and/or Street Location)

DESCRIPTION OF AND REASON FOR CHANGE:
(ATTACH PLOT PLAN,FLOW PLAN, PIPING SKETCHES, VESSEL DRAWINGS, SPEC SHEETS, PROCEDURES, ETC., AS REQUIRED TO FULLY DOCUMENT THE CHANGES)

APPROVALS	REQUIRED	SIGNATURE APPROVED	SIGNATURE CONDITIONAL APPROVAL*	DATE	PRINT NAME
PROCESS SECTION SUPERVISOR/ SHIFT SUPERVISOR (OFF HOURS)	X				
FCA SYSTEM MANAGER**	X				
SOC DESIGN SUBCOMMITTEE CHAIRPERSON**					
MECHANICAL SECTION SUPERVISOR					
IMS SECTION SUPERVISOR					
☐ I-H ☐ OM ☐ OTHER					
☐ ENV ☐ PCE ☐ OTHER					
☐ INST ☐ MCE ☐ OTHER					

** ALTERNATES: FCA SYSTEM MANAGER, SOC CHAIRPERSON, or DESIGN SUBCOMMITTEE CHAIRPERSON

* CHANGES REQUIRED FOR ACCEPTABILITY (Detailed Below)

FURTHER ENGINEERING EVALUATION REQUIRED ☐ NO ☐ YES (WHAT REQUIRES EVALUATION AND BY WHOM)

*ACKNOWLEDGEMENT OF CHANGES REQUIRED FOR ACCEPTABILTY	FCA CLOSEOUT (REQUIRED FOR ALL FCA'S)
Signature ______ Date: ______ Section Supervisor Operations Department	Signature ______ Date: ______ Section Supervisor Operations Department

Figure 7.1. Sample—Field change authorization.

provide an expedited review, but this must be done on the basis of a complete presentation of the pertinent data.

4. *Closeout of FCAs*

A closeout signature by the operations-section supervisor is be required for all FCAs. A copy of the original approved FCA cover page, stamped TO

CLOSE OUT FCA SIGN THIS COPY AND RETURN IT TO THE FCA SYSTEM MANAGER, is used for this purpose. A closeout signature will be required for temporary FCAs when the field change is no longer in use. For permanent FCAs, a closeout will be required when the change has been implemented *as described in the approved FCA*. This will trigger Design Drafting to update the appropriate drawing(s) to reflect the "as-built" change. If actual changes to equipment, piping, and so on do not exactly reflect the change(s) proposed by the FCA, the P&ID(s) in question will not be updated until the discrepancy between the proposed and the implemented changes is resolved with the FCA system manager or the operations-section supervisor. To simplify the handling of FCAs, the operations-section supervisor does not close out an FCA until it is verified that the implemented changes correspond to the changes proposed and approved on the FCA and, if there are any deviations, he/she must agree to and note it on the closeout copy of the FCA.

5. *Updating documentation*

As indicated, a copy of the completed FCA is sent to the Drafting Department to update and field check all changes to P&IDs, electrical one-line diagrams, interlock lists, equipment and layout drawings, and other plant documentation maintained by that department. In addition, the maintenance program, particularly the preventive maintenance program for critical equipment, must also be updated. To accomplish this, a copy of the completed FCA must go to the key individuals in the maintenance/inspection/instrument areas who are responsible for the maintenance function. This distribution will vary from plant to plant and company to company. As a result, each site must think through its distribution of FCAs to ensure that the correct people receive each FCA *and know what to do about it.* A partial list of documents that may require updating is given in Table 7.2.

This is a long list, but every function in the plant needs complete, up-to-date information to do its basic job and to contribute to improving

Table 7.2. A Partial List of Documents that may Require Change as a Result of a Plant/Process Change

P&IDS	Lubrication systems
Interlock lists	Battery limit isolation valve list
Start-up, shutdown, emergency procedures	Critical instrument list
Process operating specifications	Pump seals checklist
Training manuals	Vessel entry drawings
Line lists	Electrical one-lines
Equipment lists	Equipment blanking lists
Safety equipment checklists	Unit inventory checklists
Preventive maintenance checklist	HAZCOM documentation
Pump vibration checklist	Critical electrical equipment list
Computer control logic diagrams	Safety valve list
Mechanical procedures documentation	Preventive maintenance procedures

operations, safety, and profitability. There are many different interests in and uses for the same basic knowledge, albeit requiring somewhat different emphasis on details.

Fundamentally, two kinds of information are required: process and mechanical. In many cases, the two must interact, but I believe the responsibility for keeping information up to date can be split among two, or at most three, functions. A group composed of operators and technical/engineering people can be charged with maintaining the process information bank. This includes the "bibles"—that is, complete P&IDs and any supplementary computer interlock and control logic diagrams, electrical one-line diagrams and sewer diagrams, and the operating and emergency procedures, including safety relief systems.

From the above description, it should be clear that neither the operations nor the technical group alone can handle this responsibility in larger plants. In very small plants, where the two groups are one in reality, perhaps they can, but in many such, there are too few people to do the job. One of the major challenges in many plants is to get the operating folks to really "own" the process information, particularly the P&IDs and interlocks/alarms. Updates to these critical documents are prepared by others (i.e., draftsmen) and operating people often feel too busy to review them in detail. The discovery of any minor error leads to a general lack of credibility in the efforts of staff that, to my mind, is intolerable. A system to ensure that updates are done right the first time by teamwork between technical/engineering and operations people at the hands-on level is essential. Each group must work together to ensure correctness and certify the drawings, lists, and the like. So much is based on the process documents that anything else is simply unacceptable as an internal standard and likely to be viewed as criminal by outsiders.

The responsibility for updating mechanical information may have to be split between maintenance and inspection depending on plant size and organization. As will be discussed later, there are some advantages to having the inspection group independent of the line functions, but there is a price to pay. If practical, a maintenance/inspection team can be charged with maintaining mechanical records and procedures. Blending the instrumentation function into this team requires some extra effort. If a plant has the benefit of a computerized maintenance system, life is easier. If not, compatible teams need to be formed for each unit and supervision must insist on stewardship of their work product.

Wherever possible, I believe it is advantageous to have a maintenance control function through which maintenance (mechanical, electrical, and instrument) and inspection work activities flow. This function can then ensure that the appropriate records are updated as each task is finished. In larger plants, each unit may require such a focal point, but by so doing, perhaps document maintenance can be simplified to only two responsible functions, namely operations and mechanical. Fundamentally, operations management must "own" the information, but they must have help in keeping it correct.

6. *Updating operator training*

   Since the operating-section supervisor is an integral part of the FCA process, the need to update his/her operators and ensure that they understand how to accommodate the change in question should be obvious to him/her. Depending on the nature of the change, separate training sessions may be required for each operator. These may, in fact, be required by law as well. Simpler changes may be handled by documenting the changes in a logbook and requiring all operators to initial their reading (and understanding) of it. In some cases, the shift supervisor may follow-up closely on the first operation of his/her shift on the change. There is no rigid system that is universally applicable. It is very much a performance standard: The people who operate the plant must know and understand the impact of any change on how they go about their day-to-day operations.

Using the Form

To demonstrate how these aspects are captured on a one-page form (with a page of instructions on the back), let us talk through Figure 7.1, a sample FCA form. The top few lines of the form identify the location and nature of the change and the originator and provide space for a tracking number. The next block catalogues the type of change: permanent or temporary, original or renewal, with expiration dates approved and whether a maintenance system update is required.

The TCPA box is important in New Jersey and a similar label may become an important consideration throughout the United States now that proposed OSHA regulations have become law. Those who have read this book from the beginning know that TCPA is a New Jersey law that mandates a risk management plan for companies that use certain chemicals. One requirement is that every *modification* to a TCPA-covered process have a *safety review*. We use the FCA review process as our safety review, so if the TCPA box is checked, some documentation needs to be a little more thorough. Specifically, we must answer explicitly three questions:

1. Is the probability or size of a release to the environment increased by the change?
2. Is the change in accord with company engineering standards?
3. Is the technology "state of the art"?

Since "state of the art" includes economic considerations, there is seldom a problem with the concept.

Any "wrong" answer to these questions gets thorough scrutiny and complete documentation. If it will not stand the light of public scrutiny, a different solution is developed. Our first NJDEP audit of TCPA procedures endorsed this approach, but the chance remains that subsequent inspections may require more.

The next blocks cover the specific location of the change and its specific scope and logic. The attachments are often critical to this definition and help determine whether an FCA is the appropriate vehicle to document the change. If it is too big or too complex, a design specification should be prepared. This is mostly a judgment call by the FCA system manager but may be further reviewed by the safe operations committee (SOC), its chairperson, or its design subcommittee.

The next block covers approval signatures (conditional or complete) determined to be required by the FCA system manager. Because a Change Review system will only work effectively if it is rational and timely, arbitrary multiple reviews *must be avoided.* Each change must be reviewed and approved by *all* those essential to the issue, but *only* those essential to the issue. Bureaucracy will cripple the system and cause its bypass.

Timeliness is also critical. We use bright yellow envelopes to identify FCAs in the review process. Those people in the normal approval loop know they must handle their part promptly. It is not unusual for a critical FCA to be hand-carried through the approval process in one-half day. Less-critical items might take several days and problem changes often cause a meeting of essential people in an attempt to resolve the issue at one sitting.

The bottom block provides for a description of the changes required for acceptability, if any. Often this area will be used to require technical coverage on an off-shift or specific operator training, as well as specific improvements or limitations to the requested change. At the bottom left, the operations-section supervisors must commit to the stipulations before the change is implemented.

When the change has been implemented or a temporary change has been phased out, the section supervisor again signs the form and distributes the closed-out FCA to the required functions. This is important because some of the appropriate changes in documentation are triggered by this step. For example, if the FCA involved a change in plant facilities, the closeout of the FCA signals the Drafting Department to update the P&IDs, electrical one-line diagrams, and so on. It also alerts the maintenance people to implement any required updates in their system. If a temporary change is being phased out, the FCA alerts those interested that a return to normal practice has occurred, that a plant test period is over, or that a permanent repair has replaced a temporary one.

One debate that rages on is whether temporary FCAs should mandate an update of critical documentation such as P&IDs. Many temporary changes are of such short duration (approximately thirty days) or are nonprocess oriented (a temporary leak repair) that P&ID update is not thought practical. However, changes that linger for three to six months also occur, and whether to update documentation is left to the judgment of the FCA manager or SOC. NJDEP is not quite comfortable with this because it depends on a person, not a system. They have a point.

Schematic Change Flow Diagram

In another effort to save a thousand words, I shall again resort to a picture to capture what has been said so far. Refer to Figure 7.2. Two kinds of changes are identified: "complex" and "simple." The difference between the two is a judgment call by the change-system manager or SOC. Complex changes involve significant changes in facilities, a new product or process, or pervasive realignment/rework of existing equipment. Simple changes involve minor equipment changes, evolutionary changes in process conditions and/or materials, and a plethora of temporary changes, such as temporary repairs or test runs.

Complex changes require thorough process and mechanical design studies and documentation. These are usually referred to as process design and mechanical design specifications. In cases where the change is largely mechanical with no process implications, documentation may be rigging plans, foundation or structural rework, dismantling plans, piping schematics, and so on. The changes may be either capital or expense oriented but will usually involve significant cost and/or potential impact on operations and other units. A brief checklist for the kind of data needed in a specification for a modification is given in the Appendix to this chapter.

Most companies have a system in place to review such projects and manage them through to completion. It should consist of a safety and operability review of the front-end engineering, of P&ID reviews and a HAZOP, of approval of the mechanical interpretation of the process intent, of purchasing and inspection procedures, of project change authorizations, of mechanical checkout results and the resultant lists of incomplete items, and of pre-startup safety checks. The P&IDs, vendor drawings, equipment details, code calculations, and so on need to be preserved for future use; spare parts need to be collected and catalogued; maintenance procedures planned; operating procedures need to be written; and operators (and possibly mechanics) need to be trained. In major modifications (as opposed to new facilities), most of the documentation will consist of modifications to existing drawings, data, and procedures as opposed to creating new ones. The net result must be the same: a complete record of the facilities "as built" and the way they are intended to be operated.

For simple changes, the objective is also the same. The key element is that a Change Review/approval and documentation system be used. To make this happen *for all changes,* the system must be simple, timely, and efficient. It must also be strictly adhered to! Any bypassing will lead to proliferation. If your system is not working, find the root cause and fix it.

As a reminder, all changes not covered by a design specification must be captured by this system. This includes all of those changes to control systems,

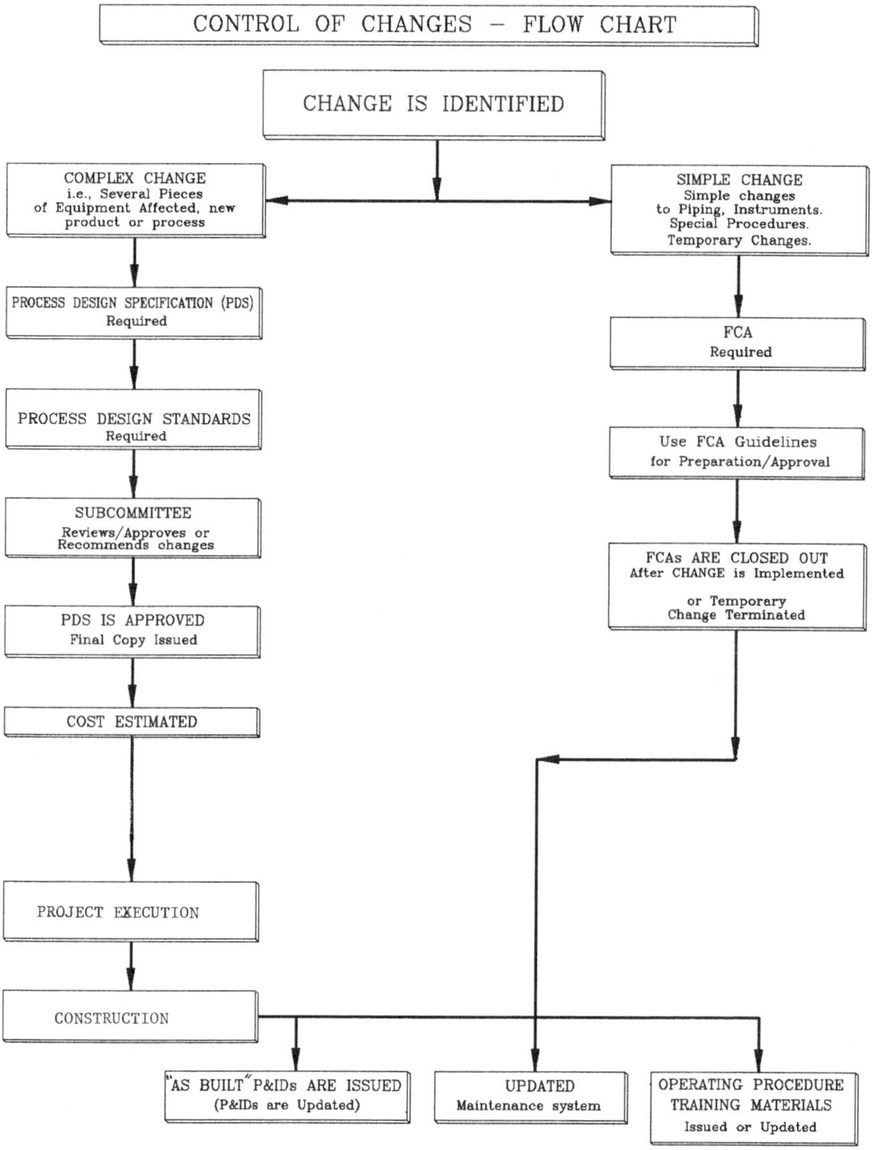

Figure 7.2. Control of changes—Flow chart.

to instrument ranges, to computer logic, to safety interlocks, to the operator intraction with the control system, to operating procedures and/or flow rates that might impact permitted effluent rates, and to all other nonhardware changes that may be initiated by process, instrument, or application types who may not be as intimately familiar with the change management system as the mechanical types.

A brief example may serve to emphasize the importance of the changes mentioned above. A seemingly routine FCA arrived requesting a range change in a particular flow instrument. At first glance, there seemed no safely implications, but a brief conversation with the instrument engineer who initiated the request revealed that this meter controlled the feed rate to the unit. Changing its range made it possible to increase the capacity of the entire unit! Further investigation revealed that the potentially higher flow rate changed the load on some safety valves and on the refrigeration system machinery and inventories. All of these implications were lost on the instrument engineer and were a surprise to the operations supervisor. The FCA review prevented blundering into some significant risks.

Work Permits

Work-permit systems are often treated as a separate entity in plants and by government agencies. To me, they are the final control on the implementation of many changes and on the verification of how the equipment performs. Thus, I have chosen to include a discussion of the essentials of a permit system here.

Many of the "standard" permit rules have evolved from the refining industry, where repair/construction work must go on amid continuously operating units that contain explosive and other hazardous materials. Everyone knows that there must be controls on sources of ignition in areas when flammable mixtures might be present. Not everyone has been so alert for other possible accidents in the workplace, especially if a unit was shut down, purged, and supposedly "safe." OSHA has recently instituted its "Lock Tag and Try" rules, covering sources of dangerous energy that greatly broaden one's perspective on the possible sources of accident; but any force, including gravity, must be respected.

The fundamental principles of a good and workable permit system including the following:

- The responsible operations person must prepare the equipment for the work and inspect the equipment and the work site to ensure that they are in the condition planned. This may include arranging for an independent analysis for flammable gases or adequate oxygen.
- The permit must specify the work to be done, the permitted time period, the risks associated with the job, the required personal protective equipment, and any possible interactions with other work groups.
- The operator and the mechanic must verify that all appropriate lockouts, tags, and tests are done.
- The mechanic must sign off when the job is complete and the site is cleaned up, and return the permit to the operator.
- The operator should verify completion.
- Management must audit the permits frequently to ensure the system is being followed.

These principles are implemented in different ways by different companies. They are just as critical in nonhydrocarbon, nontoxic environments as in those featuring more hazardous processes. Very early in my career, I was witness to the removal of two mangled bodies from a sodium bicarbonate dryer. The accident resulted when the scraper was started with the men inside because either the operator or the mechanic had locked out the wrong drive mechanism. Such issues are now covered by the OSHA rules, but OSHA rules will not ensure safety in all plant work. Thorough permit procedures implemented by knowledgeable operators and mechanics can help.

There are generally several types of permits used in a plant. Hot-work permits cover welding, burning, grinding, open flames, sparks, high temperatures, and the use of non-explosion-proof motors and motorized equipment (e.g., trucks). Cold-work permits are normally issued for work that cannot result in enough heat to ignite a flammable mixture, but care must be taken to consider the possibility of static electricity discharges. Usually separate permits are issued for entering confined spaces (e.g., a tank) in addition to the hot- or cold-work permit. Separate permits are also used sometimes for other specific cold-work activities such as excavation, fire water-system repairs, and electrical system work. The specific rules for each type of permit must be understood and followed.

Analysis of incidents often show that permit violations contribute to the causes of unwanted events.

Bypass of Critical Safety Systems

As indicated earlier, I believe this category of change needs a separate, highly visible system. Of course, there are some who say that this circumstance should never arise. The unit should be shut down rather than put a process in this situation. The fact is that there is a vast history of successful approaches to compensating for the short-term absence of a critical safety device through other safeguards. It is equally true that some organizations still just bet the odds that the double contingency of needing a safety device during the (hopefully) short time it is out of service will not occur.

Perhaps the most common critical-safety-system bypass occurs when a safety relief valve is removed for preventive maintenance while a unit is operating.

Some companies cannot tolerate the irony of removing a safety device to ensure it will operate properly. These companies install spares. Most do not, but rather substitute an additional pressure watch of some sort, or reduce throughput, or go on total recycle to reduce the risk of overpressure below what normally exists.

In these days of computer control and safety software, identifying critical safety systems and defining adequate countermeasures is more complex. This is the primary reason why I believe that separate scrutiny and daily review by high operating management is essential for this category of change.

The first step in this system is to define what is meant by a critical safety system (CSS). A CSS is a device or system that prevents damage to containment, injury to personnel, or a release of hazardous materials to the environment. As such, it is not the normal high-pressure alarm on a reactor, but rather, the safety valve that prevents the reactor from exploding or blowing a seal when the alarm pressure is exceeded. In the software area, it is not the alarm that signals that the rate of temperature rise in an exothermic reaction is higher than normal, but rather, the system that shuts off the feed, injects the catalyst poison to quench the reaction, and opens the cooling valve to maximum.

The actions that must be taken to create a CSS bypass system include the following:

1. Identify all CSS. This would normally be done as part of a HAZOP or in a separate review with SOC participation.
2. Develop a simple form to describe the desired bypass, its duration, the countermeasure to be put in place, and the reason for the bypass.
3. Develop clear and immutable requirements for the authorization level for a bypass that are operative 8,760 hours per year. These are often second-level operations supervisors and site shift superintendents.
4. Review each bypass daily for compliance.
5. Grant extensions only by authority of the operations manager.
6. Steward performance quarterly via an independent safety team such as the SOC.

Such a system has been accepted as satisfactory for compliance with TCPA requirements in New Jersey as part of an overall risk management plan that met all other requirements of the law. It is independent of specific personnel, brings maximum expertise to bear on the problem, and provides for high-enough authorization authority and frequent review. The attributes that separate it from the normal change system are of degree. These changes are *more critical,* sometimes *more sophisticated,* and need to be *more visible* to operating management.

Criteria for Reviewers

The systems espoused in this chapter rely on knowledgeable and independent review and analysis by some key people. The field-change-system manager and the SOC (of which the former is a member) are critical. What kind of people are these and how many of them does a plant require to control change effectively while still remaining competitive?

The criteria for selecting these key people include some general criteria as well as expertise in certain specific areas. The general criteria involve high standards, integrity, leadership ability and the ability and willingness to be a helpful consultant. Effective consultants experience few occasions on which it is necessary to reject a proposal, but the strength of character to do this is also essential. The specific experience required includes operations, mechanical process design, HAZOP techniques, personnel safety, industrial hygiene, and risk management. Knowledge of the plant's operations and inherent hazards are also important.

Obviously all of the people cannot process all of this experience. The FCA system manager and the chairperson of the SOC require the broadest range of skills. These are key consultants, risk assessors, leaders, and question-raisers. Depending on other duties, they could be one and the same person. Background in process design and operations or mechanical is critical. So is stature in the organization. Ten-plus years of experience is probably the minimum here.

Other members of the SOC are chosen to round out the expertise of the committee. Often the personnel safety and industrial hygiene input is by career specialists. In many plants, people circulate through mechanical and operations assignments, so one member may bring more than one skill. If people are available who have had experience at another company or at another plant site, these often bring added value to the committee. In larger plants, it sometimes is efficient to set up a separate subcommittee of senior process designers (say, three) to review all designs and refer only critical questions to the main SOC.

Some plants choose an operations manager as chairperson of the SOC. I happen to think this is a mistake. Often, operations is the initiator of the change that is being reviewed. The inherent conflict of interest can lead to less independence than I believe optimum. If the plant has an independent safety or risk manager or a senior technical person expert in process design, he or she would be the preferable choice. In this role, the SOC chairperson *must* report directly to the plant manager and not to a line manager. After all, it is the plant manager whom the authorities will try to convict if an accident occurs.

The truly critical skill for SOC leadership (and membership) is the ability to be a helpful consultant. This means the ability to listen and understand, the ability to help solve problems and give guidance. Many plants structure their SOCs like traffic cops. They know all the rules, refuse to discuss preliminary ideas, and reserve judgment until the final step of the process. Such SOCs are viewed as roadblocks. No champion of change volunteers his or her worries to such a group. If the SOC does not ask the correct questions, any sort of potential danger can pass undetected, unanalyzed, and, therefore, unprevented and unprepared for.

Consider the consultant approach. In ideal circumstances, when someone first gets an inkling of an idea for a change, he or she talks to his/her favorite member of the SOC or to the Change System manager to get some input, to learn of issues of possible concern, to solicit guidance toward the safest approach and sources of expertise related to his idea, and so on. Hallway conversations ensue, calls are made to sources of expertise, and solutions coalesce with SOC input. There is no hill to climb at the end and the full range of possible risks has been addressed. In addition, all of the participants have learned something that will help on the next question. The total involves a better solution, less recycling, and far better relationships.

The possibility exists that even this process does not result in an acceptable solution. At the very least, the initiator knows that everybody tried to get one and he or she should have nothing to complain about.

Summary

Having developed a sound low-risk configuration for a plant, the first step in risk control is to prevent changes from undermining that configuration. An effective control of change system is required to accomplish this goal.

Most plants have effective systems to control large changes such as new units, new products, and new processes. Even if driven by classic economic consideration, most project management practices are effective in providing appropriate safety reviews and documentation of the changes brought about.

Managing smaller changes requires a timely and effective system to accomplish three goals:

- An effective safety/environmental review
- An update of critical documentation
- Appropriate training of operators and supervision to implement the change and an update of operating manuals and so on

To accomplish this, a field-change-authorization system is proposed that relies on the judgment of two key individuals, the operating-section supervisor and a risk management advisor, to achieve a timely, knowledgeable review and approval of *any* change, permanent or temporary. Others are drawn into the review process by these two as appropriate. Inherent in the system are ways to update process (P&IDs) and mechanical (maintenance and inspection) data bases so that plant documentation, upon which so much else depends, is kept up to date.

While the vast majority of changes are handled by the above procedures, a separate system is recommended for the special cases in which critical safety systems (CSS) are temporarily bypassed. This system provides more sophisticated analysis, more timely review, and more visibility to the management of these critical changes.

The implementation end of any change often involves a work permit. The principles of such a system involve careful preparation by the operator and effective communication to and from the mechanical group involved. Musclemen are not required to manage change effectively. Leaders with consulting skills are. A "helpful consultant" approach is the key element to obtaining the most comprehensive safety review and most effective configuration for any change in the shortest time. Safety reviews cannot be roadblocks in a competitive business. Nevertheless, some changes may not be approved if a plant is to meet all its responsibilities to its workers and the neighboring community. With the right attitude and stature, a plant's SOC can develop this kind of role.

The appendix to this chapter lists some minimum requirements for specifications involving changes to existing units.

APPENDIX TO CHAPTER

7

Minimum Requirements for Process Design Specifications (PDS) Involving Modification to Existing Unit

A. General Requirements

- Provide overall index with page number references if specification is larger than twenty pages.
- Include simplified flow diagram if PDS contains more than three onsite process flow plans.
- In a large PDS, format specification with sectionalized divisions to aid in reviewing (i.e., safety section, plot plan section, flow plan section) unless there is a good reason for not following.
- Include PDS sign-off sheet and environmental sheets for air, water, and solid wastes, if applicable.
- Clearly spell out in text exceptions to standards.
- In a large PDS, provide summary lists in each section of all new mechanical equipment in that section.
- State in text required project completion date if critical.
- Specify temporary start-up or interim operation equipment or piping that is required for project, and instructions on whether or not it will be dismantled, and timing for dismantling.
- The designer must field check all existing facilities and should not rely solely on P&IDs.

B. Spec Sheets and Checklist Requirements

- If safety valves are installed or modified by the project, the design spec must contain a safety valve spec sheet and a safety valve contingency sheet in the safety section of the PDS.
- Specification sheets must be provided for all new or modified instrumentation with all applicable physical data on flowing fluid filled in. Sheets must show type of alarms, special shutdown-system requirements, and instrument relocation requirements. If the instrument is critical, then the critical class, alarm priority, and operator response must be indicated.
- Specification sheets must be filled in with all process data on all new or modified equipment.
- Vessels shall be specified by a detailed sketch sheet that provides all process information required to start mechanical design.

C. Process Flow Plans—General Requirements

- Flow plans must show all new piping, block and check valves (instrument valves not required), blinds, instruments, major and minor equipment, and tie-ins to existing lines or equipment).
- Flow plans must be neat, legible, and understandable by engineers not familiar with process.
- All flow plans should be as complete as possible in themselves so that continuous reference to written material is not necessary for clarification. If flow plan cannot clearly show intent of design, include additional data on sketches or spec sheets.
- All equipment on flow plan shall be identified by information block at top or bottom of flow plan, with name, number, and pertinent design conditions and information. See Attachment 7.1 for design information required.
- Actual operating pressure/temperature of all pressure vessels should be shown inside of flow plan symbols to aid in review.
- Use standard symbols/connections for equipment.
- Flow plans shall be arranged to show the interrelationship of several pieces of equipment, including existing equipment and instrumentation that interacts with design. Tie-ins may be detailed on separate sketches for clarification but should not take place of overall flow plan.
- Be sure final drawings are legible (i.e., adequate spacing around lines, lettering and symbols easy to read, etc.).

D. Process Flow-Plan Piping, Instrumentation, and Safety Valve Requirements

- Each process line "entering" and "leaving" each separate drawing shall be identified in its own information block, at left or right edge of drawing, as

ATTACHMENT 1

(Information to be included on Flow Plans)

EQUIPMENT	PRIMARY IDENTIFICATION	DETAILS REQUIRED (1)
Vessels	1. Size (overall dimensions) 2. Design temperature (min/max) 3. Design pressure	1. Elevation of bottom tangent line above grade 2. Vessel internals: a. Type and number of trays in towers b. Type of internals in drums
Heat Exchangers	1. Surface area 2. Heat transferred 3. Design temperature and pressure (both sides)	1. For condensers and reboilers show elevation above grade
Fired Heaters	1. Heat fired 2. Heat absorbed 3. Coil outlet temperature	---
Boilers	1. Steam capacity 2. Design pressure 3. Superheater outlet temperature	---
Cooling Towers	1. Waterside throughput 2. (Operating) inlet and outlet water temperatures	---
Tankage	1. Size (diam. x height) 2. Net capacity 3. Type (floating roof, cone roof, etc.)	1. Surface are of any heaters 2. Number and type of agitators or other mixing devices
Pumps	1. Rated capacity 2. Delta pressure 3. Specific gravity	1. Warm-up (or cool-down) piping
Steam Turbines	1. Steam inlet and discharge temperature and pressure	1. Warm-up piping 2. Washing facilities 3. Strainers, traps and drains
Compressors	1. Power rating 2. Volumetric flow at inlet conditions 3. Discharge pressure and temperature	1. Flushing and drain facilities
Instrumentation	1. Type (symbol) 2. Local or board mounted (symbol)	1. All primary instruments
Control Valves	1. Size 2. Flashing service 3. Valve position with actuating medium failure (symbol)	1. Block valve and bypass piping
Pressure Relief Valves	1. Set pressure	1. Inlet and discharge piping including drains and vents 2. Snuffing steam to discharge 3. Interlocks 4. Car sealed valves
Piping	1. Line size and rating (flange rating) 2. Line number 3. Line material 4. Valve size if other than line size 5. Each drawing shall include: a. Description of process or utility media entering or leaving b. continuation reference to subsequent drawings	1. Bleeds associated with valves or between double block valves 2. Chemical cleaning and flushing connections; sample connections (indicate any heaters or coolers) 3. Process drain and vent piping 4. Temporary strainers 5. Utility stations 6. Location of Figure 8, Hamer, or other permanent blinds 7. Line slope 8. Car sealed valves 9. Identification of valves for emergency isolatio or shutdown, liquid pulldown or vapor blowdown services

(1) HEAT TRACING, INSULATION AND OTHER PROVISIONS FOR WINTERIZING AND PROCESS PROTECTION AGAINST AMBIENT TEMPERATURES SHALL BE INDICATED FOR ALL FACILITIES TO BE TREATED.

to source or destination and drawing on which line continues.

- Identify flow plan lines by showing size/flange/rating/material/corrosion allowance, insulation requirements (hot/cold), and steam tracing requirements. Existing lines should have similar data plus tie-in location. Show all piping specification breaks.
- Show new piping in heavy lines with flow direction arrows; existing piping should be shown light.
- On flow plans, show failure position of control valve, size of control valve bypass line and valve, safety valve size, safety valve or R/D set pressures, and restriction orifice sizes.
- Clearly show destination on flow plans of water streams to sewer.
- Clearly show extent of dismantling on flow plans (or on isometrics if complex piping arrangement). Field-verify that existing piping that is dismantled or modified agrees with P&I drawings.
- New facility instrumentation symbols on flow plans should be compatible with symbols used on existing unit flow plans.

E. Plot Plan and Civil/Structural Requirements

- Any special information that is available that would affect the cost estimate, such as fireproofing requirements, must be included in the PDS.
- Plot plan(s) showing actual location and spacing distances of new/existing equipment and plan(s) showing location of off-site tie-ins should be included.

F. Design Considerations and Design Calculation Requirements

- Design considerations must be issued for each process design. They can be issued later as a separate document or included with the PDS and must explain the intent of the design where not obvious.
- PDS comments received during the PDS review period, together with responses to the comments, should be retained in the designer's file system. On large projects, PDS review meeting minutes should be documented and saved for future reference.

G. Design Spec Review Requirements

- A safety consistency review should be considered before the final spec is issued. This review will be done by the designer(s) with one or more of the senior technical staff members present.
- An SOC design subcommittee review is currently required before SOC signature for all PDSs. Operations approval is required before PDS is submitted to SOC subcommittee.

CHAPTER

8

Risk Control/Reduction I: Mechanical Aspects

Introduction

A great deal has been said so far about the more process aspects of hazards and risk. We have talked about eliminating/minimizing hazards, about risk identification and assessment, and about systems to prevent the infiltration of new risks. All of these aspects are very important to understanding the nature of the risks associated with a given operation and defining the problems to be solved. We must know the limits of both normal and abnormal process conditions to be able to deal effectively and economically with them.

We deal with all process requirements in three ways:

1. First and foremost: containment
2. Safe handling and disposal
3. Accident mitigation and emergency response

The next several chapters will focus on these steps. The aspects the are primarily mechanical will be discussed in this chapter, the human/mechanical interface (procedures, training and ergonomics) in Chapter 9, and mitigation/emergency response in Chapter 10. Ensuring compliance with all of the internal standards, systems, and procedures set up to deal with risks will be the subject of Chapter 11. Once all these aspects have been explored, we shall attempt to unite them into a cohesive risk management system in chapter 12.

Before discussing specifics, a word about the expecations of the public is in order. The public expects that every company will do its best to prevent accidents, especially those that might affect the health of workers and the community. The public includes a company's employees and the various agencies charged with regulating it.

Please notice that I have said nothing about deliberate releases of harmful materials. These are either officially permitted by a regulatory agency (and, therefore, a matter of public record and scrutiny) or illegal—that is, criminal. Nothing in this book is intended to deal with criminal activities.

Inherent in both public expectations and public policy is the term *state of the art* (SOA). It is expected that equipment design, operating procedures and controls, accident mitigation provisions, and emergency response capability will be state of the art—that is, the best available. Various agencies may have different terminology, but the expectation is the same: Do the best there is!

At first glance, this is a little scary. NJ TCPA makes the term SOA an official part of its regulations. Its definition, however, is reasonable. It includes technology that is commercially available or at least proven at a scale that would be readily adaptable to commercial use. Economics also plays a part. The technology must be available at reasonable cost and any technology considered must provide benefits in risk reduction that justify its incremental cost. At this writing, NJDEP is willing to listen to (and learn from) reasoned engineering analysis about what constitutes state of the art. They do compare various companies' solutions to similar problems and will play one off against the other to rachet up standards, but, as yet, they have set no absolute technology standards. This may come as their data banks collect more information on various companies' views of what is economic, however. Setting such standards would make their job easier, but individual companies often know far more about what is possible and effective technically in their process areas than any agency is likely to be able to assimilate for a wide range of processes.

Design Standards

That is not to say that there is not a base technology position that must be attained by everyone. Various building codes, boiler codes, fire codes, fugitive emissions regulations, and so on have the force of law and require certain technology in anything built. In some states, the ASME codes (in Europe, the DIN standards), the ANSI standards, and the NFPA standards also have the force of law. Other standards indirectly assume the force of law via overall endorcement by industry. An example would be the Chlorine Institute Standards, which discuss in detail proven ways of dealing with this extremely hazardous material. Thus, various API standards and those espoused by industrial insurers, testing laboratories, and others form a part of a technology base that is at least expected and often required as a method of doing business.

There are also, of course, some common-sense principles which should be applied in designing containment systems for hazardous materials. A simple one is: Provide as few opportunities for leakage as possible. This means mostly all welded piping with minimum flanges (for maintenance); double mechanical seals on pumps and compressors, or canned pumps; accounting for thermal expansion possibilities in liquid piping and heat exchangers; reinforcing small piping to prevent breakage; enclosing sample-taking arrangements; using sophisticated valve packing; eliminating gauge glasses on certain vessels; using plugs on drain valves; using tight shutoff valves at critical points; back welding screwed fittings; supplying proper pipe supports; and so on.

Thus, knowledge of much of this information is important to establishing an acceptable technology base for any business. This is not to say that blind adherence to some standard is either required or the answer to all problems. Thorough, documented engineering analysis goes a long way toward establishing both credibility and competitiveness. This might be as simple as a one-hour brainstorming session that concludes, "There is no reason why we should not adhere to the Chlorine Institute Standards." Or, it might be a brief engineering study to show that the remote activated shutoff valves recommended by the Institute at tank-car unloading stations are not needed in the system because other methods of dealing with a piping failure are in place. In either event, proper documentation is important. Such studies, in effect, set internal company standards.

Sooner or later, a manager (or company lawyer) is bound to ask whether setting internal company standards does not imply some future liability if they are not universally followed. This discussion will probably go on for some time, but the direction of the regulatory agencies is pretty clear. They will either set standards for you (if they do not accept yours) or hold you to your own. Either way, you are damned if you do not adhere, sooner or later. In my view, it is better to assume control of your own situation, if possible, and enjoy the benefits of that initiative in your competitive business situation. Having someone else's standards imposed upon you may or may not result in a level playing field in the business sense at any point in time.

Conclusion: Decide what is right for your situation and take the initiative in establishing it as the standard for you (and your competitors). Then live with it and carefully document any deviation.

Layout/Arrangement Considerations

The factors that go into plant layout/arrangement decisions are so complex that few standards exist except for some regarding storage of flammable materials. Yet space is one of the major safeguards around any hazard. Some ideas about plant location were discussed in Chapter 4. Similar ideas apply to the arrangement of equipment inside a given site. Many of these considerations impact both the safety of the community and the continuity of operations.

Fundamentally hazardous materials should be kept a minimum of 200 feet from plant property lines, regardless of the nature of activity across the fence. If a residential area or other sensitive population abuts the property line, 200 feet is inadequate spacing for any but the tiniest toxic leaks or flammable gas clouds unless winds are *always* blowing from the sensitive area into the plant. The Chlorine Institute calculations cited earlier and the Pemex LPG incident in Mexico are ample evidence of the truth of this statement.

So what is one to do if he or she must store/process hazardous materials close to a sensitive population? There is no set answer, but the logical approach would be to process the least-hazardous materials close to the fence, provide accident mitigation facilities (sprinklers, foam, remote shutoff valves, enclosures, scrubbers, drainage, fire walls, foliage, alarms, detectors, etc.), and then assess the risk to decide whether you really want (or can afford) to do what you say you must. Recent history says that the cost of killing some birds and/or sea otters is horrendous. The current cost of a Bhopal is prohibitive. Only a very, very low probability of such an event would be acceptable now.

Inside the plant, spacing is also critical. For units containing flammable materials, there must be enough space between units both to prevent proliferation of a fire and to allow adequate response to it. Generally, a fifty-foot separation is specified for the former and twenty-five feet with twenty-foot overhead clearance for the latter. In addition, access from two sides is recommended for firefighting, and if significant light ends inventories are involved, a distance of seventy-five feet is recommended between separate fire-risk areas. Stacking equipment more than three high is never recommended, nor is placing other equipment over pumps or compressors handling flammable materials, over air fins, or over any equipment running above 600°F or above the autoignition temperature of the contained materials.

Remote isolation valves can do much to minimize the fuel to a fire or the release of toxic material resulting from a fire or other accident. If these valves cannot be located a minimum of twenty-five feet from the most likely fire point, they should be remotely operated with the actuator and control cables properly fireproofed. Remote activation points should be fifty feet from the most probable fire point (e.g., a pump seal), as should manual activation valves for water sprays or a snuffing steam.

Spacing of control rooms merits special consideration. Depending on the nature of local hazards, the construction of these buildings is also critical. There are two purposes for special standards for control rooms: (1) safe shutdown of the unit(s) and (2) a safe haven for unit personnel. A control center for a single hazardous unit should be at least fifty feet from the unit and have substantial fire resistance if appropriate. A multiunit control center needs to be significantly further away (perhaps 200 feet) and blast resistant. The latter is consistent with both objectives above if fire/explosion is a major hazard. If toxic releases are possible, the control center should be capable of isolation from outdoor air and of harboring all unit workers safely for a significant period of time. Perhaps a supply of safe breathing air will be necessary for critical operating posts to affect safe shutdown.

Pipe bands also need special consideration. Those carrying interunit connecting pipes should be located at the circumference of any on-site unit, and should be free of internal unit piping. All piping, except that connected directly to a storage vessel, should be outside its fire banks. These are ways to prevent involvement of pipes external to a unit in any unit accident.

Three Worrisome Elements

Three elements are sometimes overlooked as sites evolve and/or units are modified. This is especially true if a number of small equipment or process changes have gone on over the years:

1. Electrical area classification
2. Design bases for fire water and sewer systems
3. Bases for safety relief systems

Electrical area classifications are traditionally important to those who deal with flammable materials, and are covered by a number of codes and definitions: Often, these are audited by insurance carriers and state agencies, but sometimes, the details are hard for outside organizations to follow. When a plant is first built, the division and group decisions are generally straightforward. These can change in small areas if a lower flash material is put in an existing tank, if a trench or sump is added, or if a pressure/vacuum vent is added or moved on a tank roof. Periodically, it pays to audit facilities in detail and update the drawings if needed. More will be said about effective audit processes in a later chapter.

As indicated earlier, plants are generally divided into separate fire-risk areas by providing isolation spacing. Depending on the size of equipment and the inventory of flammable material in the unit, firewater requirements are calculated. Company standards and NFPA or local fire codes generally set the basis for fire water needs. Once set, these rates are rarely reviewed unless a periodic audit and test of the whole fire-fighting system is carried out.

There are many reasons to reevaluate both the fire water loads and the capability of the system both to deliver *and remove the water safely!* A three-year cycle is often practical. Design rates can change because equipment or inventory is added. Pressure requirements might also be increased by configuration changes. The capability of the system might also change as fire pumps wear, as fouling accumulates in piping, or as pump driver efficiency changes. The pressure limitations on the system might also change as piping or joints deteriorate.

The ability to safely carry away the fire water cannot be ignored either. If sewers become partially plugged with dirt or process materials, there may not be the capability to remove large volumes of fire water quickly. Since many flammable materials float on water, the specter of floating fires spreading uncontrollably through the plant is raised. Various specialty contractors are available to inspect and test a plant fire system. This should not be ignored. The potential consequences are too great.

There are two sources of difficulty in maintaining up-to-date sizing for safety relief devices.

The first is the gradual debottlenecking of equipment that might change safety relief loads. The second relates to improvements in the technical understanding of how to calculate the correct relieving capacity on any particular device.

The management of change systems should flag those changes in facilities and processes that require a reevaluation of an individual relief device size. The part of the relief system that is often ignored is the blowdown or flare headers. These are often assumed to be adequate because they were oversized originally. But over the course of time, headers can foul and cause excessive back pressure. Also, the addition of new units to the plant can use up available spare capacity. Reevaluating safety header capacity can be a complex task, but it is required periodically to ensure proper operation of relief devices.

A great deal has been learned in recent years about how to size relief devices correctly. Older methods of sizing devices in flashing or reactive services have been proven to give undersized flow-area requirements. This does not mean that the valve will actually be undersized, because it may have been purchased as a standard size containing more than the required area. However, the potential for vessel rupture cannot be ignored. In a review of two recent designs, one in six safety relief devices was undersized in one, and one in four was undersized in the other.

Valves in flashing or two-phase service that have been sized by the API method should be reevaluated by the homogenous equilibrium model methodology developed by the AIChE Design Institute for Emergency Relief Systems (DIERS). This is the current state-of-the-art technology for this vital calculation. Please note that valves that are not in two-phase service should have been adequately sized by the older methods, provided the relief loads were properly calculated.

The reason for the final caveat is discussed in Chapter 4. In reactive systems, the acceleration of reaction rates can lead to much higher relieving capacity than that calculated by linear methods. To recapitulate briefly, the vent sizing package (VSP) will measure the correct relief area to ensure the correct capacity and phase conditions are used for a design basis. The simpler RSST (Reactive Systems Sizing Test) will provide similar data, but neither system can be operated accurately by novices. Some major companies have supported groups of experts to meet in-house needs. Others use outside consultants, such as Fauske Associates, to do the tests for them. This technology, too, is fast becoming the SOA for reactive systems.

Ensuring Equipment Is Fit for Service

Accomplishing the above objective involves a combination of thorough inspection and maintenance. The objective of these programs must be "no surprises" that is, no unanticipated failures of containment.

Clearly, there are advantages also to having no unanticipated failures of processing capability, but this objective is a lower level of priority. The big payoff is in the elimination of safety and environmental incidents, not only because the cost of such incidents is becoming horrendous, but also because credibility with the community and regulatory agencies is becoming a valuable asset. The inspection program should be structured to uncover potential problems before they are realized and bring them to management's attention. The maintenance program must then remedy them in timely fashion. In today's cost-conscious business environment, inspection and maintenance are often considered costs to be minimized short term, at least until some accident occurs. Thus, these functions need to be staffed with technically competent, assertive, confident champions who can communicate the physical needs of the facilities effectively.

Obviously, management must support the actions identified with the necessary resources of money and manpower. This is often difficult in a cost-squeezed business environment. Important decisions can be made at levels that are too low because of short-term budget pressures. One way to minimize this problem is to have the inspection function report to the plant management through the safety/environmental path rather than through the operations or mechanical manager (if both such exist). This establishes a separate "conscience" path to a high level in the organization. Presumably, at this level, a broader management perspective can be brought to bear and more balanced risk management decisions can be made.

Techniques

A. S. Krisher presented a good summary of the essentials of a good inspection/preventive maintenance program in *Chemical Engineering Progress* magazine in May 1987 in an article entitled "Plant Integrity Programs." Krisher separated inspection techniques into two classes: global and local. Global techniques allow testing of an entire system at one time. Among the few examples of global technique are hydrostatic or pneumatic testing and acoustic emission (AE) testing. The purpose of these tests is to identify areas of potential deficiency before a failure can occur. The typical hydrostatic test (150% of design pressure) is a relatively crude technique that requires process shutdown and will detect only those areas weak enough to fail under the applied stress.

Recent developments in acoustic methods have made it possible to extract a great deal more information from a controlled overstress. Acoustic sensors are placed at a number of locations on the vessel to be tested; the spacing of sensors depends on the sound-transmission capability of the material. When the system is overstressed, events such as crack formation and yielding generate acoustic waves in the metal that are detected by the sensors. With computer data processing, it is possible to identify potential areas of defects. Other tools are then used to evaluate small areas in detail and focus on areas of major concern.

AE can frequently be used on-line by imposing a small increase in normal operation pressure (five to ten percent).

Once a global approach has indicated areas of weakness, local techniques are used to examine these areas in more depth. Considerable advancement in technology has occurred in recent years, making possible more detailed inspections with less interference with operations. The following excerpts from Krisher's article describe several techniques in order of increasing complexity and cost.

- Dye penetrant techniques involve the use of a penetrating dye that can bleed out to form a visible stain. Dye penetrants only detect defects open to the surface and are less sensitive than the magnetic particle methods. However, they will detect surface discontinuities (such as cracks or porosity) that are not detectable by visual inspection. Dye penetrant techniques can be used on either magnetic or nonmagnetic materials.
- Magnetic particle techniques are used to detect cracks in, or very near, the surface of magnetic materials. The defects cause discontinuities in the magnetic field that are made visible by the grouping of the magnetic particles spread over the surface.
- Radiography may or may not require a shutdown. X-rays or nuclear radiation penetrates a component and an image is developed on photographic film on the other side. The image is a shadow picture of discontinuities within the component. Frequently, these shadow pictures can define the nature and seriousness of the defects.
- Ultrasonic thickness techniques are used to determine the thickness of a piece of metal by making measurements from one side of a component. A high-frequency sound wave is injected into the metal and the time to rebound from the other side is measured. Knowing the velocity of sound and the nature of the material, the thickness can be calculated. In many cases, thickness measurements can be made without interruption of an operating process.
- Ultrasonic flaw detection operates in a manner similar to ultrasonic thickness. However, the direction of the high-frequency sound wave can be varied and a greater time span is available. With proper use and experienced personnel, a material can be inspected thoroughly for internal defects such as porosity, slag, lack of fusion in welds, lack of penetration, and cracks. Any feature that can reflect the sound wave can be detected. By looking at the measurements from several different angles and different sound frequencies, it is often possible to characterize the size and nature of the defect. Ultrasonic flaw detection is an extremely powerful tool. However, the technique depends on the operator's skill, thoroughness, and alertness.
- Computer-enhanced ultrasonic techniques address some of the limitations of the manual techniques. The scanning is done mechanically at a fixed rate and the data are collected for later analysis. Computer-based techniques permit analysis of the data in several different ways, including generation of three-dimensional hard-copy images of the reflecting sources.

For some problems, current inspection tools are not adequate for the task. A major example of this (and a major plant problem) is the detection of corrosion under insulation without removing the insulation. While the Materials Technology Institute is researching this problem, commercial techniques are not yet available. Careful design can sometimes make inspection possible when the conventional approach would have precluded it.

Corrosion Monitoring

In many units, corrosion rates can vary with time, with the phase of a batch cycle, as a result of a process upset, or local velocity condition. Corrosion probes, some of which can be monitored continuously, can provide data between shutdowns.

For cases of fairly general metal loss, the remaining metal thickness can be determined and life to retirement projected. Clearly, the retirement thickness needs to be determined on a system-by-system basis and contain an adequate safety margin to allow for variations in corrosion rates.

Often, localized corrosion controls repair frequency. Local pitting, corrosion, or cracking at welds or in the heat-affected zones around welds often defeats attempts to predict equipment or piping lifetimes from uniform corrosion rate data. Sometimes, adding an additional safety margin on calculated retirement thicknesses will be helpful in preventing localized failures. If particularly vulnerable areas have been identified, then ultrasound or AE techniques may be suitable for monitoring specific points. Sometimes, fracture mechanics can be applied to calculate the potential for cracking in a localized stress situation. Enhanced design approaches can sometimes be used (e.g., 100% weld radiography) to strengthen potential failure areas. However, until either a great deal of historic data are accumulated or other inspection techniques are developed, a degree of subjectivity will remain.

Assurance Programs

An effective inspection program requires an inspection group (or groups) staffed with qualified people armed with proven, documented inspection procedures and the tools to monitor the integrity of equipment, capable of continuous analysis of the data, and placed in the organization so that their recommendations can be implemented in timely fashion. The following twelve fundamental requirements define in more detail the elements of a comprehensive inspection program and its relationship to effective maintenance activities.

1. There must be systems in place to manage the basic inspection concerns for critical containment systems, such as:
   a. Integrity of containment equipment, such as pressure vessels, piping, and tanks
   b. Safety valves and rupture discs

   c. PV vents (atmospheric tanks)
   d. Temporary repairs, including cold weld compounds
   e. Corrosion under insulation and fireproofing
   f. PMI (positive material identification)

   These systems must apply equally to existing equipment and the quality control of new construction.

2. Responsibilities for inspection and the timely follow-up to correct any deficiencies identified must be well defined and clearly understood by the whole organization.
3. There should be a periodic reassessment of the qualifications of the inspection work force (company or contract). Planning for inspection should include the need for replacements or reassignments such that a qualified work force is always available to complete necessary inspections.
4. Those responsible for metal inspection should have, as a minimum, a basic understanding of metallurgy, corrosion and detection, NDE (nondestructive examination), welding, and codes and standards. There should be training plans in place to maintain the appropriate knowledge level as required to remain proficient.
5. There should be a program in place that defines the required knowledge and capabilities of noncompany support groups that do inspection for the company, e.g., NDE, ultrasonic testing, safety valve repair, radiographing, and so on. Contract pipe and vessel fabrication shops should understand company and code requirements; have appropriate facilities, quality control, and inspection; and be certified by company representatives.
6. A secured central file must be available, composed of a comprehensive data base, which documents information on all fixed equipment. This file should include:
   a. Basic data—original and as-built specifications
   b. History—inspection, repair, process service
   c. All correspondence pertaining to the piece of equipment
   d. Inspection frequencies
   e. Identification and history of critical piping systems, including repairs and thickness history
7. A formal system must be in place for documenting recommendations and for effectively communicating them to the appropriate level within the organization.
8. A system must be in place to provide closure on outstanding inspection recommendations. This could be either inclusion into the maintenance program for completion or documentation of the rationale (with appropriate approval) for not accepting the inspection recommendation.
9. An effective system should be in place to monitor corrosion and other aspects of equipment performance, especially items that have a recognized vulnerability, and to update the frequency and type of inspection as appropriate.
10. "Unusual items" such as underground piping systems, nonpressurized

infrastructures, and third-party systems within the plant must be identified and their integrity must be monitored.

11. Welding procedures and welder qualifications must be reviewed and approved by a qualified representative before work commences. The welding should also be inspected according to job requirements.
12. Governmental inspection requirements must be well understood and met by operations and inspection.

Quality control monitors work as it is being done to ensure its quality. The essence of quality control is "doing it right the first time." From an execution standpoint, quality control has a short-term orientation.

The quality-control inspector monitors the execution of mechanical work in progress and assures its compliance with applicable codes and standards. The quality-control inspector verifies that appropriate procedures, proper materials, acceptable work conditions, and qualified people are used for the job. The quality-control inspector evaluates the completed mechanical work. NDT techniques are used in that evaluation when required by specific codes and standards.

On the other hand, quality assurance monitors the quality-control function and ensures that the system is operating properly. In addition, the quality-assurance function defines and executes programs that ensure the long-term safety and reliability of equipment. In short, quality assurance is directed at the long-term safety and reliability of fixed equipment.

The primary duty of the quality-assurance inspector is to implement and administer programs that monitor the long-term serviceability of plant equipment. This is done through routine monitoring programs and periodic scheduled internal and external inspections. The aim of the quality-assurance inspector is to provide the necessary information so that repair and replacement work is done on a planned basis rather than a failure basis. The intent of this mode of action is the prevention of defects rather than the detection of defects. The quality-assurance inspector also ensures that quality-control programs are functioning properly. This position provides assistance in evaluating the impact of new or abnormal operating conditions on plant equipment. In addition, the quality-assurance inspector reviews the design of new facilities and consults on unusual failures and repairs.

Both quality-control and quality-assurance activities can be managed within the job scope of each inspector. Given the dual role of the inspectors, particular attention must be paid to ensure that one function is not given preference over the other.

Clearly, these roles involve different viewpoints but require many of the same skills, experience, and certification. In addition to the fundamental skills of the quality-control function, quality assurance adds greater demands for process and mechanical standards knowledge, some formal training in metallurgy and NDE, and stronger leadership and communication skills. While most fully independent inspection people are quite experienced (approximately ten years), those more involved in quality assurance generally have a few more years at their calling.

Clearly, the supervision of contract inspectors and inspection services is a serious activity. The same quality standards must apply to the equipment they inspect and the procedures they use. Thus, the quality-assurance overview is critical to successfully employing outside people as part of the normal quality-control work force, in fabrication shops, or as specialists in NDE, radiography, and so on.

An area often overlooked is coatings and insulation. The primary duty of the coating specialist is to provide technical support for maintenance and project activities. The coating specialist contributes the technical expertise necessary to plan, evaluate, perform, monitor, and document all coating-related activities. Included is the responsibility for quality-control inspection of coating-related work, and for dealing through the purchasing function with the selection and supervision of contractors.

In plants small enough, the coatings specialist can also be charged with materials responsibilities for insulation. This is critical in the selection of new materials, and is increasingly important in managing asbestos removal. The specialized skills needed in these jobs are related, as is the need to work with an industrial hygiene specialist on a number of personal protection questions.

As an afterthought, it should be abundantly clear that appropriate metallurgical expertise is a critical component of all the activities discussed here. It would be unusual for many of the inspectors themselves to be graduate metallurgists. Their practical knowledge must be supplemented at the professional level somewhere in the organization. In my judgment, at least part of this should be located in the inspection group itself.

The implementation of the various requirements at a given plant will vary according to the needs and resources at the plant. Computer data management systems and procedure banks are obviously a great asset. A data base for each piece of equipment should include (as a base case subject to specific additions) the following information:

- Engineering data
- Operating parameters
- Design conditions
- Materials
- Manufacturer
- Test reports
- Drawings (or reference to hard-copy file)
- NDE reports
- Maintenance and repair procedures
- Preventive maintenance schedule and record of results

All of this information can also be processed manually if the manpower and will to do it are available. With the increased interest of government agencies in maintenance records, the advantages of computer use are increasing.

The detailed discussion of an effective inspection/maintenance program could easily fill a volume this size. We have already touched on the highlights of organization placement, personnel qualifications and approach, and data management needs. Table 8.1 offers a brief checklist of factors to be considered in meeting the other requirements. Most of these items are fairly self-explanatory. However, several items merit further discussion and will be elaborated upon in the following pages. Specific sections will be devoted to procedures, contractor quality control, inspection plans, critical-equipment maintenance plans, safety relief devices, and getting closure on inspection/maintenance recommendations.

Procedures

Both general and equipment or system-specific inspection procedures will be required in most plants. The list can get quite long even in the general area alone. (See Table 8.2). Control of these procedures and any additions/changes will be an important task.

Table 8.1. Checklist of Inspection Activities

A. Inspection Procedures
- Applicable codes and standards
- Inspection procedures—location and control
- Inspection procedures—preparation and maintenance
- Documentation of nondestructive testing activities

B. Routine Maintenance Inspection
- Routing maintenance overview
- Welding quality control
- Replacement and repairs of pressure vessels
- Replacement and repairs of tanks
- Replacement and repairs of piping
- Safety relief devices
- Nondestructive testing of routine maintenance
- Strength tests
- Input to maintenance plans (preventive and turnaround)

C. New Construction Inspection
- New construction overview
- Review and approval of process design specifications
- Review of fabrication and construction drawings
- Rerating and reuse of fixed equipment
- Source inspections
- Construction contractor quality control plan
- Inspection of project construction

D. Equipment Monitoring
- Design and development of inspection plans
- Inspection plan upkeep, maintenance, and revision

E. Corrective Action

System corrective action

Closure on inspection recommendations

Table 8.2. Typical List of Site Inspection Procedures

- Positive material identification (PMI)
- Weld quality assurance
- Certification practices for NDT personnel
- Visual inspection
- Magnetic particle examination
- Radiographic examination
- Dye penetrant examination
- Ultrasonic examination
- Inspection reporting
- Handling of nonconformances
- Inspection of data management
- Pipeline numbering system
- Use of radiography for routine pipe inspections
- Unit ultrasonic testing documentation
- Ultrasonic testing locations and intervals
- Pressure vessel inspection intervals
- Tube bundle inspection intervals
- Tank inspection intervals
- External visual inspection of pressure vessels
- Internal inspection of pressure vessels
- Exchanger tube bundle in-service inspection
- Inspection of internal floating roof tanks
- Tank safety equipment inspection
- External visual inspection of piping
- Hydrostatic pressure testing of piping
- Tightness testing of piping and equipment
- Inspection/testing of underground equipment and piping
- In-Service inspection of safety relief valves
- In-service inspection of rupture discs
- Inspection for corrosion under fireproofing
- Inspection for corrosion under insulation
  a. Piping
  b. Tankage
  c. Equipment
- Temporary repair inspection
- Inspection of structural steel

Construction-Contractor Quality Control

Just as in-house construction and repair work must undergo quality-control inspection, so must that done by contractors, both on- and off-site. An often-used general approach is to require the contractor to be responsible for the inspection function and use in-house talent to approve his or her plan and verify compliance with it. Such requirements are included in contract/purchasing terms. The key elements of such a program are as follows:

- The contractor provides a quality-control plan for field construction work. This plan provides procedures, instructions, and systematic actions necessary to achieve adequate confidence that fieldwork complies with the owner's requirements and standards. The quality-control plan needs to specify the means for controlling the quality of materials and services that the contractor proposes to purchase, fabricate, and install.

- The quality-control plan is submitted in a preconstruction meeting, and is reviewed and approved by the project inspector before fieldwork begins. The plan is modified as necessary to comply with owner requirements.
- The contractor's ability to implement his plan will be reviewed by the project inspector. An integral part of this ability is the designation of the responsible quality-control personnel in the plan. These personnel ensure that the quality-control activities are carried out. The contractor shall specify the qualifications, certification, and duties of quality-control personnel to accomplish the scope of the work anticipated.
- Contractors perform their own inspection and acceptance of work prior to submitting to the owner for acceptance.
- The quality-control plan of the contractor obviously must meet the owner's requirements. For example, the following items should be included in the welding quality-control program:
  Welder qualifications and testing
  Welder performance tracking
  Weld tracking
  Welding material control
  Welding procedures
  Nondestructive testing procedures and plans
  Hydrostatic testing procedures and plans
  Postweld heat treatment procedures, records, and charts
  Record keeping and reporting
  Follow-up procedures to nonconformances
- While the welding quality-control program is critical to containment, the contractor plan must address a number of other items. It needs to list the personnel responsible for carrying out these functions and their qualifications. Likewise, the plan should address the procedures for execution, the follow-up activities, and the reporting of the results for the following areas:
  Soil
  Concrete
  Structural steel
  Painting and protective coatings
  Insulation and linings
  Piping
  Material control
  Control of subcontractors and fabricators
  Nondestructive testing
  Mechanical equipment installation procedures
  Fireproofing including anchors
- The contractor establishes and maintains files of the results of the quality-control activities listed here. All quality-control files must be available for review by the project inspector.
- The contractor supplies routine quality-control reports of activities to the

contractor coordinator and the project inspector as identified and defined prior to the start of the project.

- In his review of the quality-control plan, the project inspector specifies those files that will be retained permanently. At the completion of the project, these records are surrendered to the project inspector.
- Nonconformances to job specifications and standards shall be described in detail and reported to the project inspector.
- Construction nonconformances shall be reviewed to determine rework requirements or disposition if rework is not feasible. Faulty welds or improper installation are examples of nonconformances requiring rework prior to acceptance. Nonconformances that involve material or specification changes require use of the field-authorization change system.
- If corrective action is agreed upon, the project inspector shall verify compliance.

The objective of this program is clearly that the company be able to expect the same assurance of quality construction from contractor activities as from those of its in-house staff.

Inspection Plans

The document that ties the inspection and maintenance activities together and gains endorsement from management for these critical activities is the inspection plan. In it, the specific inspection and repair items planned for the year are agreed upon, budgeted, and scheduled. Problem areas are also highlighted as well as critical testing issues.

One important value of the inspection plan is the discipline it imposes on inspection, maintenance, and operating people. Each item must be thoroughly evaluated and a consensus must be reached. If the latter is impossible, inspection must have a path through which potentially serious problems can be escalated. However, in the vast majority of problem areas, the three functions can agree on a satisfactory plan. This done, endorsement by management is usually forthcoming, although if the cost or interruption of operations for a specific item is particularly painful, a demand for rethinking or an improved technique is not all unusual.

An excerpt from a typical plan is given in Table 8.3, along with a definition of the type of inspections planned and appropriate comments. Some samples of plans for exchangers and vessels are shown. Some of these require standard inspection procedures (e.g., EXBB); some require specialized maintenance explained in the attachments to the table (e.g., EXPM). With this information in hand, maintenance planning, scheduling, and budgeting can go forward unimpeded.

Critical-Equipment Maintenance Plans

In any plant, there are a number of instrument, computer, and electrical systems that are critical to the safe/environmentally sound operation of the plant. There is often confusion about what is truly "critical." In a first listing, many

Table 8.3. Typical Inspection Plan

Equip. ID. #	Title	Hydro Shell	(Psi) Tube	Last Insp.	Next Date	Insp. Type	Comments (History, Problems, Work Items, etc.)
7160	10R-1 O.H. condenser	175	225	1991	1993	EXBB	• Channel and outlet piping are subject to corrosion due to hot salt water
7162	Filter feed preheater	225	210	1988	1991	EXPM	• Steam leak at front end of exchanger is a constant problem; weld repair needed (same is true for spare)
9158	10D-20 reboiler	—	300	1990	1993	EXPM	• Inspect this SS bundle when 10D-20 needs to be cleaned out
9179	Precondenser	143	150	1990	1991	EXBB	• Need to repair or replace shell
9183	Condenser for 10R-3	143	113	1991	1991	EXPM	• Shell nozzles wrapped; dome gasket area corroded; exchanger needs major overhaul
OD-4X	Sour gas K.O. drum	117		1990	1995	VSPM	• Drum has shallow pitting internally; internal inspection needed to check the progress of pitting
OD-7	B/D K.O. drum	75		1991	1993	VSPM	• U/T readings taken, internal pitting present; next inspection will be an entry
OD-10	Phenate collection drums	172		1986	1992	VSPM	• Need an internal inspection on one of the three drums in order to extend inspection interval; shallow pitting noted in last inspection
OD-14	Glycol receiver drum	164		1990	1999	UTPM	• Replace insulation on shell and bottom head; drum looked good
OD-44	HTF surge drum	110		1991	1992	VSPM	• Corrosion in upper section of drum; need scaffold to remove insulation for U/T readings; entry next year
OJ-1	Steam ejection stripper section	—		1985	1991	X-ray	• Repairs made on ejectors in 1985; will X-ray for corrosion 3Q91; new ejector purchased in 1989

Pressure Vessel PM

VSPM (internal inspection)
- — Slip blank
- — Open
- — Clean
- — Inspect
- — Close
- — Remove slip blanks
- — Test

Reason for PM
- + 10 years of service
- + Corrosive service

CUII (corrosion under insulation inspection)
- — Identify area of insulation to be removed
- — Remove insulation
- — Clean area under insulation
- — Inspect area
- — Perform N.D.E. on area
- — Determine additional areas to inspect
- — Paint area
- — Reinsulate
- — Caulk insulation
- — Log area into history of equipment

Reason for PM
- + Damaged or worn insulation
- + Damaged tracing
- + Suspected area for external corrosion

X-ray
- — Notify I.H. if needed
- — Obtain proper permit
- — X-ray vessel or part of vessel
- — Review film

Reason for this PM
- + Area known to have had a problem (corrosion, erosion, pluggage)
- + Suspect area of having a problem

UTPM (ultrasound thickness)
- — Take U/T reading on:
    - • All nozzles
    - • Top head
    - • Midsection
    - • Bottom head
- — Compare present reading to past reading
- — Determine corrosion rate

Reason for this PM
- + To determine corrosion rate
- + Search for problem area on equipment
- + Extend inspection interval

Exchanger PM Codes

EXPM	*EXAA*
— Install blanks	— Install blanks
— Drop dome	— Clean exchanger
— Drop floating head	— Drop back cover
— Clean bundle	— Remove elements
— Inspect	— Inspect elements
— Reinstall floating head	— Inspect shell
— Reinstall dome	— Reinstall elements
— Hydrotest shell side	— Hydrotest tube side
— Reinstall channel cover	— Reinstall back cover
— Pull blanks	— Hydro shell side
	— Pull blanks
	— Tightness test
Reason for this PM	Reason for this PM
— 10 years of service	— 10 years of service
— Floating head gasket leak	— Corrosive service
— Corrosion of FH nuts and bolts	
— Tube end corrosion	
— Corrosive service	
EXBB	*EXGG*
— Take U/T reading on:	— Take U/T readings on back and front pressure plate along with their nozzles
+ Channel	— Visually inspect for external leaks when exchanger is in service
+ Channel nozzles	— When disassembled for cleaning,, inspect plates for wear
+ Shell	
+ Shell nozzles	
+ Dome	
Reason for this PM	Reason for PM on plate exchanger
+ Mild corrosion rate	+ Check for gasket leaks
+ No history of problems	+ Check for corrosion/erosion on plates
+ Extend inspection interval	

production/quality controls are often included, leading to a large, unmanageable list of items requiring critical preventive maintenance. Thus, a clear system is needed to develop a truly "critical" equipment list.

Sanders and Wood describe the PPG Industries system in *Chemical Engineering Magazine* of February 1991 (pp. 110–118) and refer briefly to a May 1987 article by G. K. McMillan in the *Technical Journal.* At PPG, three priorities are used to set preventive maintenance frequencies:

1. Items whose failure *would* cause or fail to prevent or signal a fire, explosion, or dangerous release or major property or production loss.
2. Items whose failure *could* cause or fail to prevent a serious situation as in priority 1.
3. All other alarms, cutouts, controls, and so on.

McMillan proposed a four-tier priority system.

1. Community protection items
2. Employee protection items

3. Major property and production protection items
4. Minor property and production protection items

I recommend a slightly simpler system. As described earlier, "critical safety" items are those whose failure would cause or fail to prevent (or signal) an accident or release of hazardous material. All other items are second priority. We find this split practical to manage and to provide reasonable continuity of production at our Linden, New Jersey, facility. The system applies to both on-line critical items and to standby items. Malfunction of on-line items—for example, a temperature controller, is noticed almost immediately; the temperature goes out of control. Standby items, such as a feed shutdown in the event of high temperature, only function in emergencies. Thus, the testing of these items is essential to having the degree of safety designed into the plant available when needed. In the author's approach, the temperature controller would not be safety critical if backed up by the feed shutdown device. The feed shutdown clearly would be critical.

The preventive maintenance procedures to be used for each critical item must be clearly thought through and documented. They must cover the entire system from detection through corrective/preventive action, and the test should not interfere with production if it is to be performed faithfully. Any changes/improvements to either the system or its procedures need to be controlled through the change management system as discussed in the previous chapter.

Sanders and Wood provide some data on testing frequencies that may be of interest. The PPG Lake Charles, Louisiana, facility has about 1,300 instrument loops in each of the priority 1 and priority 2 categories. About half of the testing frequencies for these have been developed using detailed reliability studies; the rest were developed on an historical basis. The breakdown of frequencies is approximately as follows:

12 months	25%
6 months	40%
4 months	15%
3 months	10%
Other	10%

Many of the high-pressure and high-temperature alarms are included in the six-month-frequency group.

The above authors also provided some data on test frequencies at DuPont's Sabine River works (Orange, TX), which has 35,000 instrument loops. DuPont reports that every safety interlock is tested regularly—mostly at a twelve-month interval—using a written procedure. At DuPont a significant review and multiple approvals are required if the frequency or content of a test procedure is to be changed.

Such large numbers of safety systems may sound intimidating to those who run smaller plants with very small staffs. While a computer tracking system and data bank is a real asset, small plants can implement a system manually. Until about five years ago, this was done at Exxon Chemical's Linden plant at a time when critical instruments, emergency power supplies, computer interlocks, remote operating valves, and so on numbered close to 1,000. In the 1986–1988 time frame, this system was converted to a computer data base, but staff requirements stayed the same.

Safety Relief Devices/Systems

The 1,000 quantity above did not include safety valves, rupture disks, tank P/V vents, and the like. These critical relief devices are usually handled separately by either the inspection group or a specialized maintenance group. Experience at a number of companies has shown the benefit of investing in a specialized group of technicians and an adequately equipped shop to maintain and test these critical devices. Inadequate test-stand facilities can give misleading results and inexpert technicians can use the wrong lubricants, not catalogue corrosion and other materials problems, not properly account for relief system back pressure, and not record data properly. At PPG's Lake Charles Louisiana plant, there were over 2,000 safety relief devices in a wide range of sizes from twenty-four different manufacturers. Until a specialized group was set up under the direction of a single professional, there was little consistency and no learning from past problems. Smaller plants may wish to have the actual work done by an outside shop, but the capabilities and expertise must be comparable.

The first step in developing a comprehensive safety relief device maintenance system is to make a philosophical decision on whether to operate the plant while the device is out for maintenance. Some plant managements will not permit even one day/year operation without protection, so they shut down or provide spares. Others require extra precautions while the device is out, such as a safety watch, reduced throughput and/or pressure, and so on. There is no clear-cut choice on process equipment, but laws are often quite specific about boiler safety-valve procedures. Where laws permit, a consciencious plant is still free to choose its approach.

A further note to this question involves whether block valves are permitted between the safety device and the equipment it protects. Obviously, these are needed if operations are to continue, but they must be sealed open when the device is in place. Some companies that shut down during device maintenance still provide block valves to help deal with emergencies—another philosophical choice for management, but most plants use valves.

With these decisions made, a systematic approach to safety device installation and maintenance can be developed. It should include the following elements:

- Identify each device. Often, specific painting schemes are employed.
- Legibly tag each device and its host connection point with a characterization legend. This will often codify set pressure, back pressure, and orifice size.

- Include the most recent test date on the tag.
- Seal all adjusting screws.
- Ensure what discharge piping is aimed at a safe location.
- Ensure that suction and discharge valves are sealed open and of adequate size to meet code pressure-drop requirements.
- Evaluate whether the relief device needs protection from corrosive process fluids (such as a rupture disc).
- Require the mechanics who remove the valve to report on the condition of the inlet nozzle and discharge piping. Inspection of the rest of the relief system is as important as the relief device itself.
- At the test shop, pop the device to see whether it would have relieved at capacity at the time of removal. If not, a change in maintenance frequency or installation arrangement may be required.
- Disassemble the valve, replace/repair/clean internal parts as required, and record the condition of such parts.
- Lightly lubricate moving and guiding surfaces with a material compatible with process materials.
- Reassemble the valve and retest at the design back pressure. Paint if appropriate.
- Install tamper-proof seals on all adjustments, cover ports, and tie-down levers for transportation back to the field.
- Complete all record keeping and highlight any problems for the inspection group. This is critical to determine whether inspection frequency is correct.

Testing frequencies can also be impacted by state laws and company policy. At PPG, Lake Charles, about thirty percent are tested annually and sixty percent more every two years. Some systems, such as instrument air manifold and machinery lube systems, are tested only every three years. In addition, devices that are forced to relieve corrosive or potentially plugging materials are removed as soon as possible after any action to clean the system of corrosives and ensure there is no downstream plugging or restriction. In many plants, most process relief devices are tested annually to avoid problems with fouling or polymerization.

The importance of proper identification of each device and host point is critical. This is especially true if spares are to be affixed either in place of a valve to be maintained or in a rotation out of the storehouse. Obviously, the pipe size of any device can have little to do with its relieving pressure or capacity. Much more data than pipe size must be provided on the capabilities of the device and the requirements of the service to ensure the correct device is installed at every position all the time.

Maintenance Action and Follow-up System

How a job gets done will vary appreciably from plant to plant. Aside from unplanned failures, most repair/test jobs will be initiated from an inspection or preventive maintenance plan. A work request is generated for the specific task.

This work request usually goes to a maintenance planning person who will recommend a schedule and manning to the operations supervisor of the unit in question. If the operations supervisor approves, a work order is written and execution is scheduled. This is a highly interactive process. In a small plant, there may be only one or two people involved. In a large one, there will be dozens of work requests on a given unit every day. These will be prioritized by operations, but some mechanism must exist to ensure that critical jobs are not ignored.

There are several checkpoints in the system to accomplish this, whether computerized or not:

- If no work order is written for a given work request, the request initiator can check with the planner.
- If the work order is not approved, interaction with the operations supervisor is indicated.
- If an approved work order is not completed, both the originator and the operations supervisor can discuss it with the mechanical group.

The key element is a paper or electronic trail from request to work order to approval to completion. Inherent in that trail is a system for tracking closure on each request, ensuring that the information is readily available to those interested. This is usually done via "read-only" access on computer systems. In a paper trail, interested parties must go to the proper mechanical file to follow-up. The importance of keeping these files constantly up to date is obvious.

On an overall basis, a monthly report is generally required to keep management (and possibly government agencies) apprised of overall performance. This is assembled by grouping each work order into categories of interest and comparing the numbers of those jobs planned to those actually completed. A sample tabulation is shown in Table 8.4. This is usually covered with a note explaining special issues such as reasons for some lack of completion, manpower considerations, and so on. From this tabulation, a list of specific overdue jobs can be generated along with the time overdue. In our company, jobs overdue by more than two months are highlighted directly to the operations manager, and he or she must approve plans for completion or a new schedule for the job.

While this system is easier to handle via computer, it is manageable (and was managed) manually. The scope at this plant was for about 1,400 jobs annually, including safety relief devices. With proper management attention, it works.

As a final closure system on problem items, the originators of all work requests review their plans monthly with operations and mechanical management. At this meeting, priorities are reassessed, outstanding concerns and work requests are reviewed, and risks/vulnerabilities are evaluated. Input from the SOC is part of the process if appropriate. The net result is a revised maintenance plan and renewed commitment to get it done.

Table 8.4. Chemical Plant Overall Zone a *Preventive Maintenance Status Report—Month: October 1989*

Code	Description	Jobs Carried Over	Carry Over not yet Done	New Jobs This Month	New Jobs Compltd In Month	Percent PMs Compltd	Jobs not Done	New Jobs Ytd	Jobs Compltd Ytd	Percent Ytd Compltd
302B	Compressors, Recripr	4	4	1	0	0.0%	5	1	0	0.0%
313A	Safety valves	11	10	16	12	75.0%	14	178	171	96.1%
313P	PV vents	1	0	2	2	100.0%	0	54	54	100.0%
313R	Rupture discs	0	0	2	0	0.0%	2	12	10	83.3%
318A	Platforms	1	1	0	0	0.0%	1	2	1	50.0%
321	Hoists & cranes	1	0	0	0	0.0%	0	2	2	100.0%
322	Instruments	4	2	59	59	100.0%	2	396	394	99.5%
322A	Analyzers	6	0	23	23	100.0%	0	193	193	100.0%
323	Computers	0	0	2	2	100.0%	0	8	8	100.0%
337	Motor	2	0	2	0	0.0%	2	13	11	84.6%
337A	Electrical circuit	4	2	2	2	100.0%	2	38	36	94.7%
338	Conveyors	0	0	1	0	0.0%	1	4	3	75.0%
343	Miscellaneous machinery	1	0	15	15	100.0%	0	142	142	100.0%
400L	Scales	1	1	0	0	0.0%	1	2	1	50.0%
	Total zone A	36	20	125	115	92.0%	30	1,045	1,026	98.2%

Work Permit System

The ultimate control on any risk or change is to see to it that any work that goes on in potentially hazardous areas is done under conditions of maximum safety. Most plants have a work permit system to attempt to ensure this. Most of these systems are sound, but sometimes the implementation of the system suffers from shortcutting. Therefore, a work permit system should be the least cumbersome possible and be strictly enforced.

The purpose of the work permit system is to ensure that jobs are safely planned, equipment is properly prepared, employees are adequately informed, and work is safely executed. All those included in the activity covered by a work permit share in the responsibility for the safety of the job.

The primary elements of a good work permit system are:

- A formal training and qualification testing program, with periodic refreshers, should be provided for all site personnel involved in the work permit system. Qualification test results should be retained.
- The work permit system should provide for a work permit document that is used as a tool to:
  a. Describe the type of work and specifically identify the area and equipment where work is to be performed.
  b. Confirm that the work area has been properly prepared and inspected for the specific task before the work permit is issued.
  c. Specify time limits of the permit.
  d. Designate if an atmospheric test is needed, the type of test (flammability, oxygen, toxics), the time, and the results.
  e. Specify standby equipment, personnel, and any special precautions, such as protective equipment and hazardous materials monitoring. If special hazards exist, attachments may be necessary.
  f. Provide space for signatures/initials of the responsible parties:
     — Atmospheric tester (if applicable)
     — Permit issuer: individual(s) issuing or approving the work permit
     — Permit accepter: Maintenance person(s) responsible for performing the work, indicating that the task has been reviewed and is clearly understood by all (in some cases, language may be a consideration)
     — Cosigner: If the work assignment should overlap into another operating area or area of responsibility
     — If procedures require it, the responsible operations supervisor, indicating that he is aware of the work being done
     — Completion/sign-off by both permit acceptor and permit issuer indicating that the area has been inspected and the work is complete, and the equipment is properly prepared for return to service
- Work permit systems should separately describe conditions, procedures, and approval responsibility for different categories of work. As a minimum, a permitting system should deal separately and directly with cold work, hot

work, confined-space entry, electrical lockout, and atmospheric testing. Other types of permits may be appropriate for a given site.

- Atmospheric testing should be done only by employees who have been formally trained, tested, and approved by management. Refresher training for these employees should be held on a regular basis to maintain an acceptable level of proficiency. Site management should maintain up-to-date records of individuals qualified to make tests.
- Appropriate atmospheric tests should be performed before a work permit is approved for all appropriate conditions, including hot work, confined-space entry or situations where toxic or combustible materials may be present.
  — Specific criteria for acceptable conditions should be defined before the test.
- Plant management should routinely audit various parts of the work permit process to ensure compliance with procedures and to see that the procedures in place are meeting the needs of operations.

Special considerations are necessary when working with sources of potentially hazardous energy. This used to be thought of as only electricity but recent OSHA rules have recognized what was inherent in many existing plant practices: that pressure, flammables, asphyxiants also require a lockout procedure to ensure worker safety. The OSHA lockout/tagout procedures usually require two separate locks on any source of energy that might impact on a job. Often the operations lock is placed to control the process aspect and the mechanical lock is placed for personal safety.

The recent OSHA rules have caused some discussion about how they would apply to process plant operations. Various examples have been discussed and written up to help clarify the correct procedures. The example given here is taken from a proposed draft appendix to OSHA's instructions for compliance with its lockout/tagout rules. It is meant only to illustrate the principles that need to be applied and not to specify exactly what procedures must be employed. Thus, actions are labeled "Possible Approach." It also gives an example of the implementation of a normal permit procedure.

Lockout/Tagout Example Pump Repair—Small Job

Background

A pump is located at a refinery unit and is under the operating control of a process post manned at all times by a process employee. It is driven by an electrical motor and has one suction and one discharge block valve.

The work on the pump will be performed by mechanical employees. The refinery process operator who "owns" the pump must specifically authorize, by a written permit system, the refinery mechanical employees to perform maintenance on the pump. The permit system will not allow mechanical to touch the pump without process authorization.

In this example, the two block valves provide two blinding points. The pump motor has one electrical circuit breaker.

Procedure

The existing procedure is presented in steps 1–10 below. New implementation strategies being considered are highlighted by the phrase "Possible Approach".

Step 1:

The process operator turns off the electrical switch to shut down the pump.

Step 2:

The process operator closes the two block valves and tags them with a white DO NOT OPEN tag. Process attaches a pink INSTALL BLIND HERE tag to each of the two blinding points.

Possible Approach:

A new tagout device could be developed to replace the Do Not Open tag. This new tagout device would be a durable, standardized, substantial device similar to a car-seal. In addition, it could be color-coded to identify it as a process tag that could only be removed under authorization of the process post operator for that location. The device would prevent valve operation. It would also indicate the identity of the process post employee on duty at any time.

Step 3:

The process operator locks out the electrical circuit breaker with a specially designed hasp. This hasp can accept a number of separate locks. The process operator installs a process lock on the hasp.

Step 4:

The process operator clears and drains the pump. At this point, process has prepared the pump for mechanical. However, neither process nor mechanical knows who the actual employees will be who will perform the work on the pump. Those employees will be identified at a later time and will be working under their existing training and knowledge to perform the task assigned.

Step 5:

Sometime later (perhaps hours or days), a mechanical crew of three mechanical employees will arrive to "pull" the pump so that it can be taken away for servicing and maintenance. The process operator must first issue a permit to mechanical. This permit requires both process and mechanical to make a safety tour of the job site. During this safety tour, mechanical verified that the valves have been closed and tagged and process also identified for mechanical each location to be blinded. Mechanical is given a written Blinding List on which the location of each blind to be installed is recorded. Both process and mechanical sign the permit at the job site, and a copy of the permit is posted in a conspicuous place at the job site. Process is not permitted to open the valves while the mechanical permit is active.

Possible Approach:

An Isolation Blocking List identifying the location of the closed valves could be prepared by process and attached to the permit issued to mechanical.

The permit itself could serve as the device to ensure mechanical that the energy isolating devices (the two valves that keep process fluids back) are in the appropriate position. It would function as the group lockout/tagout device. Once mechanical has signed for the permit, process cannot unblock (i.e., open) a valve on the attached blocking list. The group procedures could be further standardized by either requiring the authorized mechanical employee signing the permit to be responsible to the other mechanical employees for the permit, or, for small jobs, allowing each mechanical employee to sign the permit. Each mechanical employee could be trained to understand the use of the work permit as the group lockout/tagout device.

Step 6:

Mechanical applies a mechanical lock to the electrical lockout hasp. At each of the blinding locations, mechanical unbolts the flanges, inserts the blind, and rebolts the flanges, locking the blind in place. Each blind has a metal piece that extends beyond the flange, providing visual observation that a blind is in place.

Possible Approach:

New color-coded tagout devices, similar to the process tagout devices described in the proposed change to step 2, could be developed for mechanical. After locking each blind in place, mechanical could attach the new mechanical tagout device to the blind extension.

Step 7:

Mechanical pulls the pump and sends it for repair, removes the mechanical lock from the electrical-circuit hasp, signs and returns the permit (with the completed blinding list) to process, and leaves the job site.

Possible Approach:

Before the permit could be turned in, standardized group procedures could require the designated authorized mechanical employee to verify that no mechanical employee is on the job and dependent on any energy isolating device. Alternatively the plant might require each employee to sign-off the permit. The mechanical tagout devices would remain on the blinds. The process tagout devices would remain on the closed valves.

Step 8:

Sometime later (perhaps days or weeks), a completely different mechanical crew of three mechanical employees will arrive to reinstall the repaired pump. Process and mechanical will conduct a job-site safety tour with process identifying each location to be unblinded, and process will issue a new permit with the blinding list to mechanical. Once the permit is signed by mechanical, the valves and blinds may not be moved by process.

Possible Approach:

Process could include an Isolation Blocking List (identifying the location of the closed valves) with the permit that is hung at the job site. The permit could be signed by either the designated mechanical employee or each mechanical employee.

Step 9:

Mechanical places their lock on the electrical-circuit hasp, reinstalls the pump, removes the blinds, removes their electrical lock, signs and returns the permit, and leaves the job site.

Possible Approach:

As blinds are removed, mechanical would remove tags from the blinds. Before this second permit could be returned to process, the designated mechanical employee or each mechanical employee could be required to sign the work permit.

Step 10:

Process removes their seals from valves and starts up the pump.

As you can see, certain small changes will likely be required to "normal" procedures to accommodate the OSHA lockout/tagout rule to this small and relatively simple task. Larger tasks and more subtle sources of potentially dangerous energy will require some additional analysis and planning to ensure compliance. In most cases, the fundamental precautions already in place (blinding, car-seals, locks, etc.) will be adequate. However, some additional tags will probably be required as well as some signatures on permits. In addition, OSHA was quite emphatic about *only one key* being available for any lock: Each lock is personal. Thus, if a job is to extend over a shift change, locks must be changed out or some other acceptable procedure must be developed to meet the "personal" lock principle.

Summary

Public expectations are that companies will do everything within reason to prevent accidents, especially those that could cause an impact on the public. As a minimum, the following mechanical elements of this accident-prevention program will be scrutinized:

- Appropriate design standards
- Qualifications of inspectors and mechanics
- Quality-control/assurance procedures on mechanical work and construction
- Inspection of operating equipment to prevent unexpected failures of containment
- Follow-up system to ensure implementation of required work and recording the necessary data
- Procedures to ensure any work is done safely

Safety critical equipment is defined here as any equipment whose failure would cause or fail to prevent (or signal) an accident—primarily loss of containment. Some companies break down their definition further by separating safety from environmental, major from minor. The appropriate design standards are usually embodied in legal or industrial codes (e.g., ASME VIII for vessel design) or standards. Increasingly, however, the concept of state-of-the-art technology is being applied for safety critical equipment.

Plant layout, equipment spacing, fire protection, and sewer systems and the proper sizing of safety relief devices are areas of technology that often are not completely covered by codes, laws, and industry standards. These will require the development of company policies and procedures appropriate for the specific risks involved.

Required personnel qualifications will vary with the type of operations in the plant. However, those techniques relevant to containment are likely to be universal for inspection and fabrication operations. Thus, welder qualification, certification under API 510 (pressure vessel inspection), and knowledge of API 620 and 650, ASME VIII and IX, ANSI B31.3, and company mechanical standards will be essential. Knowledge of specific testing techniques, such as radiography and ultrasound, may or may not be needed depending on the scope of plant operations and the use of outside resources.

The objective of the inspection function is to monitor the condition of equipment and warn of any impending failures. It consists of two roles: quality control and quality assurance. Quality control involves monitoring the ongoing work to ensure quality. Quality assurance monitors the control system and carries out longer-term programs to ensure that operating equipment is fit for service. In most plants, the same people divide their time between the two rules.

Inspection technology is becoming more sophisticated as new techniques are brought into commercial practice. Simple global techniques are now supplemented by more precise local techniques to pinpoint problems. There are areas, such as corrosion under insulation, where sophisticated, more cost-effective techniques are still lacking, however.

Preventive maintenance is also critical to containment and safe handling. In many cases, accident prevention relies on the proper operation of instrumentation, remote-operated valves, safety relief devices, emergency power systems, and the like when needed. An effective preventive maintenance program will not only proof test and service critical items at regular intervals but evaluate their performance and the appropriateness of the preventive maintenance program itself. Safety relief devices usually merit a separate emphasis and specialized personnel because of their criticality.

Both inspection and preventive maintenance require a complete, accurate, and continually growing data base to be effective. Typical requirements are listed in the text and, clearly, computerization can be a real advantage even in small plants.

A system to ensure that the required repairs and testing are carried out in a timely and proper manner is also essential. As part of this system, a certain independence for the inspection function from the operations line management is an advantage to avoid undue budgetary pressures. In addition to tracking the implementation of work requests and identifying those overdue, the tracking system should receive top management review at frequent intervals to ensure the work is getting done.

The final element of the mechanical phase of an accident prevention program is an effective work permit system. In the United States, this system must comply with OSHA rules. Its objectives should be that the risks associated with every job are understood by the workers, that the equipment and job site are properly prepared and in safe condition, that the job requirements and those for personnel protection are clear, that there are no conflicting jobs going on simultaneously, that operating people know what is going on and when, and that acknowledgement be given that the job has been completed.

In such a broad subject there are many details to be worked out. This chapter presents some examples of how this might be done, but these are not recipes for success. Each plant must develop procedures and systems that will be effective at its particular site. As we shall see in the next chapter, these may well differ from plant to plant because of the culture of the work force and the history of the site.

CHAPTER 9

Risk Control/Reduction II: Procedures/Behavior Aspects

Introduction

Many accident and near-miss investigations tend to focus on human errors as the cause of the incident in question. Often, these incidents are attributed to procedures not followed or to procedures that were not up to date, were inaccurate, or were nonexistent. Investigators tend to be content to end their investigation by pointing to human error in general or to a specific individual as the cause.

As we have seen in the foregoing chapters of this book, many risks are controlled by procedures. Many are also controlled by the knowledge of the people charged with plant operation. In addition, some risks are related to the interfaces between individuals and the hardware and software used to control plant operations. Since management is relying on human beings for these critical services, how can it afford to be content with less than optimum performance of this resource?

It is the contention of many writers in the safety area that management's responsibility for ensuring the best possible human response goes far beyond the usual approach to procedure-writing and employee-training. Management has a responsibility for the effectiveness with which it communicates information to these employees and for providing an environment in which optimum human performance is possible. This implies a recognition of the factors that influence behavior, including the general job environment, the interface with facilities, the level of stress associated with normal and emergency operations, personal issues, and the many other performance-shaping factors that influence human response. According to many authors, the most important performance-shaping factor is the commitment shown by management to avoiding incidents. Under normal behavioral patterns, dealing with all these factors is difficult. Increasingly, concern about the fitness for duty of critical plant workers is growing. Without fitness, the task becomes much more formidable. We already have federal regulations about drug and alcohol testing of transportation workers, and the implication is that employers should assure that critical plant employees are at full capability at all times. At the same time, the legalities of random drug and alcohol testing for existing employees are complex at best, making it difficult to assure that all plant operators are at full capability all the time.

When we think about how certain humans have either excelled brilliantly or overcome tremendous handicaps to function normally we still get only a glimpse of the ultimate human capability. Aside from the relatively few Albert Einsteins and Olympic athletes of the world these are many, many others who have demonstrated intellectual and physical performance that is a step beyond the average. Perhaps blind people are the most frequent example of those who have overcome a very serious physical disability to function almost normally in society. We have a big-league baseball pitcher who has only one hand. By providing ramps and other special facilities society has created an environment in which many physically disabled people can work and contribute. The combination of an enabling environment and individual motivation has been shown to be very powerful. Industry (and society in general) needs to work with these forces to provide maximum assurance that the human side of the incident prevention effort is the best it can be.

This chapter will discuss some of the behavioral aspects of human response as well as some of the criteria for effective transfer of information to those who need to use it.

Behavioral Aspects

Let us start with some definitions. When I refer to *behavior,* I mean an observable act. A *consequence* is the event that follows the specified behavior—results from it. Behavior and attitudes are interactive. Behavior that is perceived as successful or unsuccessful produces positive or negative attitudes that, in turn, result in additional behaviors that reinforce the attitudes.

Perhaps a simple-minded example will help. The telephone rings. If one's attitude toward the telephone is negative—that is, the ringing is an interruption, it can only mean more work—the behavior associated with the telephone ringing might be to leave the office or just not answer it. The consequence of that behavior might be that the person's superior becomes angry as a result of not being able to reach the person and, at their next encounter, criticizes him or her for being inaccessible. This will only reinforce the employee's negative attitude toward the telephone. On the other hand, one might consider the ringing of the telephone as a new and exciting opportunity. Such a person will pick up the telephone before the first ring is complete out of curiosity to find out what wonderful opportunity awaits him or her. If it turns out to be a nasty job, this will have some impact on his or her eagerness to answer the telephone the next time.

Thus, consequences of a given behavior tend to influence future behavior. Social approval from co-workers and superiors forms part of the perceived success of any given behavior. For example, if an employee jumps onto a moving forklift to hitch a ride, he or she commits an unsafe act. However, if co-workers express admiration for his/her agility and creative approach this attitude will reinforce similar behavior from the worker in the future. In effect, the worker will be saying, "They like me; I'm one of the boys."

Similarly, if an employee carries too much to the job site in an effort to make one trip instead of two, his supervisor might pat him on the back for being a good worker, or a hustler, and a good example for others. The worker is likely to repeat such behavior in hope of further rewards, perhaps even a raise or promotion, at least until an injury occurs. Observation of these interactions led to the observation that Dan Peterson, noted safety consultant, uses as a philosophy statement: "What gets measured, what gets rewarded, gets done." In short, the way to control the behavior of employees is to recognize both positive and negative behavior appropriately. Whether or not intended, vague and conflicting management signals are perceived by the work force and responded to. The stress is particularly high on those in the first level of supervision who must provide management's directions but at the same time often have been promoted from the very group they must supervise.

Influencing worker behavior can be visualized to be a bit like a triangle. At one junction is the behavior; at another is the organization environment; and at the third are the attitudes, awareness, and perceptions of both the workers and the first-line supervisors.

Both positive and negative feedback are important influences on behavior. It is just as important to recognize and reward proper behavior as it is to correct any unsafe behavior immediately. It seems obvious that if you see one of your co-workers put a hand inside a piece of equipment to unclog it, you should say something immediately in hopes of preventing an injury. But it is equally important to recognize the worker who takes the appropriate safety precautions before attempting to unclog the equipment, so as to reinforce safe and prudent behavior. Waiting for the next safety meeting to comment on either behavior may lead to the perception by the workers that safety is only important during safety meetings. You may be telling them it is acceptable to perform unsafe acts as long as they do not get hurt, or that taking the time to work safely is frowned upon because it was not recognized positively immediately. In both alternatives described above, it was not the technology that was critical but rather the discipline to perform and reinforce the appropriate behavior.

Just as behavior influences attitudes, attitudes influence behavior. It is management's task to create an environment that breeds positive attitudes. Fundamentally, workers will attempt to perform up to the standards set by their leaders.

Leadership needs to participate rather than merely direct and support safe operating practices. Workers examine their supervisors' attitudes continually to determine what their true values are. They want to know what pleases and displeases their supervisors. They get very good at determining and analyzing this information, so every level of supervision must have a consistent positive attitude about preventing accidents if worker behavior is to be optimum. If something is important to the supervisor, it will be important to the worker. If supervision is doing something to cause the workers to make unsafe decisions, this needs to be ferreted out and corrected. Management's attitudes, and the environment they create, influence the behavior of workers.

Thus, it is critical to avoid confusing signals and/or body language in the organizational approach to safe operations. Some of the typical mistakes are as follows:

- Nonconformance with safe practices is considered to be the result of not being tough enough on the workers (i.e., the workers do not want to work safely).
- It is thought that it is okay to rely on a remote staff function to ensure safe practices are followed (i.e., safety is not really important).
- An emotional response of the management transforms investigations into "Who caused the accident?" rather than "What management system failed, which resulted in the accident?" (i.e., it is somebody else's fault).
- Tacit approval is given to possibly dangerous shortcuts in order to speed production (i.e., safety is second to production).
- Stress is placed on external boundary conditions or barriers rather than the opportunities to correct the situation using resources under a supervisor's immediate control (i.e., it is not my fault).

To address many of these questions, an orientation toward the future is critical. If the supervisors and workers have a joint belief that safe operations are a common interest (which can be fostered by recognizing behaviors immediately and consistently), the level of trust needed for open communications and unequivocal feedback in both directions can be created.

Numerous consultants (see reading list at end of chapter for a few of the studies in the area) have attempted to distill these principles into a more systematic approach for influencing human behavior to be more sensitive to avoiding incidents. Some of these specific approaches have been and will be touched upon in this chapter. Most of these programs tend to be formulated in generic packages that (hopefully) can be adapted to any site. However, it is my observation that any hopes that an outside consultant can parachute in and be effective in changing worker behavior are futile. I firmly believe that full involvement of local people (both management and wage) is essential for any lasting cultural impact. So is considerable patience. While the outside programs available have similar elements, each comes from a slightly different point of view. Each, therefore, must be adapted to the local situation. The common elements of all behavior modification approaches are:

- Data gathering (interviews, questionnaires, accident data, etc.)
- Focusing attention (identifies factors for improvement)
- Action items (e.g., train supervisors in behavior principles)
- Measure and trace progress over time (focus on action items, not statistics): enforcement
- Feedback (progress charted publicly, adjustments to action items): reinforcement
- Cultural impact (recognition is consistent, communication enhanced, etc.)

Some examples of criteria to use in adapting these elements may be useful. If employee reporting of incidents and near-incidents is important for data gathering, then reporting must be made easy, barriers (real or perceived) must be eliminated, and rewards must be provided. Interpreting the data to provide focus and identify action items must include participation by all groups. Priority setting can include any number of factors agreed upon by the group as important: chances of successful implementation, cost/benefit effects, time for implementation. Tracking progress requires a system to collect, analyze, and display results against scheduled objectives. Adjustments to these schedules need to be kept current, complete with the reasons for any changes. Problem areas are faced and reevaluated. Feedback on problems must result in positive efforts to redefine the solution and implementation plan, not in criticism.

A project to add remotely operated shutoff valves (ROV) to our chlorine tank-car unloading system may serve to illustrate the dynamics of the interaction process. The Chlorine Institute recommends such valves be located on the piping very near the tank car to provide rapid shutoff if a leak develops in the piping between the car and the consuming operation, and NJDEP required this improvement of us. Small leaks are a particular problem since the cars have excess-flow check valves sized for larger flows. An installation was designed and installed with little input from the operating people.

After a few days operation, some small leaks were found in the new valves even though they had been tested both at the supplier and in the field. The valves were removed for evaluation and the system was returned to the previous configuration. At that point, the operator's difficulty in handling the ROV arrangement became important. The valve with its operator represented a heavy, clumsy end fitting to the land piping, which needed a certain degree of flexibility to match up with the car fittings. The operators were having difficulty in maneuvering the combination to match up the flanges, and were worried about possible leaks. Muscle strains and lower productivity also seemed possible. Since other valves of this type were used successfully in the plant, we concluded the complaints were real, not just antagonism against a new component. A joint design/operations group was formed to rethink the design, do some field observations, assess the cost/benefit ratio, and resolve the issue. In the meantime, the safety engineer explained the problem and the ultimate solution to the state agency to whom the improvement had been promised. While the project went forward in less than ideal fashion, it went forward to a successful solution once all groups had become involved and the full extent of the mechanical and operating problems became clear. Ultimately, clear communication and joint effort provided a practical answer.

Finding the True Cause of an Incident

Accident prevention means anticipating what is most likely to happen in the future and devoting the time and other resources to preventing all foreseeable incidents. There are many unknowns in predicting the future, but many things are known and history has a tendency to repeat itself. From careful analysis of what has happened, we can anticipate a great deal about what might happen in the future. Thus, accident investigation is often required by law and is recognized as a valuable preventive tool if done well.

To learn from history, we need an energetic, penetrating attitude. An accepted approach to achieving this goal is to ask "Why?" at least five times. The rationale here is that the first causes identified are generally not related to the fundamental problem. Cures that are developed from the first-, and perhaps second-, level analysis will probably prevent the same exact problem from recurring, but will not prevent other related problems from being dealt with inadequately. In addition, history shows that cures at this level often have a negative impact on productivity, product quality, and costs. They simply do not reach deep enough to deal with the fundamental problem.

As the layers of the onion are peeled back and more fundamental causes are revealed, the impact of solutions to problems is more positive and more pervasive. Related problems are covered by the solutions, the solutions are simpler and easier to understand and implement, the results are less costly. At this level, history shows that cures generally have a favorable impact not only on safety but also on productivity, quality, and cost. If this is not true, then the investigation has probably not been penetrating enough. Often times, these penetrating solutions require changes in the management systems in place to deal with a process or a function. Stopping at the person involved with the incident being investigated is usually a sign of an inadequate analysis: It has been suggested that the following two questions are a good starting point for a penetrating analysis.

- What was the person trying to accomplish—that is, what was his immediate objective?
- What was the real reason that this objective was important?

Knowing these specific goals and the true reasons why people were trying to accomplish them in a given time frame are the keys to selectively asking questions to uncover the root cause of any incident. The only accident prevention system that has a chance to work comprehensively is one that is designed, developed, and implemented by the people that work within the system. History shows that progress results primarily from ideas that flow up from the working line to impact the management systems in use. Management's task is to make this upward flow of information easy and to listen carefully to what is presented. Its incentive is that when the business is being managed in the correct way, safety is present and productivity, quality, and cost advantages accrue.

A less-positive aspect of incident investigation is that workers being questioned about the events surrounding an incident are often on the defensive and sometimes do not relate the facts accurately.

This could be the result of the effect of the excitement on their powers of observation, but also it could be a simple lie. The technique of the interviewers is very important. Their prejudice must be controlled. Starting the interview with broad questions and gradually narrowing the focus is helpful. Listening carefully and being patient to give the interviewee time to formulate an answer is very important. Paraphrasing what the interviewer heard as a way to check what was said is also a valuable technique.

A critical factor will be to resolve the differences in testimony obtained from separate sources without angering the people involved. The paraphrasing process is one way to test conflicting data. Obviously, asking several witnesses the same questions will reveal any discrepancies and identify additional questions that may clarify the truth. It is not unusual to have to return to some witnesses with a second round of questions and perhaps a new strategy to get all the available facts. Incidentally, electronic tape machines usually heighten defensiveness and distrust. Take careful notes instead.

The Work Environment

As indicated earlier, management creates the environment in which its people work. Whether real or perceived, this environment contributes to the mental and physical state of the employees. We all have days in which we feel more capable and energetic than others. The "state" of the employees will influence their ability to work safely and productively. It is in management's interest not only to avoid creating systems that negatively impact the mental and physical state of its employees, but also to seek consciously those systems that will enhance the state of its employees during the workday. Thus, programs to help with child care for dual-career families may well enhance the state of employees because worries are relieved and self-esteem is enhanced. Wellness programs involving improvements in the physical condition or preparation for heavy work can also have a positive effect on the state of employees. It is my thought that physical warm-up before the day's (or night's) work not only prepares the body for the tasks ahead, but also helps to focus the mind. Amateur athletes warm up and practice skills before a game. Why should not industrial workers benefit from a similar approach? There are many, many other approaches that have been discussed in the literature and/or tried on an ad hoc basis at a given installation. None of these is a panacea. Each must be evaluated for effectiveness at the specific site.

One mechanism that has been found to be useful in communicating the importance of avoiding incidents to everyone is the execution of frequent field surveys by a management/worker team. These surveys are focused on the behavior of the workers that are observed. The objective of the team is to eliminate unsafe behavior immediately by forestalling unsafe acts, *and* to provide on-the-spot reinforcement for safe, responsible behavior. For example, if members of the team find an employee without a hardhat or safety glasses where required, the situation should be corrected immediately. On the other hand, those who are wearing the proper protective equipment should be complimented immediately. If they come upon an operator taking a sample wearing protective glasses and face shield or goggles, they should stop and congratulate him for working safely. Enforcement and reinforcement are the key processes. The participation of the management and the hourly employees in such surveys has made possible an important two-way communication about critical safety elements. The four elements of an effective communication program are:

- The ability to listen
- The ability to observe what is going on and interpret the information being received
- The ability to consistently present the "no-incident" message
- The ability to accept, and provide, feedback to ensure that the message has been properly understood

Consider this example and the alternatives for dealing with the employee in question. An employee suggests modifying a small section of tank truck loading piping to eliminate spillage on the roadway that ultimately flowed into the process sewer. The idea would save some product, eliminate a slipping hazard, and reduce waste-water treatment costs. The supervisor of the employee told him where he might find the materials he needed. The employee got what was needed and changed the pipe himself. No safety reviews or change authorizations were carried out.

Here we have the classic "nonideal" situation. We want to applaud the employee's idea and initiative, but equipment-change procedures were ignored. To applaud or even ignore the act will enable other bypassing of the management of change procedures. To make a major issue of failure to follow these procedures could stifle future creativity and initiative. Clearly, a way needs to be found to reinforce the positive and discourage bypassing established procedures.

Our first-line supervisors felt the best approach would be to have the employee process the change authorization after the fact. The purpose was to emphasize the importance of managing change and familiarize the employee with the system. Their logic (to be used in discussion with the employee) was that recognized change procedures provided for after-the-fact review for emergency changes—why not for surprisingly rapid implementation as well? I believe this reflects a bias toward encouragement of initiative and motivation, which was a wish at this plant.

In addition, the judgment was that change authorization procedures were well respected at the plant, and the incident/improvement conflict presented a good topic for reinforcing the right procedure at a safety meeting. Under different circumstances, a different balance might have been appropriate. The reader may wish to take a moment to analyze what the appropriate response should be at his/her plant.

By taking a thorough, active, and open approach to communication, the confusion associated with saying one thing, but behaving in a way that is perceived to mean another, can be addressed.

The Impact of Stress

A number of people have explored stress as the cause of human error. A. D. Swain et al. of Sandia Laboratories (see reading list) did considerable work on the causes of human error in nuclear power plant operations. C. J. Latino of Reliability Center, Inc., published an article in *Chemical Engineering Progress* in May of 1987 that attempted to apply modern psychology to process-plant operating errors. Both authors provide insights into ways to improve the chances that operators will respond correctly to all situations.

In high-stress situations, humans are much more prone to make errors than under normal circumstances. When the Three Mile Island Nuclear Power Plant went out of control, over 100 alarms were sounded and, in some cases, the plant operators responded by doing exactly the wrong thing. For example, they actually shut off the main reactor cooling system, which was desperately needed to help prevent the emergency because of pump vibration. Swain's study of the general nuclear-power-plant situation indicated that if plant operators have only one minute to react to an out-of-control situation, the chances are greater than ninety-nine percent that they will do the wrong thing. If they have ten minutes to react, the odds are only one in ten that they will do the wrong thing when faced with a single alarm, which improve to one in 1,000 with a half hour in which to make a decision. An interesting factor is that with two hours to analyze the problem and choose an action, operators will still make the wrong move one percent of the time.

These figures apply when there is no error in detecting and understanding the alarm. During emergencies, when corrective action is urgently required to prevent a minor upset from becoming a major accident, time factors have a large influence on the performance of the operator. In these situations, the operator may be faced with a barrage of alarms and data from which he or she must select the important information, ignore trivial data, and make an accurate diagnosis. All these actions must be completed within possibly tight timing constraints in order to prevent escalation of the incident. Thus, the time available to identify and perform the task will, if not sufficient, lead to extremely high levels of stress. As described earlier, these high levels of stress will tend to cause an increase in the error rate.

In addition, errors may also arise as a result of the quantity of information the operator receives in a short period of time. At the moment when an upset occurs, the operator may be unable to interpret and assimilate the information being received due to the large quantity of information available in a short period of time. This is normally referred to as *information overload.* Thus, a critical reading or parameter may be lost among the clutter of irrelevant data and valuable time may be lost in finding the relevant information. This situation is more likely to be the case as the magnitude of the upset increases since the number of systems affected will generally increase. The level at which information overload occurs varies from individual to individual, and it is a strong function of short-term memory. Typically, short-term memory has a capacity of only seven items and can only retain information for a few seconds. As these limits are exceeded, the error rate increases rapidly.

Researchers have shown that a prerequisite for disaster is an error chain. They have noted that a single mistake rarely causes a major accident. Indeed, they find that it takes about fourteen errors in series to result in a bona fide disaster. Since some of the chain of error will flow from the same misunderstanding of the situation that led to the initial mistakes, it is clear that a chain interrupter must be operative very early in the process if disasters are to be averted.

A widely used approach to preventing initial errors in a high-stress situation is to carry out various simulated emergency drills. In these drills, a realistic hypothetical emergency is presented to the operators, and they respond in a time frame comparable to that of the real emergency. They are forced to communicate to the others with whom they would be involved in a real case such as government agencies, the local fire department, plant emergency response organizations, and so on. A very important part of these drills is a meaningful critique of exactly what transpired. The thorough investigation techniques described earlier need to be applied here to ensure that the true causes of any weaknesses in response are identified.

Inherent in this approach is another technique that is of value. This is the anticipation of potential accidents so that remedial action can be considered before any such event takes place. Obviously, the drill scenario represents the anticipation of a particular incident and reduces these anticipated emergencies to a more routine upset, thus reducing both the stress and the strangeness of the situation.

Clearly, the automated control system of many sophisticated process plants addresses and attempts to forestall many incidents. Computers are often programmed to shut down a system safely if certain combinations of measurements occur. Such action is often programmed when system characteristics are such that an accident could occur in a time so short as to place the operator in a very high-stress category as discussed earlier. Again, eliminating the need for unfamiliar decision in high-stress situations is a valuable step. The costs of this safeguard include an effective maintenance program and the risk that spurious trips will increase costs and impugn trust in the system.

The design of information displays and instrumentation readouts can go a long way toward ensuring that the information on which an operator must base decisions is clearly understood. The size of the print on instrumentation or digital readouts, the location of adjacent buttons and switches, the color coding of these buttons and switches, and even the tactile perception of one switch vs. another can be used as secondary indicators to the operator about whether he/she has his/her hands on the right control lever. Swain points out that any action that an operator must take that is counterintuitive is much more likely to result in an error. For example, if an operator in the United States is asked to push a switch down to turn on the lights as opposed to the more conventional upward movement, he/she is 100 times more likely to push it the wrong way because of his/her intuitive programming. Add stress to the equation and he is 1000 times more likely to make the mistake. Consequently, any switch that has the "on" position in the down direction is likely to be subject to the same errors.

What this says is that the human/control-panel interface can also contribute to the stress on the operator, and when combined with other external stimuli such as a fullblown emergency situation, can often lead to bad decisions. It is imperative that the control station be as foolproof as possible, which can only be done with significant input from the operator who will use it.

Errors occur even in low-stress situations. The most remarkable of these to strike the author in recent days was the transplantation of the wrong heart into a living person. This, of course, made great headlines but represents a dramatic example of the potential for error even among highly skilled, highly experienced professionals. What happened in this instance was that a heart donated by a person with type A blood was transplanted into a person who had type O blood. This increased the chances of rejection of the organ and would appear to have been totally preventable. If the highly trained professionals involved in a life-or-death heart transplant process can make mistakes of this nature, even though they are presumably very confident of their abilities, is it any surprise that our operating personnel can make similar unthinking mistakes? One accident victim characterized this as "losing respect for the risks" through familiarity.

Some low-stress errors may be minor, even inconsequential, but the potential for, let us say, introducing the wrong reactant into a vessel from a manifold of similar-looking charge valves, or the introduction of the right reactant in the wrong order, could have serious consequences. Thus, the causes of low-stress error are worthy of study. We referred earlier to the human/control-panel interface as a potential source for a mistake. This may be critical in a high-stress situation but can also contribute in low-stress situations.

Distraction or lack of attention is another obvious cause. Plant management has control over some of this, and here again, we return to the plant environment as a fundamental cause. If the morale in the workplace is low, if there is a threat of shutdown or layoff, if working conditions on very hot days verge on intolerable, the mental and physical environment provided by management may well be conducive to a distracted state for one or more employees.

Some distractions are not controllable by management, but recognition of these factors and applying effective countermeasures may still be possible. During the stock-market crash of 1987, we had a rash of minor incidents in our plant. One might have thought that those associated with the financial markets would have been much more prone to distress during this particular time period than blue-collar workers. In retrospect, I believe that many of our employees were concerned enough either about the drop in stock values, or with the opportunity to buy stock at unusually low prices, so as to be distracted from the job at hand. Perhaps we could have addressed this situation in a safety meeting and avoided some of these accidents.

Even very personal distractions are sometimes known to plant supervision. It is rare that an employee gets divorced, becomes a parent, or suffers major financial stress without some co-worker knowing about it. On a less-dramatic note, some employees are less effective on the night shift, or on the first day of any new shift. In an environment in which co-workers watch out for each other and care about preventing incidents of any sort, it might be possible to deal with even these very personal distractions in one way or another. Clearly, in the best-of-all-work-situations, there would be enough trust between supervisors and workers to air out these concerns and see whether there could be positive (temporary) changes in the work situation aimed at minimizing the chances for accident.

A more serious cause of human unreliability is alienation. The recent downsizing of many companies toward a "leaner and meaner" configuration have been perceived by workers as a repudiation of long-standing, but unofficial, contracts with them. In other cases, work reorganization has removed some or all of the challenge from the task and/or degraded work conditions to an intolerable level. Latino writes of an interesting example. In a plant that had fifty employees working in a 100°F environment, three robots were purchased that displaced eight of the people. In order to protect the electronic components in the robots, the room temperature was then reduced to much more comfortable levels for the humans. The clear message received by the remaining forty-two workers was that the machines were much more important than they were. Would anyone be surprised if those workers were less diligent, less concerned with their productivity, in short, much less loyal to the company and less attentive to their tasks under these circumstances?

In some cases, lack of knowledge or skill must be addressed to prevent human errors. Psychologists call this human deficiency.

It is difficult to reveal deficiencies in our current society, particularly in the industrial area. People generally compensate for a deficiency by forming mental models that they attempt to follow. Based on what they know and what they think they know, they emulate the behaviors of others who appear to be successful at their job. In many cases, these people get by and even become comfortable while functioning without understanding the operation they are trying to control. This lack of understanding can cascade when these apparently competent workers are asked to train others in their job.

When jobs are new or changed, a high potential for deficiencies exists. This is particularly true if tasks have been made more complex and involve high sensory loads. This introduces the subject of effective information transfer to the workers, commonly referred to as training, and the evaluation of the effectiveness of this information transfer, often referred to as qualification or certification. They will be discussed in the next section of this chapter.

Before going forward with information transfer, let us consider the following list of items drawn from Latino's article, which may signal potential for human error.

- Any action planned but not carried out according to the plan.
- The potential for infrequent events that must be controlled promptly.
- An event that can easily be avoided by the action of the operator, but said action must occur within the first half hour after the commencement of the event.
- New or changed technology, or equipment, has recently been introduced.
- New people are involved.
- Logistics and human/control interfaces are less than optimum.
- The required task lacks imagination or challenge.
- Machines, not people, are the focus of management's attempts to improve productivity.
- Working conditions are at the extremes of human tolerance.
- Management is task oriented and appears to lack consistency of purpose.
- No mechanism exists for exchange of views and gathering the expertise of the workers in solving problems.
- Some work tasks are socially unacceptable.
- No means exist for recognizing unusual contributions.
- Only certain "star-quality" people get the assignments perceived as the best.

This list is not meant as a comprehensive checklist for identifying stressful situations or work activities that are conducive to error. Clearly, there are many other situations, such as labor/management strife, which may form such an environment. The list does present some examples of the types of things for which management should be alert, and not afraid to address, if the best possible human behavior environment is to be achieved.

Transmitting Information

In more normal situations, the level of performance is still tied to the time available, but it is more strongly a function of the amount (and quality) of the training received and the quality of written procedures. The well-known learning curve principles apply to this situation. It is clear that an operator who lacks experience or skills will commit errors at a higher rate. As the operator gains experience, the error rate decreases. A particularly dangerous phase is when the operator has sufficient skills to run the plant yet lacks the in-depth knowledge that will enable him/her to deal with unusual operations or with rare upsets. These, if mishandled, could lead to an accident. In terms of written operating procedures, the operator will refer to them only if he or she finds them useful and easy to use. Thus, if the written procedures are not used, incorrect, unofficial procedures will become ingrained in the workplace. These are likely to eventually lead to an incident.

It is important to recognize that everything that is going on presently in any plant is derived from the current knowledge base of the workers. This base includes what is actually known plus what people think they know. This is the basis for all decisions. If current knowledge is lacking, decisions on the best way to manage the process and deal with abnormal operations can be in error.

Demming's observation that if an organization continues doing the same thing, it will continue getting the same results, is very pertinent to the process of gathering and transmitting information. Decisions should be based on proven knowledge, not assumptions or hypotheses. Continuous improvement in the knowledge base for decision-makers is essential. If something good is discovered and not incorporated into the knowledge base, it will soon disappear. Preservation and growth of the knowledge base upon which decisions are made should be the objective of plant management.

In developing a management code for process safety, CMA set out some specific objectives for training of personnel. Training, to me, is a loosely defined term bandied about frequently by management and government agencies. In this book it means the transferring of information necessary for expert plant operation and maintenance.

The objective set out by CMA is that all personnel reach and maintain proficiency in safe work practices and in the skill and knowledge necessary to perform each job safely and productively. To ensure that these skills and knowledge are transmitted and properly applied, companies must not only provide thorough training but also periodically test employee proficiency.

The initial training of employees, whether new to the firm or new to the assignment, must include both general work practices in the area and specific job skills. To accomplish this, both the appropriate training materials and qualified instructors are required. Included in the subject matter must be not only the specific process and equipment information required to operate and maintain the equipment properly, but also the information on the hazards and risks associated with the operation. In addition, general safe operating procedures, such as the work permit system, the general approach to starting up a pump, and procedures for taking samples, using tools, and lifting, must also be part of the training syllabus.

Refresher training is also essential. Part of this refresher is simply repetitive: a review to ensure that what has been previously taught continues to be understood. A second part of the refresher training exercise updates employees on changes in materials, controls, equipment, and so on and their impact on safe operating procedures. Both of these aspects of refresher training are required under NJ TCPA and are also required by upcoming federal programs on process safety.

To many, training means a formal classroom environment. This is often counterproductive. By the same token, complete reliance on on-the-job training techniques in the field often leads to the proliferation of shortcuts or bad habits developed over the years by experienced operators and mechanics, and perhaps even endorsed by first-line supervision by failing to correct these developments. A proper balance between classroom and hands-on training experiences is important.

A third phase of training, which is particularly important in dealing with emergencies, is the periodic exercise of the employees in response to simulated accidents. In many cases, these emergency drills can be broadened to include emergency response forces from other units in the plant or from outside response organizations. This form of training is exceedingly valuable and increasingly required by regulation.

Documentation of training activities is also an essential. This has many positive values as well as the obvious one of proving to an agency that specific employees had been adequately trained even though some sort of incident had occurred. Internally, it is of value to know which employees have been trained in specific jobs and perhaps have taken additional training to increase their value and flexibility to the company. It is also of value to know the timing and content of the required refresher training and the deadlines for renewal of various operating licenses required in many states.

The training of contract employees has received a great deal of attention in the days since the Phillips polyethylene explosion in Houston. The realization that contract employees must be trained to the same levels of accident prevention proficiency as the company's own employees is gradually being acknowledged. Corporate lawyers who had advised their employers to leave the training of contract employees to the contractors are revising their advice in the face of major fines and other punishments levied against host companies, even though a contractor was found to be the cause of an incident. This development is consistent with public expectations that the company should do everything possible to prevent an accident. It is also consistent with the CMA's forthcoming code of product stewardship, which will impose an obligation on the manufacturer of a chemical to attempt to ensure that its customers know how to use and dispose of its product safely.

A large variety of materials and methods can be used in an effective training program. Many consultants are available who specialize in tailoring training programs to the intended audience. One factor that is becoming increasingly important in effective training is recognition of the language proficiency of the trainees. In our plant, some crafts are predominantly made up of those for whom English is not the native tongue. In many plants, Spanish is a language that must be considered in training employees, but for our painting contractor, Polish was the critical language. We worked with the contractor, and a number of our training materials were translated into Polish, and proficiency tests were developed in that language. This notwithstanding, it was necessary to develop certain specific procedures to ensure that an English-speaking supervisor was available on the specific job location of the contractors in case an emergency occurred. No plant can assume that everyone is proficient in English unless this has been established. Nor can any plant ignore its obligations to provide for training and proficiency testing in other languages if a need is identified.

The certification or qualification of employees is becoming an increasingly important aspect of the information transfer process. Not only is it part of the expectations of the public to which I have frequently referred, it is often required by law. In response, the Synthetic Organic Chemical Manufacturers Association (SOCMA), an association of mostly smaller chemical manufacturers, has developed a pilot program to assist its members, and anyone else interested, in employee certification. To SOCMA, workers' certification is "the process of training, testing, documenting and continuing performance evaluation to assure that workers in critical jobs (those that can have an impact on the public) have the necessary knowledge and skills to perform their work safely and effectively."

The program is specifically designed for a small location. This could be either a small company or a relatively small installation of a major company. It consists of two levels: the first level is the basic operating skills and the second relates to specific training in a list of unit operations. It is suitable for a wide array of training situations. The program was developed by a mix of small-, medium-, and large-size companies. Additional criteria used as a basis for its development were as follows:

- It was designed with in-house training in mind, but suitable for taking advantage of courses available in local adult education facilities or community colleges.
- It was cost-effective.
- It did not require a large staff.
- It was aimed at meeting existing and anticipated regulatory requirements.
- It provided a positive approach to certification.
- It could be endorsed by the industry.

The fundamental components of the Level 1 Program are shown in Table 9.1. These include some fundamental skills and concepts appropriate to understanding and safely working in a process plant. The components of the Level 2 Program involve unit operations such as heat transfer and would be added to by the plant doing the training to provide the specifics of its own operations within the general framework (Table 9.2).

SOCMA is supporting the development and implementation of this program by developing the fundamental training materials, helping to train the trainers, and providing support to those who are using it. One of the fundamental contributions will be a testing program to establish that the fundamentals of the material have been understood by the trainees. In addition, remedial training

Table 9.1. Outline of Level 1 Program

I. Basic Skills
- Mathematics
- Chemical concepts
- Physical concepts

II. Basic Safety
- Hazardous substances
- Fire safety
- Physical and mechanical safety procedures
- Personal protective equipment

III. Applied Unit Operations
- Reading instrument drawings
- Basic instrumentation
- Utilities
- Piping
- Pumps
- Valves
- Fire & emergency response
- Emergency planning

IV. Overview of Federally Regulated Training
- Respiratory protection
- Hazard communication
- Carcinogen training
- Asbestos standards
- Hazardous waste operations
- Powered forklift truck operation
- Fire systems
- Portable fire extinguisher use
- Hazardous materials transportation
- Hearing conservation

Table 9.2. Outline of Level 2 Program

- Reactors
- Separation systems
- Material transfer
- Utilities
- Controls and instrumentation
- Heat transfer
- Environmental
- Size reduction
- Quality
- Process safety
- Mixing & blending
- Maintenance

materials will be developed for those found in need in the initial testing procedures. As of the second quarter of 1991, approximately eighty companies had purchased the fundamental program and eighty potential trainers had attended seminars put on by the association.

It is the general content of this certification program that is of interest.

SOCMA's program is not asserted to be the best or the only approach that is useful. The discussion here is intended to show that help is available, that at least part of the industry recognizes that help is required, and that it can be obtained at a reasonable cost through a group effort. It is expected that effective training and certification programs would result in safer and more profitable plants through the efforts of more-effective workers. It is also expected that a thorough training and testing program would deliver a message to the public and legislators that the plant cares about doing its best and is taking this step prior to any regulation. It is the author's opinion that if the industry is not successful in convincing the public of its sincerity in these areas, it will receive more mandated assistance than it can deal with in the accident prevention area. The legislature of the state of New Jersey is already grappling with a bill with the euphemistic acronym HELP. It would give committees of local citizens the right to inspect any operation and shut it down if it felt the situation was unsafe.

Procedures

The CMA Responsible Care Code of Management Practices addresses safe work practices from a very comprehensive point of view. Their scope includes both mechanical and operating procedures, as well as general safe work practices and specific unit operating procedures. The objective of written procedures in these areas, according to the Responsible Care view, is to "equip knowledgeable people to perform operating and maintenance tasks in a safe manner." Because knowledgeable people are the audience, procedures should be aimed at bridging the gap between training and the actual "hands-on" experience. They should include critical information so that every employee can protect him/herself, co-workers, and the public by avoiding accidents.

The Responsible Care guidelines go on to point out that procedures can erode with time. Minor changes and shortcuts can be introduced over time and passed on from veteran to neophyte unless several things happen to prevent this.

The first of these steps involves involving the workers themselves in procedure writing. Not only does this promote realism and "buy-in" by the people who actually do the work; it also exposes what really goes on in the plant to critical safety review. In addition, the language and terminology used in the procedures will more accurately reflect what is clearly understood by the workers.

The second critical step is to be sure that *all* procedures are written down and reviewed. Frequent surveys are then needed to ensure that only the written procedures are used. If a truly better way to do something evolves, the written procedures should be updated to reflect it. This gives explicit approval to the new variation and avoids tacit approval of irregular procedures that can then proliferate.

A periodic job safety analysis (JSA) has been found a useful tool in verifying and improving procedures on a specific job task. A team observes an operator or mechanic at a task and breaks it down into a step-by-step procedure. Problems and potential improvements are then discussed with the workers to promote joint analysis and gain endorsement of the end result. The documentation serves to ensure consistency from shift to shift and provides the basis for updating procedure manuals.

The third step involves providing a mechanism for feedback about problems. Exxon Chemical's Bayway plant provides this at weekly safety meetings, but other companies provide for written suggestions or critiques. Still others develop feedback as part of their periodic review and updating of procedures.

The last step in maintaining the integrity of all procedures is review and control. It is easy to say (and government agencies often do) that procedures should be continually updated. In fact, this happens periodically, often annually, with interim modifications captured in the management-of-change system. This is a practical approach but may not meet all of the requirements of current and possible future regulations for proper documentation. Inherent in the review process is the concept that procedures may only be changed with specific authorization. Certain locations for the *true word* are established and only certain positions (usually the second-line operations and/or mechanical supervisor) are able to authorize changes. These rules must apply whether procedures are maintained in hard copy or computer format.

Dow Chemical Canada's Murry Trask has written on what he calls "operating discipline" (*Plant/Operations Progress,* July 1990, Vol. 9, No. 3). Much of his thrust involves the documentation and objectives of plant procedures. He sees what I call procedures as providing:

- A way to help manufacturing workers understand their process (and its hazards) thoroughly
- The ability to operate 100% of the time in accordance with that understanding
- A system to preserve that knowledge and understanding in documented and approved form

Dow's goal is to have up-to-date information immediately available to workers and supervisors. This includes not only critical operating parameters but also both unusual and routine mechanical procedures and process hazard data. Clearly, this can be successful only if an effective management-of-change system is in place and frequent vertification of both documentation and practice takes place.

Procedure Content

Writing clear procedures for operating and maintenance activities is often a task without clear boundary conditions. There are those who wish to include every known bit of information about each task/process for the sake of completeness. There are those who recognize that a four-foot-long shelf of procedure manuals will never get used. I find myself in the camp of the concise. There have been too many experiences in which even simple procedures were not followed because they were relatively hard to find. Detailed backup information must be readily available, but procedures—especially operating and emergency procedures—need to be concise without resorting to a mindless cookbook approach.

The regulations implementing the New Jersey Toxic Catastrophe Preventive Act (1989) deal explicitly with the content of operating procedures for those units that handle EHS (extraordinarily hazardous substances). In addition, they require that the procedures be "readily available" to the operators and be written in English "in a manner understandable by the EHS operators and [that they] shall also be written in the language of fluency of EHS operators not fluent in English."

The content of the procedures "shall include but not be limited to":

1) Simplified process flow sheets and a process description defining the operation and showing flows, temperatures and pressures.
2) Procedures and conditions for normal operations.
3) A description of abnormal conditions, including the control and mitigating procedures to be followed to return to normal conditions.
4) A description of emergency conditions which could occur, including the control and mitigating procedures to be followed to reduce the impact of the emergency conditions.
5) Pre-startup procedures covering, at a minimum, testing for leak tightness prior to placing into EHS service.
6) Start-up procedures, including conditions to be maintained during start-up.
7) Shutdown procedures, including provisions for normal and emergency shutdown and details on the conditions to be maintained after shutdown.
8) A description of the type, location and purpose of safety relief devices, interlocks and alarms with their respective activation points indicated.
9) Sampling procedures addressing apparatus and specific steps involved in the taking of samples.
10) Safety procedures related to each specific operation in the standard operating procedures.
11) Procedures to prepare EHS equipment for maintenance and inspection of maintenance work upon completion and prior to placement of equipment in EHS service.
12) Material safety data sheets or fact sheets for each EHS used in the operation.
13) Logsheets and checklists where appropriate to the operation.

In addition, the regulations require that all changes to procedures be incorporated into the documentation before use and that an index with the latest revision dates be maintained.

All of the above is essentially consistent with the general objectives of procedures laid out in the previous section of this book.

The publication by Mr. Trask of Dow Chemical Canada referred to earlier lists fourteen steps for operating discipline of which twelve speak to this author of the content required in operating procedures for reactive systems. These are:

1) Define the important control parameters, including reactive chemicals data and why they are important.

2) Define interfunctional relationships of process variables.
3) Establish limits of control parameters.
4) Ensure that each control parameter has a field sensor with adequate redundancy and identify each.
5) Define the control strategy and system.
6) Monitor important process variables and alarms and report exceptions.
7) Specify remedial actions for problems that can be anticipated.
8) List appropriate safety precautions.
9) Document detailed operating procedures.
10) Provide graphic flowsheets where required.
11) Create energy balances.
12) Create material balances.

Translating these generalities into a practical system poses a number of questions. Does all of the information have to be in the same format (i.e., hard copy or computer file)? What does "readily available" mean? How many copies of procedures need to be maintained and where? How should plant units be segregated? Should critical mechanical procedures be included in operating procedures? How much should be covered in general safe-work-practice manuals and how much in unit specific procedure manuals? A host of other questions will arise as the specifics of your situation are dealt with.

First, I think a distinction should be drawn between practices and procedures. Truly plant-wide safety practices, such as when a self-contained breathing apparatus is required, sounding of emergency alarms, lockout of hazardous energy rules, work permit rules, and general evacuation and "safe havens" procedures should go into a plant-wide safe practices manual. While this manual forms part of the accident prevention program, it will not include the details of the procedures used to run a unit or do a specific mechanical job. It should be available in each place that unit-specific operating/maintenance procedures are kept (and possibly in many other places such as mechanical work sites).

The unit-specific procedures need to be clear, up to date, and readily available. To me, this means they are used and, therefore, useful. How this is managed in any given unit may be a function of the culture of the plant or the specific unit. In a highly computerized operation, data files accessible from the console may be most appropriate. In a highly dispersed manual operation, perhaps laminated cards would be appropriate. Between this extreme lies the more conventional approach of having some hard-copy manuals and perhaps some computer files in the unit control house. The point here is that while generalizations can be made about content, a very wide range exists for specific implementation plans, and this should be recognized.

A further point needs to be made about style and clarity. There are a number of approaches that might be chosen and a number of specialist consulting companies able to assist if needed. Choosing the style and format that fits your people is a critical first step. Choosing the complexity of the language used is part of that decision. It is necessary to consult with the workers on what approach will be most effective at each specific unit. The primary factor here is to recognize that procedure manuals are being prepared to help the user do the job. The style and format that help them the most are what they should get.

A starting point I can recommend involves three separate manuals: A unit procedures manual, a unit information manual, and a plant-wide safe practices/emergency manual. The unit procedures manual may include MSDS sheets for the chemicals involved or increasingly, this is provided via some form of computer data base. Typical contents of each of these three manuals are shown in Tables 9.3 for procedures, 9.4 for information, and 9.5 for plant-wide safe practices. As long as any government agency agrees to meeting its procedure requirements in separate volumes and formats, this approach should be found acceptable.

In addition, it provides manageably sized information and training packages for a medium-sized plant.

As mentioned earlier, dealing with contractors is becoming increasingly important. If contract employees regularly perform certain mechanical or

Table 9.3. Guidelines for Content of Unit Procedures Manual

General

- Pages numbered
- Table of conents
- Tabs for major sections, dated for latest revision
- Plot plan of unit
- Simplified flow plan
- Cover sheet showing date of review and revision number
- Overall process description, including process control philosophy

Procedures

- Start-up
- Shutdown for T/A
- Pre S/U and S/D checklists
- Basic operating procedures
- Short S/D including cold-weather concerns
- Checklist for putting equipment in/out of service—specific procedure for unusual equipment
- Assistant operator rounds (task books)
- Examples of standard logsheets, sample tags
- Product quality considerations and specifications
- Operating targets, batch sheets
- Typical material balance

Emergency Procedures

- Power failure, power dip
- Cooling water failure
- Loss of instrument air
- Loss of steam
- Major hydrocarbon leak
- Leak of toxic materials (EHS)
- Vapor cloud
- Safety valve or rupture disk release to the atmosphere

Table 9.4. Guidelines for Content of Unit Information Manual

General

- P&IDs, electrical and one-line diagrams
- Electrical area classification
- Glossary of terms
- Sewer diagrams

Procedures

- Blanking including permanent and boundary blanking lists
- Winterization lists
- Special precautions for extreme cold weather
- Rotating equipment do's and don't's
- Air freeing the unit and specific equipment pieces
- Special warm-up and cool-down procedures
- Loading/unloading tank cars/tank wagons/ships/pigging
- Equipment lubrication
- Cooling tower operations and precautions
- Drawing water from equipment containing light hydrocarbons
- Railroad operations
- Ecology upset notification

Equipment

- Listing of all control valves with failure position
- Safety valves and their set pressure, type
- Rupture disks and their set pressure
- Tankage list, including details of strapping, vents, roof type, contents
- Listing of motors and the electrical system details, including switch rack fed from and reacceleration capability
- Lists of pumps, exchangers, spheres, towers, reactors, drums, furnaces, compressors, centrifuges, filters
- Vessel drawings showing internals
- Electrical distribution system
- Technical data sheets such as pump curves, compressors data
- Utility station plot plan
- Utility flow diagrams
- Steam tracer header and tracer service
- Plot plan showing underground hydrocarbon piping

Safety

- MSD sheets or use separate PC system
- Listing of temperature-sensitive materials, including critical temperatures and what could happen if they are reached
- Loop seal systems, depth and testing
- Safe handling procedures for chemicals and catalyst
- Automatic S/D devices and protective systems
- Car-sealed open (closed) valves list
- Safety calender/task book
- List of pressurized equipment required to comply with area classification designation

Emergency Procedures

- Loss of fresh water
- Loss of nitrogen
- Use of deluge systems or other fixed fire-fighting equipment
- Fire-fighting with unusual reactive materials involved

operating procedures, perhaps specific procedure manuals should be prepared for these procedures. Surely, your contract employees need to be familiar with the plant-wide safe practices manual. In our plant, we also test contract employees on their knowledge of the latter and keep a computer data file on which individuals have been trained and tested. A contract welder, for example, may work for one contractor one week and a different one the next.

Procedure Manual Reviews

While an independent review of procedure manuals is often desirable, it often is not practical because the reviewer has insufficient knowledge of the technical information contained to pass judgment. Occasionally, an expert not currently connected with the unit is available to review procedures, but often a team approach must be taken to provide both an independent viewpoint and depth of expertise in the specific procedure. The group at the site (management committee, safe operations committee, etc.), which oversees safety and environmental policy, should ensure that these reviews take place at the appointed frequency. A somewhat generalized policy for procedure review is given in Table 9.6. We believe it is appropriate for ensuring that our procedures comply with NJ TCPA regulations.

One might conclude that this is part of a more comprehensive audit system. It is. More will be said in a later chapter about auditing approaches.

Fitness for Duty

Recently, an allegedly drunken New York City subway motorman drove his train off the tracks at high speed, killing six and injuring 145 more. This incident added a great deal of fuel to the public fire about doing something to ensure that workers, especially those in key positions, do not go about their jobs while impaired. In the United States, driving an automobile while impaired is a very serious offense and transportation workers are routinely tested after accidents. New employees are tested for drug use prior to granting employment in many industries, but current employees are usually not tested routinely unless they occupy "critical" jobs. This inconsistency is justified in this country under the First Amendment to our Constitution, which includes rights to privacy.

In this writer's view, the tide is turning on this issue. The public's expectations of those who deal with hazardous materials/machinery/diseases are becoming more demanding. More disclosure, more testing, and more scrutiny will be forthcoming in spite of union or legal objections. A recent court decision upheld the discharge of a union employee who tested positive for drugs after an industrial shipping accident.

Aside from the demand of the public, companies have a high stake in ensuring their employees are fit for duty. National statistics show that, relative to non-users at work, substance abusers:

Table 9.5. Guidelines for Content of Overall Emergency and Accident Prevention Manual

General

- Sampling including hot or toxic samples
- Tank/vessel cleaning and evacuating
- Entry into confined space
- Emergency reporting/alarm procedure
- Local evacuation alarm
- Safety valve/RD removal and reinstallation with the unit in service
- Work permit procedures
- Personal protective equipment
- Traffic safety/control
- Lockout/tagout procedures
- Use of tools and equipment
- Scaffold requirements

Equipment

- Plot plan showing location of pipe band line identification diagrams

Safety

- Plot plan showing location of: fire hydrants and monitors, extinguishers, hose reel boxes, Scott Paks, acid suits, safety showers/eyewashes, ROVs, deluge and sprinkler systems, snuffing steam, boundary blanks, hydrocarbon and toxic gas monitoring equipment
- General safety considerations
- Loop seal systems, depth and testing
- Safe handling procedures for chemicals and catalysts
- Blowdown systems, including simplified flow plan
- Sewer diagrams showing flow direction, vents, and seals
- Fire water grid system showing hydrants and grid isolation valves
- Plot plan showing defined high-noise areas
- Near-miss documentation
- Field change authorization procedures

Emergency Procedures

- Agency notification procedures
- Alarms

- Have twice as many accidents/mistakes on the job
- Are late to work twice as often
- Are sick three times as often
- Are only about two-thirds as productive

The cost of these abuses to industry in 1987 was estimated to be $140 billion. When added to the fact that half of cocaine users *sell* drugs on the job, the incentive to develop a clean workplace is tremendous.

Most companies hold that substance abuse in the workplace is grounds for dismissal. Many of the larger companies provide a confidential employee assistance program for those who ask for help in ending a substance dependency and will not dismiss an employee because he/she asked for help in this way. However, such employees are usually prohibited from "critical" jobs and may, therefore, suffer some job dislocation.

Table 9.6. Policy on Procedure Reviews

The SOC Policy on Reviewing, Operating, and Mechanical Procedures
• Each unit supervisor is responsible for updating his/her procedures on a continuing basis; for units involving Extraordinarily Hazardous Substances, there will be an internal review of procedures just prior to the legally required annual refresher training for all operators under the New Jersey Toxic Catastrophe Prevention Act
• Both internal standards and NJ TCPA require periodic HAZOPS of operating units; a formal review of operating procedures, training manuals, and critical mechanical procedures will be carried out prior to the scheduled HAZOP date to ensure that up-to-date documents are available for the HAZOP
• The review team will be nominated by the operations-section supervisor and submitted to the SOC for approval; this review team will include at least one member who is not associated with the unit in question, but who has the experience and capability to evaluate the completeness and effectiveness of the procedures
• To ensure that the procedures include correct and up-to-date technology in various critical areas such as personal protective equipment, internal sign-offs, similar to those provided by experts (e.g., industrial hygiene) in process design specifications, will be included in the procedures; it is the responsibility of the section supervisor to identify the needs for such internal sign-offs and submit this list with his/her proposed review team for SOC approval.

The contents of operating manuals have been defined both by the SOC and in NJ TCPA. Deviations from this content should be discussed with the SOC since too much, as well as too little, may interfere with the effectiveness of the procedures.

A problem occurs when different companies define "critical" differently. To my mind, any job that could cause a loss of containment of a hazardous material is "critical." Some companies have taken a much more narrow view, presumably to ensure being on safe legal and union relations ground. This clearly must be decided on a site-by-site, company-by-company, basis. Holders of critical jobs are subject to drug testing. Others are not, although periodic drug/alcohol searches (using trained dogs) may be held for all employees. The public may not be satisfied with the narrow definitions.

There are some things a plant can do to create an atmosphere that is more conducive to minimizing substance abuse in the workplace. Perhaps the most important of these is to make the work as challenging intellectually as possible. The objective is to build the self-respect of and opportunity to contribute for each employee. Related to this is creating a "Watch out for your buddy" atmosphere where workers truly care about the well-being of their co-workers and are willing to act on that belief. The two are clearly related. Both are based on mutual trust and open communication. This is different from tolerance and/or covering up of co-workers' dependence, which merely enables continued dependence and leads to inevitable disaster.

Summary

Many of a plant's accident prevention strategies rely on human activities. These include both the accurate execution of normal tasks and the rapid reaction to emergencies with effective countermeasures. A positive work environment is essential to minimizing the chance for human error and maximizing creative and productive input. Companies must live their commitment to their people in the lean years as well as the fat, and foster self-respect and trust. Primarily, this means having open communications with employees and recognizing that most accidents involve a failure in some management system rather than simply "employee error."

Management needs to ensure that the objectives its people are trying to accomplish are correct and understand the real reason for any action. Its "body language" must be consistent with its official positions, because employees will quickly interpret any differences they perceive and modify their behavior—perhaps in ways that make accidents more likely.

Incongruities contribute to worker stress and worker stress makes accidents more likely. An effective way to improve/correct employee response in high-stress emergency situations is to carry out frequent simulated emergencies so that the correct responses become more second nature. Another is to program automatic safe responses into the control sequence so as to remove the decision from the high-stress environment. Careful design of the operator/control-device interface will also reduce stress and make errors less likely.

Given a positive environment, management must also transmit the necessary information to its employees and ensure it is understood. This is done by a combination of training and written procedures, followed by some form of testing program. Many approaches are feasible and effective in these areas. The "right" approach will depend on specific plant circumstances. Some specific approaches to certification and organizing procedures are discussed.

Lastly, employee fitness for duty is becoming more critical. Not only are public expectations escalating, but also the costs of substance abuse in the workplace are becoming known, and legal roadblocks to more comprehensive testing are eroding. An appropriate policy will soon be needed everywhere.

Human Error Reading List

1. J. Bell, A. D. Swain, *A Procedure for Conducting a Human Reliability Analysis for Nuclear Power Plants,* U.S. Nuclear Regulatory Commission Report No. NUREG/CR-2254.
2. J. G. DeSteese, et al., *Human Factors Affecting the Reliability and Safety of LNG Facilities, Vols. 1 and 2,* Gas Research Institute Report Nos. GRI-81/0106-1 and -2.
3. D. E. Embrey, *Human Error: Theory and Practice,* Conference on Human Error and Its Industrial Consequences, Aston University, Birmingham.
4. D. E. Embrey, *Human Reliability in Complex Systems: An Overview,* National Centre of Systems Reliability Report No. NCSR R10, UK Atomic Energy Authority, July 1976.
5. B. A. Fecht, et al., *Human Factors Applications to Computer-Aided System Design in LNG Facilities, Vols. 1 and 2,* Gas Research Institute Report Nos. GRI-85/0183.1 and 2.
6. A. H. Howland, *Human Error Situations in Large Scale Process Plant,* Conference on Human Error and Its Industrial Consequences, Aston University, Birmingham.
7. C. J. Latino, *Chemical Engineering Progress,* (1987) 83(s) 42–45.
8. F. P. Lees, *Loss Prevention in the Process Industries, Vol. 1,* Butterworths, London.
8. A. D. Swain, H. E. Guttmann, *Handbook of Human Reliability Analysis with Emphasis on*

Nuclear Power Plant Applications, U.S. Nuclear Regulatory Commission Report No. NUREG/CR-1278.

9. R. B. Whittingham, *Human Factors in Reliability and Safety—a Pragmatic Approach, Process Econ. Int.* (1985) V(3), 49–52.
10. *Fault Tree Handbook,* U.S. Nuclear Regulatory Commission Report No. NUREG-0492.
11. *Guidelines for Control Room Design Reviews,* U.S. Nuclear Regulatory Commission Report No. NUREG-0700.
12. *PRA Procedures Guide,* U.S. Nuclear Regulatory Commission Report No. NUREG/CR-2300.

CHAPTER

10

Emergency Preparedness and Response

Introduction

In the aftermath of Bhopal it is easy to endorse emergency planning but not so easy to ensure its effectiveness. For example, the state of New Jersey chose to organize its SARA III local emergency response committees (LEPC) by community. Over 500 LEPCs were to be set up and each was to develop an emergency plan for approval by the state police. Many LEPCs were set up. A somewhat smaller number of emergency plans were submitted and fewer were approved.

Those that were developed used very few data about the real industrial hazards to the community because very few data on those were available. MSDS sheets were available but the specifics of such things as equipment locations, configurations, and operating conditions were not. As a result, emergency plans were based on general considerations for Industrial Risks.

This is not to say that these plans are of no value. Many aspects of them are extremely important in defining the core capabilities of emergency response. For example, the responsibilities of the various emergency forces were defined. Communications centers were equipped and staffers identified and trained. Fundamental facilities and equipment needs were identified and contacts with key industrial and mutual aid groups were established. In addition, joint industry/community emergency response drills were now possible to add some realism to the evolution of the plans. In addition, plans for handling natural disasters were strengthened.

However, the plans remained lacking in specific realistic scenarios that might represent something close to the actual emergency with which the community (and its neighboring communities) might be faced. In addition, there was little coordination among the plans of neighboring communities.

Some communities and industries have resorted to the "worst possible scenario" as a basis on which to plan. This was a fallout from the nuclear approach. The Atomic Energy Commission felt that exploring the consequences of the worst case would "bound" all other accidents. Since such accidents (core meltdowns, etc.) were inherently intolerable, many safeguards were imposed to prevent them. The WASH-1400 report (also known was the Rasmussen report) was a pioneering effort in the quantitative risk assessment of a complex system, but was faulted for underestimating the uncertainty of its results and for the inscrutibility of its analysis. Gradually, the industry and its regulators began to see that the "worst case scenario" (which became standardized) enveloped all others as far as consequences went, but not as far as the probabilities of much smaller accidents and human errors went. It then turned its attention to these more likely, less disasterous events, but probably too late to resurrect the credibility of the industry in the public's eyes.

This is not to say that the worst case scenario should be ignored. Rather, it should be prevented, almost at all costs. For example, if a site uses ninety-ton tank cars of chlorine in its process, the worst case scenario would be the complete rupture of a full car with the wind blowing toward the nearest house under the worst meterological conditions. Unless the plant is literally miles from its nearest neighbor, this would be a major disaster, possibly equal to Bhopal.

There are many things that are routinely done to prevent Cl_2 car rupture. The cars are designed for extreme conditions and are carefully inspected and tested. They are equipped with excess flow valves lest connecting piping be severed. Guards and derails on the tracks prevent other cars from being run into the Cl_2 cars. The cars are usually located away from the plant boundaries and away from sources of fire. No heat is applied to the cars and the pressure of the inert gas used to force the Cl_2 out of the car is carefully controlled. In some plants, cars are even enclosed in a building with a scrubber, although this is unusual. Some installations are monitored by TV cameras at all times. Proper plant security is maintained so that vandals are denied access. Almost any idea that has been conjured up by man to prevent a car failure has been employed.

Some things are not provided for, however. These usually include a plane falling from the sky and smashing into the car. In addition, serious sabotage, such as arson or the use of artillery, is not usually considered. Realistically, there is some residual probability that a car could still rupture, but using scarce resources to prepare for this scenario does not seem reasonable to many industrial people. They must still convince their local planning committees and regulatory agencies that the people will be better served if smaller, more probable accidents are used as a basis for planning.

We must also realize that emergency preparedness is not a one-time exercise. Not only do scenarios and communities change, but so do technology, accident data, people, the economy, community attitudes, and laws. Emergency preparedness is truly an evolutionary process, as are the standards against which success is measured.

General Preparations

Planning

In the United States, OSHA has promulgated the HAZWOPER regulation. HAZWOPER stands for Hazardous Waste Operations and Emergency Response. This regulation specifies requirements applicable to (1) emergency responses to hazardous substance releases and (2) corrective action cleanups at hazardous waste sites.

The regulation is found at 29 CFR 1910.120 in the OSHA regulations. Because the HAZWOPER regulation addresses two largely separate activities, the requirements are often inconsistent and difficult to apply.

HAZWOPER imposes three basic emergency response requirements:

1. A written emergency response plan
2. Documented emergency response training
3. Requirements for post–emergency response operations

Emergency response plans are required to be in writing and to be maintained for OSHA inspection at each work site where the employees may be involved in an emergency response. The only exception is for sites where the employees will be evacuated when an emergency occurs and will not be permitted to assist in handling the emergency.

The written worksite plan is required to address anticipated emergency responses, which are defined as:

1. uncontrolled releases of hazardous substances, including petroleum and petroleum products
2. Those requiring response by employees from outside the immediate release area or by organized responders

This is the HAZWOPER definition of emergency response, and it is uniquely limited to "releases" requiring response by "outside employees."

Examples of emergency responses covered by HAZWOPER and requiring a plan are:

1. Larger oil spills requiring involvement of an oil spill response team
2. Fires (beyond the incipient stage) releasing hazardous substances and requiring suppression by a fire brigade

The written worksite emergency response plan must address the following plan elements as a minimum:

1. Preemergency planning and coordination with outside parties
2. Personnel roles, line of authority, training, and communication

3. Emergency recognition and prevention
4. Safe distances and places of refuge
5. Site security and control
6. Evacuation routes and procedures
7. Decontamination
8. Emergency medical treatment and first aid
9. Emergency alerting and response procedures, including an incident command system (ICS)
10. Critique of response and follow-up
11. Personal protective equipment (PPE) and emergency equipment

It should be noted that all these plan elements focus on the involvement of the employees in the emergency response: the roles the employees will play, and the measures to be employed to protect the safety and health of the employees. HAZWOPER does not regulate the type of emergency response to be undertaken (e.g., dispersants vs. containment booms, bulldozers vs. vacuum trucks). Rather, HAZWOPER mandates a plan to ensure employees participating in an emergency response know how to act and are adequately protected. Like OSHA, HAZWOPER is employee-safety directed, not operation-control directed.

Existing plans can be incorporated by reference to avoid duplication and to satisfy the HAZWOPER plan elements listed earlier. Examples of existing plans are:

1. Oil spill contingency plans
2. Spill prevention, control, and countermeasure (SPCC) plans
3. Emergency procedures
4. SARA title III local or state emergency response plan

In fact, where existing plans already adequately cover the HAZWOPER plan elements, it may simply be a matter of referencing these sources in the HAZWOPER plan. A reference folder may suffice for the site HAZWOPER emergency response plan.

In meeting these requirements, there are two general considerations of value in the emergency preparedness process: a disciplined approach and a comprehensive scope for the core capability. These are the skeleton of preparedness. The flesh comes with the development, over considerable time, of specific credible accidents, and then with the study of their prevention, detection, and mitigation.

A Disciplined Approach to Planning and Response

In many areas, the Canadian process industries and government have taken the lead in developing industry/community interaction. In this case, the chemical and petroleum industries have developed a videotape presentation that describes a *Disciplined Approach to Emergency Response.* To this author, the approach also provides a disciplined approach by which to plan for emergencies. One of the major thrusts of the video is that hasty action may well be ill-advised. Time must be taken to analyze the situation, marshal the needed resources, and coordinate the response. Obviously, this is most easily done long before the incident occurs, in a planning mode, so that many resources and actions are prescreened prior to a real emergency.

The approach has the objective of protecting lives, property, and the environment in an efficient manner and is broken into five parts. These are:

1. A complete definition of the event
2. The prioritization of critical issues
3. The development of response strategies
4. The implementation of these strategies
5. Recycling to determine if the nature of the problem has changed

Defining the Event

The first stage, problem evaluation, involves four separate phases. These include the fundamentals of the problem, any modifying conditions, the potential losses, and available control measures. Each of these phases is broken down into elements.

Fundamentals

The fundamentals of the problem involve the quantity and nature of the material involved, the type and condition of the container, and the state of the incident. The properties of the material should be available from the MSDS. The nature of the container introduces more complexity. It might be a process vessel, a tank, a transportation vehicle, or a shipping drum. There really are only two possible incident states: stable and unstable. The potential for escalation is an important point to be evaluated.

Modifying Conditions

The possible modifying conditions include the location, all time factors, and the weather. Location factors are most complex. The physical location of the accident is important, but so are its position relative to other potential hazards, and the community's available response teams, sensitive areas, and access routes. Any of these might significantly change the nature of the problem.

Time factors are also numerous. The season of the year, the day of the week, the time of day all impact the nature of the problem. Will water freeze? Will there be mud? Are people available in the plant? Is there commuter traffic and/or crowded roads? How long will it take for response teams to arrive? These and other questions need to be considered both in preparation and in the actual response.

Weather conditions can also impact the nature of the problem. Clearly, they will affect how far and how fast any vapor cloud will travel. They will also affect the effectiveness of some response efforts, the safe direction of approach, the impact on the community, and so on.

The sum of these factors will define the potential impact (the potential losses) on people and things both inside and outside the plant. In the concern for people, various environmental factors can be overlooked. The ultimate disposition of anything that enters the sewers or runs off the property cannot be forgotten.

The resources available (control measures) to deal with the problem need also to be considered. These may be internal or external. In the planning phase, any gaps can be identified and addressed. Internal needs such as training, equipment, and control agents need to be addressed early in the planning process so that internal forces can practice and so they can offer leadership to the community. Obviously, external resources need to be cultivated and included in drills and other cooperative exercises. Joint provision of resources is an effective way to reduce the cost of providing for common needs and serves to promote teamwork. Obviously, the definition of specific potential accidents is needed during the planning phase to identify resources and appropriate training exercises.

Critical Issues

Once the full extent of the problem is identified, critical issues need to be identified and prioritized. This includes determining which potential losses are most important at any given time and which resources might be deployed to protect them. For example, if an LPG leak has occurred under a storage sphere, the critical issues might be to prevent ignition, to water-flood the sphere, to have the spilled LPG flow from under the storage vessel by gravity, and to promote dispersion of the vapor. If the pool ignites, the critical issue of preventing sphere failure emerges as high priority. Clearly, the planning process must consider various escalation paths for the incident and identify the strategies and resources needed to combat each.

Response Strategies

Response strategies fall into two classes: preventive and corrective. Preventive strategies include:

- Rescue people
- Protect exposures
- Deal with the incident
- Extinguish fire
- Prevent container failure

These are fairly obvious strategies aimed at preventing escalation of the incident. Without planning and patience, some important items may be overlooked. Often, response teams are not cognizant of their own need for protection in dealing with the incident. Unless properly trained, they can become part of the problem needlessly. A knowledge of environmentally sensitive areas and ways to protect them are essential. The Alaskan oil spill clearly demonstrates the high cost in money and critical skills of being unable to contain the hazard in a timely manner. It also points out the need to have equipment and trained people available on very short notice and to have had enough practice runs involving government agencies so that decision-making on priorities and technology can occur quickly.

Questions like, "Can chemical dispersants be used?" "Is controlled burning a viable approach?" and "Is bioremediation practical?" should be answered in the planning and practice stage. The same is true of the choice of extinguishing agents in case of fire. If combustibles are present that require unique extinguishers or neutralizing chemicals if ignited, this must be known and prepared for. These are only some of the reasons why specific scenarios must be considered in addition to the general case in planning for emergencies.

Preventing container failure is extremely important. This might be done with fixed or mobile facilities. It may also be done simply by sloping the ground so that pools of flammable materials do not collect around storage or process vessels. In some cases, flammable or toxic materials can be moved or displaced with water to get them out of the danger zone. Again, planning is essential for such steps, since to improvise during a real emergency may be disastrous.

Finally, we come to corrective strategies. These include cleanup, restoration, and disposal. These are complex questions not usually subject to time pressures unless seasonal factors or adverse weather conditions are important. Environmental impact is likely to be of primary importance, but worker safety must be carefully considered. In many cases, more people are hurt during the cleanup effort than in the incident itself. History shows this especially to be true in oil spills when large numbers of volunteers become involved in cleanup.

In most cases, disposal of any hazardous material cleaned up will be governed by laws and agency decisions. The industrial partner in such efforts often needs to worry only about the cost. As a result, it may well be profitable to explore remedial technology and disposal sites in advance. Be advised that most outlets will also want to know specifics before being able to commit either to acceptance or a cost structure in the United States.

Implementation

The last step in the first iteration is to implement the specific strategies selected. This implies a management structure, the availability of the needed resources, and the necessary authority and authorization to proceed. Company authority can be established in advance through internal plans and the communication of those plans to responding agencies and cleanup contractors. This may be as simple as telling the local fire department that the shift superintendent will be in charge in any incident and he/she will be the only person with a white hard hat. It may require the introduction of your incident commanders to a number of government officials and to the management of several cleanup contractors. It pays to do this in advance and keep these contacts current.

Authorization is a separate matter. Obtaining permission to implement a given strategy in a given situation may be an issue.

Clearly, this step can be facilitated through planning and joint simulated emergencies. However, in a real emergency, public scrutiny is much higher than in any simulated situation. What seems a clear course of action in a training exercise may become cloudy as various pressures are brought to bear by public groups with differing priorities. The decisive fire captain of your simulated emergency may find his/her decision to use chemical dispersants questioned by the mayor or the environmental agency when the time comes to use them in real life. Some of this can be anticipated, but I doubt it all can. In some cases, a company may be faced with the choice of seizing an opportunity to contain an incident without formal authorization, or of watching its cleanup obligation escalate while government agencies debate. If one opts for action and is correct, one is likely to be praised as a concerned corporate citizen. If the choice is wrong, some fines are likely to be added to the cleanup cost, but much more importantly, criminal charges may also be lodged. Such decisions are clearly for top management to make, and a system to get those folks involved quickly is essential.

Recycling

As implementation progresses, reevaluation must start. This is no different than the practices used in dealing with minor plant upsets or the start-up of a new process or business venture. As new data are collected, a new iteration of problem definition, prioritization, strategy selection, and implementation is carried out. There is no perfect plan. It must evolve as the incident does and become tuned to the microspecifics of the situation. While the general capabilities discussed in the next section will be employed in each response, their use will be tailored to these microspecifics.

A further element of reevaluation that is not explicitly discussed in the *Disciplined Approach* video is a formal critique. This is an accepted part of any actual response or simulated emergency drill and should be planned for. While appreciation is generally in order for all efforts in any response, the critique focus must be on ways to do it better the next time and/or prevent the type of incident altogether. Too often the need for praise outweighs the need for analysis. Too often a company will try to keep the real cause private (on advice of counsel). The latter course not only deprives the company of the insight from outside people, which might help avoid another incident, but also deprives the industry as a whole of the learning opportunity. To their credit, some companies that recently experienced serious incidents have shared their analysis of the real causes with other companies and the public to help prevent similar problems elsewhere.

Management System

The first general necessity in the emergency response area is also required in other parts of any business: an *effective* management structure and philosophy. This structure must have clearly defined objectives and responsibilities. It must also provide for trained primary contacts and backups for each role and a plan to draw in additional people for very large or prolonged emergencies.

Some reasonable objectives for an in-plant emergency response organization include:

1. Minimizing impact upon the public and facilitating communications with governmental agencies, public officials, outside emergency response organizations, the media, and within the corporation
2. Facilitating ongoing operations of unaffected process units
3. Organizing incident data collection, investigations, and reporting
4. Supplementing the existing emergency procedures as required for the proper handling of releases affecting the public
5. Dealing with transportation incidents close to the plant, or in an assigned region

Clearly, others may also be appropriate depending on plant location, size, process or materials used, distance from sensitive populations, and so on. For example, if a highly toxic release is possible (such as at Bhopal), a specific objective might be to provide early and unequivocal warning to some or all of the community as soon as such an event is detected.

A key philosophical component of the general emergency response organization separates the doers from the communicators. In my experience, those whose job it is to deal with the incident not only lack interest in communicating with the media, government agencies and so on, but do a bad job if forced into this role. These action-oriented folks are much more capable and comfortable fighting a fire or wading knee-deep through a flammable material to close a valve than in talking to anyone, including the plant manager. As a result, I recommend minimizing the communication responsibilities of the line operations people and in-plant emergency supervision.

In our plant, the shift supervisor must make only two communications decisions:

- Whether to call out the emergency response team
- Whether to report the incident to the authorities

Once these calls are made, others will assume the communications chores and he/she is free to concentrate on dealing with the problem itself.

The separation theme continues into provisions for dealing with outsiders. The lead community-emergency-response commander is brought immediately to the command center. Government agency folks are brought to one conference room and media representatives to another. The logic is simple: the community emergency responder can actually help with the emergency if needed; he or she also serves as the eyes of the outside world by providing a credible assessment of the situation to the many other agencies that may be interested in the incident.

Media and agency representatives are separated to prevent misinterpretation of partial information that may be appropriate to share with the agencies but not yet with the public. Also, since sometimes emotions run high, inappropriate sentiments may be expressed in debate between plant and agency people that have no place in media reports.

Separate from all of this is the incident command center. Here, the analysis and command take place in private, while the field forces deal with the incident. This command center must be equipped with a means to communicate with the field, the corporation, and the outside world. If the plant is equipped with radio communication, this is an effective way to monitor action real time. (But you must recognize it will not be private!) The same receiver will also allow monitoring of police and fire department activity. Obviously, several telephones are required for internal and external communications. A number of portable telephones in the field are useful if cellular coverage is good. If the incident is prolonged, a means to monitor TV news broadcasts is often useful.

An emergency data base is also required. In addition to MSDS sheets, area maps, meteorological data, equipment data, personnel data, and a contact list are important. If community risk studies have been done, these data should also be available to help analyze exposures. Some companies have a large file of computerized community impact studies to draw on for specific accident scenarios and weather conditions.

If the incident is away from the plant site—that is, an outgoing product or incoming raw material—some transportation equipment may be required. For short distances, a HAZMAT-type vehicle, preequipped with safety and control equipment and data, may be warranted. For longer distances, the capacity to fly people or minor equipment to the scene on very short notice may be needed. In all cases, appropriate communication equipment must be included. All this takes some significant advance planning to accomplish. Management cannot ignore a plan to respond to remote incidents under the Responsible Care precepts.

The following positions are usually staffed in an emergency response organization:

- Incident commander
- Area operations manager and staff (field)
- Environmental specialist
- Risk Assessment specialist
- Industrial hygiene specialist
- Personnel safety/OSHA specialist
- Public affairs specialist
- Legal advice
- Medical advice
- Human resource staffer

- Secretarial support
- Telephone staff
- Purchasing staff

Obviously, smaller plants can combine some of these functions. There is a critical-mass effect, however. If fewer than four to seven people are available, there may not be enough communication capability, or worse, the few who are trying to manage the situation may get into a panic mode.

As indicated earlier, the responsibilities for all these people must be clearly defined. For the operations people and staff (technical and mechanical), the job is clear: Stop the incident! For the others, it pays to plan, and practice, in advance. A suggested allocation of specific responsibilities is given for each function/position below.

A. *Incident Commander*
  - Responsible manager leads the effort; provides management direction
  - Determines resource requirements and arranges for them
  - Arranges communication with upper management, with local organizations, with public officials, with the media, with aggrieved citizens
  - Authorizes expenditures, commitments
  - Organizes incident investigation and reporting

B. *Manager Responsible for Operations Involved in the Emergency (and Staff)*
  - Communicates with emergency management regarding field conditions
  - Overall responsibility for ongoing operations and emergency control activities including operational resource requirements
  - Assess business impacts and develop appropriate response plan
  - Provides appropriate process information
  - Coordinates field operational activities
  - Secures needed resources
  - Prepares plan for safe recommissioning of facilities
  - Supports risk and environmental assessment activities

C. *Environmental*
  - Ensures timely communications are maintained with state and local agencies
  - Determines quantity and hazardous nature of release; ensures activation of outside monitoring, if needed
  - Maintains communications with other environmental groups
  - Identifies need for additional environmental resourcing.
  - Notifies USEPA and other agency groups, if required, and not already accomplished

D. *Risk Assessment Coordination*
  - Determines and recommends additional actions necessary to assure protection of noninvolved plant personnel and the outside community
  - Responsible to protect all data and record emergency activities for later investigation (maintain a chronological history of the event)
  - Provide communication with other risk management organizations

E. *Public Affairs*
  - Coordinates all information releases to public
  - Releases all communications to media
  - Arranges for media entry into media room and restricts their further mobility as deemed necessary
  - Responsible for all activities and equipment in media room and arranges for meeting media's comfort needs
  - Provides feedback regarding public reactions and communications needs
  - Briefs telephone staff
  - Monitors incoming telephone complaints and recommends response actions
  - Assesses need for and secures additional resources as may be required
  - Maintains communications with corporate public affairs
  - Collects news media reports on the incident (newspaper clippings, video tapes, etc.)

F. *Industrial Hygiene*
  - Provides toxicity information needed to communicate with and protect the public and plant personnel
  - Monitors toxicity levels in the plant and at other key locations
  - Obtains input from any employee outside-tracking network

G. *Safety/OSHA*
  - Determine extent of injuries
  - Notify OSHA and government authorities of injuries as appropriate
  - Coordinate OSHA investigation

H. *Law*
  - Anticipates legal liabilities and prepares contingency action plan
  - Ensures action taken minimizes company liability
  - Maintains communication with headquarters legal and local counsel
  - Coordinates local claims activities
  - Coordinates response to any summonses, citations, criminal charges

I. *Medical*
  - Determines health effects
  - Provides on-site medical treatment
  - Provides medical consultation with outside health organizations
  - Prepares medical alert (if required) for news release
  - Arranges for additional medical resourcing, if required

J. *Human Resources*
  - Establishes telephone communications center and monitors activities
  - Arranges for emergency feeding and housing, both in-plant and possibly outside
  - Notifies employee relatives of any injuries
  - Coordinates with other company units for all additional resourcing
  - Coordinates emergency living facilities (in-plant sleeping, motels, etc.)
  - Monitors head count of all employees
  - Coordinates with plant security
  - Maintains communications with unions

K. *Secretary*
- Prepares minutes of all meetings
- Provides secretarial services to response organization
- Assesses secretarial requirements and arranges for additional resources

L. *Telephone Communications Staff*
- Receives incoming phone calls per procedure
- Assists with media and agency staff needs

Off-Hours Leadership

In most situations, it will take some time (an hour?) to assemble the full support team if the incident is during off-hours. During this time, a shift superintendent is likely to be the senior person on the site. In small plants, there may be no security force, no twenty-four-hour in-plant emergency response capability, no communications people. In such cases, close coordination with the local emergency responders is essential. In addition, management should be equipped with portable telephones so that they may be able to interact while en route.

Whatever the situation, the emergency response plan should be documented, available to all involved, and exercised frequently. This is especially so if reliance is placed on outside responders for critical portions of the response. It may be necessary to provide some equipment for these responders and some training to ensure that their abilities will meet the needs. In addition to the primary responders, mutual aid contacts should be in place to provide backups, and these arrangements should also be documented, including contact numbers.

Realistically, the probability is almost three to one that any incident will occur during off-hours in a twenty-four-hour operation. The odds may be somewhat lower in a forty-hour/week operation, but as long as hazardous material is present all the time, there is some chance that the incident will occur during the off-hours. From this, we conclude that whoever is present during these shifts must be carefully trained to respond effectively. This means that he/she must fully understand the nature of plant hazards and the general emergency response plan. He/she must also be able to assess any incident and adapt the general plan to the specific emergency—at least in the initial phases. He/she must also be knowledgeable in legal reporting requirements and know where to get advice and help on an around-the-clock basis. Sound formidable? It is. Most of these people have many years of practical experience and respond admirably to any incident. However, today's standards for outside communication and sensitivity to even trivial impacts require fleshing out these practical process and mechanical capabilities. Training and drills to ensure the system works are much less expensive in the long run than a single, even relatively minor, incident.

Media Training

Aside from the specialized emergency response training required by OSHA, which will be discussed later, specialized media training is available for designated communicators. If you have ever witnessed the difficulties a normal, conscientious manager faces when confronted with an aggressive reporter and a bank of TV cameras, you will immediately appreciate the need for this training. The best response effort in the world will be doomed to apparent failure if it cannot be communicated effectively to a hostile media. A clear knowledge of how to do this and the discipline to persevere in the right approach are essential.

Emergency Responder Training

A second type of general emergency response capability is the appropriate training of people who *may be* called upon to respond in the field. In the United States, OSHA has fined a number of companies for failing to train those who responded to a major incident properly *before* they were called upon to respond. Under OSHA 29 CFR 1910.20, various levels of training are required for various levels of response. In some places, other local laws also apply to hazard communication and emergency response personnel. Documentation of emergency response training is usually required by all agencies.

In this regard, HAZWOPER training goes beyond OSHA Hazard Communication (HAZCOM) training, although HAZCOM training may count toward satisfying certain of the requirements. In fact, OSHA has clarified its position, saying that if an incident can be handled by on-site employees and contractors using their HAZCOM training alone, it is probably not a HAZWOPER emergency response.

There are seven levels of HAZWOPER emergency response training, which are listed here, ascending from the least to the most highly trained:

1. *Skilled support personnel* are skilled equipment operators needed on a temporary basis. An initial briefing at the site is the only requirement.
2. *Specialist employees* are regular employees who provide assistance to responders at an emergency response. Refining process operators who close valves and set fire monitors, but do not fight the fire, may fit here. Appropriate training or demonstration of competency is required annually, with no set number of hours or specified training areas listed in the regulation.
3. *First responders, awareness level,* are the individuals responsible for recognizing and reporting an emergency response event. The classic example is the night watchman who calls in, but does not fight, the fire. Most operating employees probably fall into this category. Although no set number of hours is required, sufficient training or demonstration of competency is required in the following areas:
   a. An understanding of what hazardous materials are present, and the risks associated with them in an incident
   b. An understanding of the potential outcomes associated with an emergency created when hazardous materials are present
   c. The ability to identify the hazardous materials, if possible

d. An understanding of the role of the first-responder awareness individual in the employer's emergency response plan
e. The ability to realize the need for additional resources, and to make appropriate notifications to the communication center

First responders, awareness level, are trained to initiate an emergency response sequence. They would normally take no further action beyond notifying the authorities of the release.

4. *First responders, operations level,* are the individuals who respond to the emergency response event. Examples are the fire brigade members who fight the fire, the spill team members who contain the release and keep it from spreading, and the security personnel who seal the area of physical or chemical hazard. These individuals must be certified to have eight hours of training or demonstration of competency in the following specified areas:
   a. Basic hazard and risk assessment techniques
   b. Selection and use of proper personal protective equipment provided to the first-responder operational level
   c. Basic hazardous materials terms
   d. Performance of basic control, containment, and/or confinement operations within the capabilities of the resources and personal protective equipment available with their unit
   e. Implementation of basic decontamination procedures
   f. Relevant standard operating procedures

   First responders, operations level, are part of the initial response to the site for the purpose of protecting nearby persons, property, and the environment from the effects of the release. They are trained in a defensive fashion but not to actually try to stop the release.

   Please note that most process operators from the unit in question would seem to fall into this category at first glance. The phrase that calls this into question is "but not to actually try to stop the release." It seems incongruous for an operator to stand idly by while waiting for a HAZMAT technician to arrive when closing a valve, stopping a pump, or other *normal operating action,* which he/she performs in his day-to-day activities, would stop the leak. As will be seen below, HAZMAT technicians are trained to do much more than perform normal operating actions. Of course, the alternative is to train all operators to the technician level, but this may seem wasteful to companies with limited resources.
5. *Hazardous materials technicians* are the individuals who actually approach the point of release to plug, patch, or otherwise stop the release. These are the leak stabbers on the fire brigade, and the members of the hazardous materials (HAZMAT) team. These specialized employees must be certified to have at least twenty-four hours of training equal to the first-responder operations level and in addition be certified to have competency in certain listed areas.
6. *The hazardous materials specialist* is the HAZMAT reference expert, the specialist in a particular hazardous chemical. This is probably a contract

expert, sometimes a university professor. The training level is twenty-four hours of training equal to the technician level and, in addition, competency in specified areas.

7. *The on-scene incident commander,* who will assume control of the incident scene, is required to have twenty-four hours of training plus competency in specified areas.

Annual refresher training or documentation of competency is required to be maintained for each of the above training levels.

Existing training programs can be relied upon to satisfy the specified training requirements. Examples of such programs include:

1. HAZCOM training
2. Emergency training
3. Fire brigade training

In fact, most employees will probably have documentable training sufficient to satisfy the first responders awareness level, and most fire brigade and spill-team members will likely be able to demonstrate training more than adequate to qualify as first responders, operations level. Overall, the OSHA training levels appear reasonably related to being trained to do the job an employee is asked to do in an emergency response.

In summary, training satisfying the HAZWOPER training requirements for the applicable training level should be documented for each operating employee. To accomplish this, existing training should be evaluated against the applicable HAZWOPER requirements and documented. It should be recognized that, depending on their involvement in an emergency response, operating employees may qualify as skilled support personnel; specialist employees; first responders, awareness level; and possibly first responders, operations level. Fire brigade members and spill-response team members may be first responders, operations level, and, in some cases, HAZMAT technicians or the on-scene commander. Where multiple training levels apply, training should be to the highest level and should be adequate to ensure the employee is trained to do the job he or she is asked to do.

Establishing a Credible Planning Basis

Having argued that "the worst possible scenario" is not an effective basis for emergency planning, one has the obligation to define a better approach.

In my view, this can only flow from a full and detailed understanding of one's plant and all the people who work in it (employees and contractors). This implies that both the historical experience at the plant and in the industry must be analyzed in detail so that likely failures are identified. This done, possible escalations need to be evaluated to flesh out the credible scenarios. Finally, the potential impact on the plant employees and surrounding community need to be assessed. The net result will be a series of realistic possible accidents and an assessment of their potential impact.

The next step is to present these to plant management to ensure that they are fully understood, *and accepted,* by that group. Invariably, acceptance is a multistep process. Questions like, "Isn't there something we can do to reduce the fenceline concentration?" and "Can the city fire department back us up on that?" usually require answers before an optimum risk set is defined.

This risk set may be quite extensive. It will include the usual cases of power failure, cooling-water failure, flow interruption, pump seal fire, and so on as well as specific releases from relief devices, packing leaks, corrosion failures, routine maintenance and operations, and so on.

To return to our chlorine-case example for a moment, a credible scenario surrounds the changing of a car when empty. Normal procedures would involve purging the connecting piping with nitrogen prior to decoupling the car. Even so, a little Cl_2 is usually released as the piping is opened. As a result, there are usually two operators. They try to stand upwind and will often use a respirator. An understanding of how much Cl_2 is released and the local concentration is important even for this very routine scenario. A reasonable escalation should also be considered. For example, what would be the amount released if the operator failed to purge the piping well (or at all) prior to disconnecting the car? This could happen more easily than most people would care to admit, and I know of one case where a release of eight pounds of Cl_2 put six people in need of a medical checkup. Another possible escalation would involve the valve at the car not shutting off quite tightly. This would expose a rather large reservoir of Cl_2 vapor to the atmosphere, albeit at a very slow rate.

Both of the above might cause a very nonroutine situation to exist at a time when a mere puff of Cl_2 would be expected. Obviously, prevention is the objective, but preparation of a response plan is both appropriate and necessary. The plan should know how much Cl_2 would be released and at what rate. It should also know the consequences of such an accident and be prepared to deal with it.

For widely used hazardous materials like Cl_2, various industrial groups (e.g., the Chlorine Institute) have already done dispersion-model calculations and made them available to users to assist in evaluating the consequences of many typical releases. These data can be used as a starting point to determine whether any release is serious, thus reducing the time and expertise needed to evaluate many typical accidents. At the very least, these generic calculations will serve to focus in-house activities on incidents that might be critical.

In more specific scenarios, in-house analysis of consequences is required. An example worth discussing is a study done on an anhydrous HCl system in response to an overpressure incident. A pressure control system failed, resulting in a safety valve release to the atmosphere. There were no untoward consequences from the real incident, but a consequence analysis was carried out to determine the full range of possibilities of what might have happened. In short, was the current safety relief system design adequate? What other accidents should be considered to ensure a "safe" system?

Anhydrous HCl is a very dangerous material. The ACGIH threshold limit valve for an eight-hour workday is five parts per million and the odor threshold is less than one part per million. At twenty-five to thirty-five parts per million, extreme respiratory discomfort is experienced, and concentrations of 100–150 ppm are life-threatening.

In the contingency studied, the maximum release rate was only 130 lb/hr, which would be vented through a stack fifty-seven feet above-grade. The questions of interest concerned:

- The highest ground-level concentration inside the plant
- The concentration at the nearest residence (1,600 feet)
- The concentration at a nearby highway exit (5,000 feet)
- The extent of the area subject to an odor
- The impact of increasing stack height
- The in-plant concentrations of a leak at lower elevation

All dispersion calculations were made at the worst meteorological conditions for the site (low wind velocity, an "F" atmospheric stability) with a vertical release.

The results are shown in Figure 10.1. The maximum ground-level concentration was calculated to be about eight parts per million at about 0.1 miles from the release point (inside the plant). The very steep falloff in concentrations on each side of the maximum indicated a concentration close to the TLV at the nearest residence and much lower than the TLV at the highway exit. The potential for odor existed well into the community, however, up to 1¼ miles from the release. Repeating these calculations at a hypothetical stack height of 75 feet showed that this potential change would reduce the fenceline concentration by about thirty percent but have little impact on the potential odor area.

The fundamental question for plant management was whether this scenario—which was entirely realistic but of low probability (about once in ten years by experience)—represented an acceptable risk. The chance of health impact was very low, but if the release occurred, a wide area of odor impact was likely. This is not a black-and-white choice. There is no morality to a potential for an odor. The situation is simply a choice for management to make with whatever input it chooses to obtain from headquarters, the community, the legal staff. What is essential is that management must understand and decide. Such choices cannot be delegated to staffers or be made by default.

The second phase of the study involved considering what else might go wrong with the system to cause a release. An obvious risk was a small leak in piping or a valve in the high-pressure system. Upstream of the pressure letdown valve mentioned earlier, pressures of 1,000 psig are observed. Under these conditions, even a small hole results in a fairly sizable release. A one-millimeter-diameter

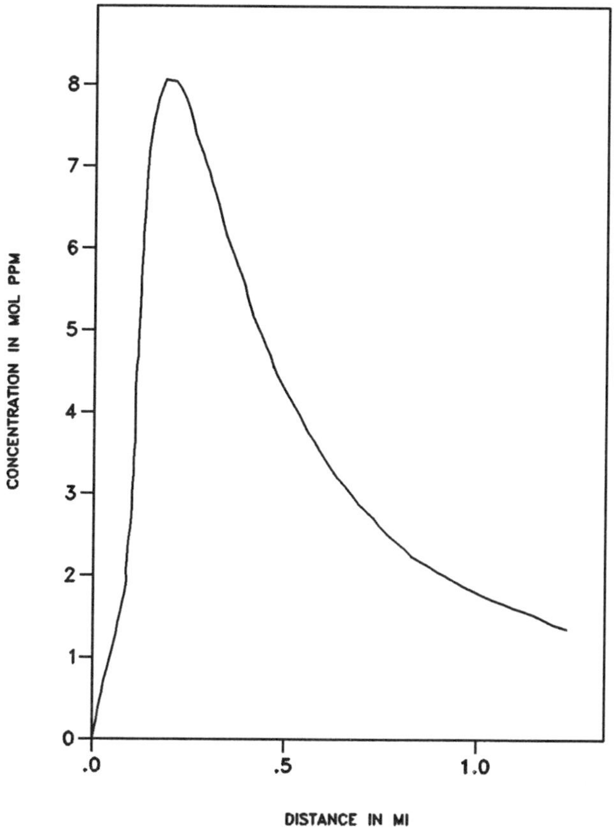

Figure 10.1. Safety valve release ground-level centerline concentration of HCl vs. distance.

hole was calculated to release 80 lb/hr. Since this release would be close to ground level, the HCl concentrations, which might be experienced by nearby workers, had the potential to be very high. Figure 10.2 shows the calculated variation of ground-level concentration with distance from such a leak. At fifty feet, concentrations of over 600 ppm could be expected. These would fall off very rapidly so that the fenceline concentrations would be comparable to those from a safety valve release (6.5 ppm at the nearest house and 1.4 ppm at the highway exit). Thus, a small leak posed no additional community problems but raised a serious worker safety issue.

Planning to Deal With Specific Emergencies

The logical course of action to deal with this possibility parallels that which should be followed in planning for any emergency:

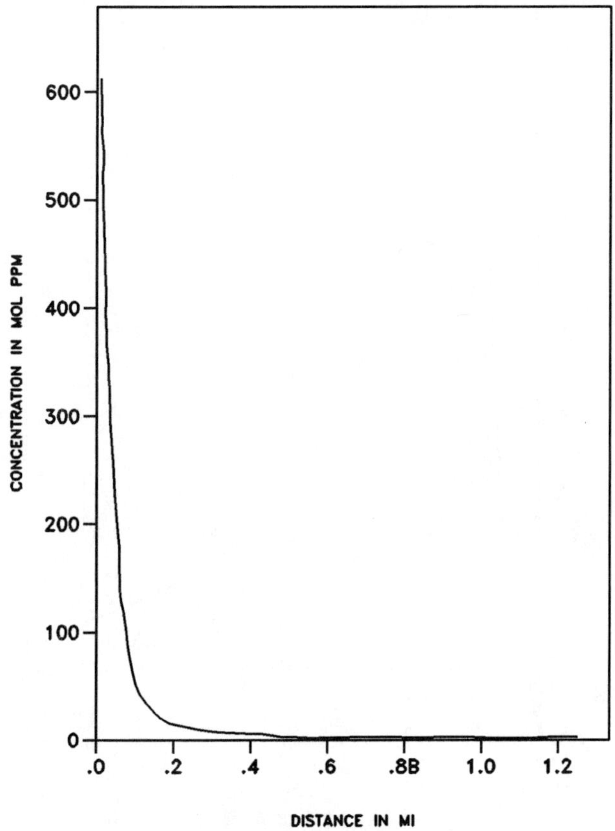

Figure 10.2. Piping Leak: ground-level centerline concentration of HCl vs. distance.

- Prevention
- Provide for early detection
- Provide for suppression or mitigation of the release
- Develop a response plan

Prevention involves good design, appropriate inspection, and effective preventive maintenance. These topics have been discussed at length in previous chapters.

Early Detection

There are a number of approaches to early detection of dangerous releases. These include process instrumentation, which signals the approach of conditions that might cause a release (e.g., high pressure), human surveillance, and atmospheric detectors.

Process Instrumentation

In theory, process instrumentation can be used to signal conditions in which a release may be imminent. In practice, many process instrumentation systems have so many alarms active at any given time that operator recognition and response may be slow. Nevertheless, prevention (or immediate detection) of a release through the process instrumentation system is an objective that should be pursued vigorously.

The problems with accomplishing this have little to do with instrument maintenance or operator training or motivation. They do have to do with the complexity of the man/instrument system interface and with barriers to recognition. As documented earlier, these barriers are much more severe in times of stress.

The first step in fostering quick recognition is good ergonomics at the panel. A lot of work on this subject has been done and previously mentioned. Much of what has been learned from nuclear (A. D. Swain, H. E. Guttman, "Handbook of Human Reliability Analysis with Emphasis on Nuclear Power Plant Applications," USNRC Report NUREG/CR-1270) and aviation industry studies is pertinent to the process industries as well. Some of the principles include:

- Primary work stations should provide full view of console and keyboard even when manuals are in use.
- There should be adequate work space around console. The console design should consider local human measurements.
- Things should be distinctively identified—legible labels.
- Ventilation, climate control, and lighting, should be adequate
- Buttons/switches should be "fail-safe" (i.e., cannot be accidentally activated).
- No counterintuitive actions should be required (i.e., do not require that a switch be pushed down for "on" rather than up in the United States).
- Group information on panel and computer screens should provide everything needed for an emergency at one time.
- Grouping and location of readouts and switches should be optimal
- CRTs should be large; a minimum number of keystrokes should be required for key information; no-glare screens should be used.
- Data should be presented in graphical form to facilitate trend identification.
- Distinguish emergency messages. Do not fill screen area >75%.
- Inoperative instruments should read off-scale (not zero) so they can be identified immediately.
- Alarm annunciators should contain only single, critical readouts with signals distinct for the unit in question.
- Prealarms and alarms should be segregated (by color or sound).
- If multiple alarms can result from a single incident, the initiating event should be identified.
- Emergency system controls should be protected from accidental activation and give immediate feedback indicating that they have been activated.

Much of this has been written about by others and will not be discussed here. These factors constitute a growing science area, in which multiple research institutes and a number of private companies offering services. Many older control panels/rooms do not comply with the best current thinking and could benefit from a review.

Segregation of critical and noncritical alarms and cutouts/ins is an area that holds significant potential. The challenge is to identify the minimum list of "critical" instruments. Extreme discipline is often required to limit the list to the last item in a chain of instruments *before* an incident occurs. These "last-chance" instruments and alarms not only receive more extensive testing and maintenance; they also are identified separately and distinctively on the control panel by light and sound. Procedures are also put in place to emphasize the need to respond immediately to these critical signals and to prevent their bypass without management approval and standby scrutiny.

Experience shows that this "critical" list tends to be diluted by the addition of other items that are important (to someone) but not critical to incident prevention signals. If a particular console has hardware or a limit of sixty-four "critical" inputs, innovative operations and instrument folks will find ways to circumvent it by ganging alarm signals together so that the original alarm intended to signal a high-level pump cutout on tank 123, for example, will be used also to signal high levels in tanks 124, 125, 126, and 127. Not only does this dilute the importance of the particular alarm condition, it also obscures the specific source of trouble. Further, this approach reduces the operator's commitment to immediate action because he or she knows the odds are about four to one that solving the situation is not truly critical to preventing an incident. This type of proliferation often goes unnoticed by management because it is never reviewed at levels high enough to be cognizant of the potential impact on overall plant security vs. the short-term improvement in operating convenience. If all changes are processed through the management-of-change system discussed in Chapter 7, an independent viewpoint should be obtained.

A second approach to ensuring critical signals are identified is to absolutely minimize bypasses and lockouts in alarm circuits. This, too, is an exercise in discipline and diligence. Experience shows that the number of locked-out, disabled, bypassed alarms can be reduced to a few if management wishes. To fuel managers' desire for this, have them sit for a day at a console with forty or fifty meaningless alarm lights lit to see if they can find a new alarm and discern whether it is important. One of the interesting aspects of the Bhopal disaster was that the storage tank that overheated to cause the incident had a high-temperature alarm. However, since the tank was normally refrigerated, it was set for a subambient temperature. The refrigerator system was subsequently shut off, but the alarm set point was never reset, so its "normal" condition was in alarm. Obviously, the operators learned to ignore it. When the tank actually began to heat up to above-ambient temperatures during the incident, they had no alarm and never looked at any other temperature indicator until it was too late. With this anecdote, I shall rest my case for purging spurious alarm signals.

Personnel Awareness

Plant personnel constitute the second means of early detection of an incident. On their usual rounds, they will develop an "all senses" base case of what is normal. This will include sights, sounds, smells, and even vibrations. These people need to be trained and encouraged to explore or report *any* abnormality promptly.

In exploring a strange noise or odor, care must also be taken to do so safely. It is not unusual for an emergency responder to blunder into becoming part of the problem. Hence, the OSHA requirement for training for all people who potentially may be exposed to emergencies is logical.

Some clues to impending or actual trouble are subtle. Operators can log temperatures that are seen to be slowly rising over a shift and ignore the implication. Engineers can look at a temperature in a vessel that is seen to be higher than the temperature of a heating medium and still try to deal with the problem by only turning off the heating medium. Some people will believe that closing a valve will *always* cut off the steam supply and never check back (even though they know that sometimes valves leak). Visual inspection of tank cars and trucks can be cursory, leading to loose fittings or missing caps on vehicles released to the road. The potential list is endless. The only solution to these issues is a continued high level of awareness fostered by management.

A major role in fostering this awareness falls to the first-line supervisor or shift superintendent. How he or she responds to questions or reported abnormalities under the press of time or production schedules is often critical. How the possibility for abnormalities is discussed in safety meetings and training sessions is very important. Developing and carrying out meaningful simulated emergency drills constitute a very effective way to promote awareness, and this often is the responsibility of the first-line supervisor. The quality of these drills and of the subsequent critiques will give poignant testimony to the sincerity of management's efforts to involve the operators in detecting and dealing with incipient emergencies. First-line supervisors need to lead the way in responding intelligently to the first perception of abnormality in operators or those performing maintenance activities.

Area Detectors

Detectors of various sorts are playing an increasing role in early warning of incidents. Those in fairly widespread use include:

- Furnace flame detectors
- Flare pilot flame detectors
- Pipeline leak detection systems
- Chemical sensors (e.g., Cl_2, H_2S, BR, HF)
- Hydrocarbon detectors
- Fire and smoke detectors in enclosed spaces
- Detectors for oil sheens on water

- Personnel monitors on individuals who might be exposed to small amounts of radiation, benzene, H_2S, and so on

Clearly, all of these detectors require periodic preventive maintenance and testing by carefully trained crews. In addition, if automatic actions are triggered by the detectors, the entire response system should be included in the testing plan. Finally, the required response from operators, mechanics, and contractors must be exercised periodically to ensure that the people part of the response system is prepared to carry out its role.

The technology of incident detectors is undergoing rapid development. Reliable Cl_2 detectors have been available for some time. These are now being applied to detecting leaks of related substances such as HCl and SCl_2.

Until a few years ago, the operation of most flare-stack pilot burners was verified visually. In more sophisticated plants, TV surveillance was performed continuously in the control room. On bright days, the TV picture could not always provide assurance of a pilot flame, so specially designed thermocouple assemblies were designed. As might be imagined, the materials, design, and location of these probes were critical to their operation and longevity. Now, some of this technology is being applied to flame detection on pilot lights inside furnaces to ensure safe start-up operations. The point is that there are no universal solutions. Each installation must be evaluated specifically. This is especially true if a major part of the response plan is dependent on the detector and a system of automatic responses.

A major worry in any such system is spurious trips. These erode operator awareness and lead to the evolution of shortcuts and bypasses. The 1990 Arthur Kill, New Jersey, oil-spill story is ample evidence of the potential impact of this erosion. In many cases, two-out-of-three voting logic is employed in detection/shutdown systems to minimize spurious trips. This adds to the short-term hardware cost and to the long-term maintenance cost, but may prevent the enormous cost of an incident. In any case, detection systems must be carefully engineered, maintained, and tested if they are to contribute full value to emergency response.

A few organizations have coupled detectors, meterological data, and dispersion-model calculations into an automatic mitigation/community alert system. DuPont's "SAFER" organization is one of those to offer such a service. Although advertised as "real-time emergency response" systems, all such approaches involve a very heavy planning component. Part of this is because a huge amount of data must be input to the computer to map the community and its sensitive locations. This then must be discussed with local officials and some alerting/response plans must be established. Another reason is that, once a leak/incident is detected, appreciable time relative to the duration of many incidents is needed to assess the situation via calculation even when precalculated scenarios are available as a basis. As a result, much of what is labeled "real time" is based on previous calculations of various scenarios, wind conditions, atmospheric stabilities, corrected for the specifics of the instant. In short, if the planning is not well done, neither is the real-time analysis.

Suppression/Mitigation

One of the advantages of early detection is that it is sometimes possible to prevent escalation of the incident. This is sometimes done directly—for example, with an automatic explosion-suppression system. More often, it is done indirectly by providing a safe/environmentally acceptable vent system, by using any number of systems to shut off the feed/energy supply to the incident, or by supplying rapid cooling at critical points such as water deluge or sprinkler systems.

With more sophisticated understanding of the dynamic behavior of a system, it is sometimes possible to design it to be self-limiting. For example, some nuclear reactors were designed such that if the core temperature rose above certain limits (due to a coolant loss, perhaps), the density of the moderator decreased to such a point that the nuclear reaction was slowed down because of a reduced availability of thermal neutrons. Similarly, systems might be designed to withstand the overpressures built up during an explosion. These same systems would have to be designed to *prevent* a detonation. Such an approach was taken in a number of acetylene handling and chlorine liquifaction systems. Obviously, this approach can only be taken with confidence when a very thorough understanding of the kinds of mishaps possible in a system is available. Some of the ways to acquire such an understanding were mentioned in Chapter 3.

The three approaches mentioned in the first paragraph of this section are the most commonly used. They are generally a major factor in the design of a hazardous materials plant. Typical examples include:

- High-pressure sensors that automatically vent material to a flare or scrubber system designed for the load.
- Excess flow valves, remote shutoff valves, emergency blowdown facilities, water displacement systems, emergency power supplies, automatic equipment or power-supply shutdown, automatic start-up of substitute equipment, and other ways to either shut off the flow of fuel or reactant to the incident or reinitiate the controlling mechanism that failed to start the incident. For example, if the piping between a Cl_2 tankcar and the land piping is broken, flow rates will increase and an excess flow valve on the car will automatically shut. If an LPG vessel leaks, it might be blown down (emptied) to another vessel to minimize the leak or water might be injected into the bottom of the vessel to substitute water for LPG at the leak point. If loss of cooling initiates a reaction runaway, the flow of reactants can be shut down and substitute cooling water pumps can be started up.
- The automatic sprinklers in hotels and offices are familiar sights. Similar systems apply in industrial settings. LPG vessels often have automatic cooling deluges installed; so do large hydrocarbon pumps. Floating roof tanks (which usually handle more volatile materials) often have installed foam systems to cool and extinguish a fire at the seal (the potential leak point) without having to force firemen to approach an elevated small fire that might spread. Halon

flooding is a common way to both extinguish fires and render the atmosphere inert safely in enclosed spaces. (With the current concern about ozone depletion, Halon will be phased out by the year 2000.)

- Recent work on mitigating gas explosions on offshore platforms has shown that rows of water sprayers of very fine droplets can reduce flame speed and minimize overpressure in an explosion that results from a confined gas/air mixture. Detonations in propane/air mixtures were also prevented. The process was found to be complex, with the turbulence introduced by the sprays having an accelerating effect as well as a decelerating effect, particularly in areas in which obstacles were few (pipe, structural beams, etc.). Acton, Sutton, and Wickens of the British Gas Midland Research Station theorized that the decelerating effect of water sprays was due to interfering with both expansion ratio of the combustion gas and the rate of reaction in the flame front *(I. Chem. E. Symposium Series 122)*. The subject is complex and timing of the water sprays appears to be critical, but this approach might be useful if leaks are detected in confined spaces before ignition.

All of these approaches require a sound understanding of the design scenarios involved. In the offshore drilling industry, two classes of "accidental events" are postulated:

1. "Design" events (DAE)—those for which the platform will be designed to survive without serious loss
2. Residual events (RAE)—those extreme events that the platform will not be able to withstand

The Norwegian Petroleum Directorate sets a limit for frequency of RAE of about 10^{-3} per year per platform. Obviously, the frequency has to be determined by quantitative risk analysis. More important, the physical scenarios for all accidents have to be developed in enough detail so that design calculations can be made.

Basis for the Accident

In the offshore oil business the main risks are fire and explosion. The determination of the physical basis for such events is relatively easy. One simply has to determine how much hydrocarbon will be involved in the fire/explosion. Calculating the overpressures resulting in a confined space filled with obstacles (process equipment) is another matter. A great deal of work was done at the Christian Michelsen Institute (Norway) to define overpressures and effective explosion venting considerations. The best answers have long been known to the onshore segment of the industry: adequate spacing, open construction, and containment. Offshore production platforms have very restrictive spacing options, and weather conditions often prevent open construction. Similar constraints often exist in smaller process-industry plants, especially those in northern climes. Thus, it may well be possible that some of the lessons learned in the offshore oil business can be applied in other segments of the process industries.

Development of design criteria to deal with more complex accidents, such as runaway reactions, is more difficult. The kinetics of such runaways often cannot be characterized by general data. In addition, the physical state of reacting mixtures is often multiphase, making the design of relief devices difficult. Thus, experimental programs involving reaction calorimetry and relief sizing tests are often required to obtain a sound design basis. More chemical companies are setting up "reactive chemistry" laboratories to understand better both the safe ranges of operability and the consequences of a runaway. As mentioned earlier, such facilities are not cheap ($0.5–1 million capital) and a considerable manpower investment is required to develop the needed expertise. Appropriate manning in a large central laboratory, which also does some modeling of events, is in the $0.5–1-million-per-year area. As a result, several consultants have established themselves as providers of these services. Also, the AIChE sponsors the development of emergency relief sizing technology through the Design Institute for Emergency Relief Systems (DIERS) and conducts users' meetings to exchange technology and foster round-robin testing to help ensure consistent results. The chemical industry has been quite open to the exchange of information in this area.

Clearly, all of the above involves identifying the risks to be dealt with quite specifically. This is generally done by one of the techniques discussed in Chapter 5.

Emergency Response

Effective emergency response is heavily dependent on planning and communication during an incident. The key challenges to be dealt with include:

1. Leadership
2. Responder knowledge and experience
3. Getting the necessary resources to the scene
4. Rapid, "no fault" communication with community responders, mutual aid groups, and critical governmental agencies (e.g., state police)
5. Medical and environmental support
6. Communication with the public at large

The first five challenges are dealt with only by training and exercise. This is typically done via simulated emergency drills, involving everyone who might have to participate in a real incident in an appropriate way. It is important to have operators and mechanics participate both in the formulation of the incident scenario as well as the exercise. Their input on what is really likely to happen to cause an incident and in response to it is very valuable.

Simulated Emergencies

There are three levels of simulated incidents that need to be considered and four phases to each drill. The levels are:

- Plant unit
- Entire plant
- Community

The phases are:

- Tabletop discussion
- Communications only
- Full rollout
- Critique/investigation

Frequent unit drills are a must. A minimum of six per year per unit is recommended and twelve would be better. Often the communications-only phase can be skipped and the critique can be held immediately after the drill. These typically cover small incidents peculiar to a given unit—a small Cl_2 leak, a pump seal leak.

In the tabletop phase, the person in charge plays out the scenario one step at a time; the group sits around a table and discusses what response each person would make as each step in the incident unfolds. Typical questions are:

- Who would you call?
- What would you do?
- Where would you get the equipment to do what you say?
- Where would you go?

at each step in the development of the scenario. The tabletop discussion can usually be carried out in less than an hour and provide a dose of realism for the full-scale drill.

The communications phase involves executing the telephone calls, radio links, and so on to all (or almost all) of the full-scale drill players, but no action is undertaken. This might be broken into sections for manageability. For example, an emergency-forces call-out section might include off-hours calls to all members of the emergency response team with instructions to call certain numbers to check in rather than actually report to the plant.

For major drills, the entire communications system is exercised separately before the rollout drill is conducted. This is usually approached by marshalling a core group in a control location and activating the community agency's control centers, hospitals, mobile command posts, and so on for a test of the operability and effectiveness of the people, the hardware, the software, and the systems associated with getting all of the necessary information to all those involved in an accurate, timely fashion. The entire scenario is run through on a compressed time scale, while monitors observe the speed and accuracy of the communications.

Major drills require serious planning and can usually be staged only once per year. The public relations and political aspects of the drill cannot be underestimated. Local, and perhaps national, political figures may wish to attend kickoff press conferences and even get their picture taken in a hard hat. All of this means a good opportunity for positive press for the industrial participants that should not be ignored. Unfortunately, it will also add to the cost of the event.

Because of the extensive participation involved in major drills, they are usually held on a Saturday (in the US). The duration has to be two to four hours to allow all of the agencies to get into action. This is usually accomplished by adding certain ancillary incidents to the main one. Typically, an auto accident or fire is loosely connected with the main event to prolong the activity. Periodic press conferences are held at a place chosen by the participating municipality to exercise their capabilities and maximize positive media coverage. A critique session is usually held immediately after the incident is over, usually along with a meal provided by the industrial participants. Independent observers report their observations of the various aspects of the drill response.

Perhaps recounting an example will put all of this into a realistic perspective. Exxon was host for the annual HMAC (Hazardous Materials Advisory Council) drill in 1990 in Linden, New Jersey. The scenario was developed by a committee made up of the New Jersey State Police, New Jersey Department of Environmental Protection and Energy, Exxon, the HMAC executive director, and the Union County Office of Emergency Management. All had different political and technical objectives: the state police wanted to test the capabilities of the city of Elizabeth, New Jersey, which abutted the Exxon property. Exxon was interested in a scenario that involved a NJ TCPA substance, and NJDEPE was looking for some exercise for its HAZMAT team. All civic agencies wanted to use the event to qualify for their government budget for emergency management.

Eventually a scenario was developed in which an NH_3 delivery truck would make a wrong turn and approach the wrong Exxon gate on a little-used road. The truck would overturn on a curve and be punctured. The resulting NH_3 cloud would blow one way for awhile, but then go another in response to a hypothetical wind shift. The cities of Elizabeth, Linden, and Rahway, plus the borough of Staten Island, would be involved. The truck driver would be hurt and be rescued by the Exxon team, ambulances would be called upon to rescue victims in a downwind industrial establishment, rescue workers could have to be decontaminated, the tank hole would be plugged by the HAZMAT team, the truck offloaded to another and ultimately righted by crane. All of the emergency management centers in the area were activated as were several ambulance corps and hospital emergency rooms. The U.S. Coast Guard got involved because some of the liquid NH_3 might have gotten into New York Harbor. Several related incidents were hypothesized, the largest being a school-bus accident that occurred when the driver smelled NH_3. All of the above was translated into specific data input by a NJDEPE simulated emergency expert. In such major drills, this kind of skill is important because it provides realism, controls the timing of events, and ensures that all players eventually get something to respond to. In effect, certain pieces of data were interjected at certain times to motivate responses. No one but a chosen few knew the entire scenario, so responses to each new data input were tested. For example, at the start of the incident, the wind was from the south, blowing fumes into Elizabeth. Access to the site, command center locations, and so on was from the south (upwind) through the Exxon property. After a time, the wind changed to westerly and then northerly, forcing the relocation of command centers and access routes and involving other communities and the workers in the Exxon refinery as well. The manipulation of these data inputs and reports of NH_3 concentrations at various points in response to wind changes and leak-stopping efforts were critical to the success of the drill.

The overall technical objectives were to foster communications and exercise a major part of the region's emergency response capability. In addition, a clear goal was to have everybody involved get some good public notice.

Key Considerations

We have mentioned some of the overall objectives of simulated emergencies in the foregoing. On a more detailed level, there are some specific elements of the response plan that need to be tested in each drill. These include, in the form of questions:

- Who is really in charge of the response, especially during off-hours?
- Did he/she respond properly? Understand the correct priorities? Call for the appropriate resources in time?
- Were the resources needed available in a timely fashion?
- Did the callout system work as planned?
- Can the needed help realistically be available in time?
- Were communications handled well? Is more equipment needed? Is coverage complete for radio or cellular telephone networks?
- Were all legal requirements met?
- Were those outside agencies that might be called upon to help contacted in a timely fashion?
- Did communication needs interfere with the attempts to control the actual emergency?
- Was the cooperation with outside agencies effective?
- Was traffic control effective?
- Were agency and media people handled separately and effectively?
- Was all of the information critical to handling every phase of the incident readily available—MSDS sheets, equipment information, P&IDs, air and water samples, special response equipment, government agency tests?
- Was the management in a position to monitor outside response?
- Did the system for recording detailed chronology of the incident work? Were the data necessary for the investigation preserved?
- Were the authorizations needed for corrective action immediately available—

the permission to use chemical dispersants, the authority to pay for plywood to cover broken windows? Could cleanup repair contractors be committed for immediately when needs were identified?

- Was there a transportation need for equipment and people, including evacuation?
- Was backup manpower alerted in case needed?
- Did effective communications with the community begin early and continue to be responsive? Were the necessary trained people available in time?
- Would a likely accident impair the plant's communication ability by cutting power to the system or disabling telephone lines or radio repeater stations?
- In certain weather conditions, would smoke or fumes impair visibility to the point that either proper evacuation or proper remedial action would not be possible?
- If responders did not act in the optimum way, did your critique session find out exactly why? Did you obtain and consider the workers' input on how to respond more effectively?

The list of such elements can be quite long, especially if site-specific items such as weather or transportation issues are critical. A comprehensive list is impossible in the general sense.

I do wish to consider briefly the needs for medical, industrial hygiene, and environmental expertise and networking. It is clearly important, and in some places legally required, to discuss toxic and corrosive hazards with nearby emergency rooms and ambulance squads. MSDS sheets, recommended treatment methods, and sometimes even materials must be available to these medical responders. In my view, it is the duty of the industrial-risk manager to ensure that the hospitals to be used are trained and equipped to give speedy treatment to accident victims. This is also an ongoing process because the emergency room medical staff is often notoriously transient. For smaller companies, this might need to be accomplished via a consortium arrangement due to lack of company medical capability at the site, but the responsibility remains. Communication from the hospital to a responsible medical expert retained by the company is something that should not be overlooked.

Industrial hygiene and environmental staffs normally have fairly effective networks among their industrial neighbors. These can encompass public agencies and this is much to be desired. The objective is agreement on sensitivities and measurement techniques beforehand to avoid any public debates at the time of an incident. Both the company and the agencies will be making measurements in times of trouble. It is very important that comparable approaches be used so that data obtained by both groups are subject to comparison. You do not want the agency to use one method to measure an air contaminant and report, say, "less than ten parts per million" while you use another and report "less than one part per million". This is especially true if the odor threshold or the TLV is between one and ten parts per million.

On the environmental side, it pays to know what the agency considers a "sensitive area" well beforehand. In the Arthur Kill oil spill, Pralls Island, viewed by the industrial community as a place where people in boats dumped garbage, turned out to be crucial to certain migrant birds. Getting some oil on its shores (and the ruined auto tires sitting there) was viewed with panic by environmentalists, to the delight of the media. Now plans are in place to make sure the island can be protected quickly in the event of any new oil spill. It could have been prepared for much earlier and saved much pain.

Public Communication

I mentioned earlier the need to train responsible managers in how to deal with a hostile media/agency audience. In my view, public affairs spokespersons are not credible. Neither are visiting vice presidents from headquarters. There is just no substitute for prompt access to responsible local management and no substitute for training them to handle the situation.

I had the opportunity (not the responsibility) to sit in on the media/agency sessions at Exxon after the pipeline oil leak into the Arthur Kill in New Jersey. The head of the Refinery Safe Operations Committee and the manager responsible for the pipeline transfer operation tried to answer questions from the company's viewpoint. They tried not to speculate; they tried not to incriminate anyone; they tried to protect all legal flanks; they succeeded in self-annihilation, in producing terrible headlines, in convincing the agencies that someone should go to jail, and in being depicted as thoroughly incompetent. Fundamentally, it was not their fault. They were lambs in the lion's den who listened too long to the lawyers from headquarters. In addition, they were not even supported properly by supposedly knowledgeable public affairs experts. The mayor of Elizabeth, whose harbor was invaded by some of the oil, tried to ask a question for about five minutes, was not recognized in the hubub, and got up and stormed off with his entourage. The next day he announced a lawsuit. I can understand why the hard-pressed spokespersons did not recognize the mayor, but fault the public affairs people for not bringing their attention to him.

Such disasters are easy to achieve. Preventing them takes preparation, quick response, and some courage by the spokespeople. It is often far better to be seen as responsive and open to the public and apologize to the lawyers later than to hold the company line too long. By contrast, the manager of a Shell plant in Norco, Louisiana, was perceived as a caring person even though an explosion in the plant killed several people. He did his communication in the streets, authorizing repairs immediately, comforting families, and generally being visible. To my knowledge, no one thought he should go to jail.

These days, upper management should consider well the initiative and courage of the candidates before selecting plant managers.

Summary and Example

Emergency preparedness/response depends very much on the effectiveness of planning and communication activities—that is, on planning before, and communication during, any incident. Developing general plans and capabilities is important, but ultimately each organization must understand its own risks well enough to postulate and prepare for specific incident scenarios.

The general preparedness strategies are the framework on which specific steps are hung. Plans must be in writing and comply with OSHA and other regulations. A step-by-step approach described by the Canadian process industries is a good starting point. An effective management system is essential. Its objectives should include:

- Minimizing impact on the plant and community
- Facilitating communications with all impacted parties
- Facilitating continued operations of unaffected units
- Organizing data collection about the incident
- Providing leadership and aid in incidents that affect the public
- Dealing with related transportation incidents close to plant as well as those inside the gates.

A fundamental principle of this system should be to separate responsibility for remedial action from that for communication insofar as possible. Similarly, the separation of governmental officials from the media is advised.

The management system must also provide for the resources needed and an assignment of responsibilities. Clearly, the necessary equipment and supplies and appropriate mutual aid agreements to combat any emergency are high on the list of resource needs. So too is the necessary training for both responders and management. OSHA requirements may govern the former, and training in dealing with the media is important for the latter. Appropriate communication and transportation equipment is also important.

Concentration on the worst case scenario may not serve a plant well as a basis for a well-thought-out emergency plan. Smaller, more-probable incidents are much more likely to be the subject of the real response needs. Specific scenarios should be developed and response plans should be formulated. These plans need to include early detection, suppression/mitigation measures, response strategies and investigation procedures. Mitigation may well involve reevaluation of equipment and control/relief system designs.

Aside from the necessary training, emergency response involves six key functions. These are:

- Leadership
- Knowledge/experience of responders
- Resources
- Communication
- Medical and environmental expertise
- Effective public communication

The first five need to be exercised frequently. This is normally done via responding to simulated emergencies at several levels: individual units, plant-wide, and major community exercises. The first level can be exercised every one to two months while the last is probably possible only once/year. Unfortunately, realistic practice for public communications is difficult to obtain short of the real thing. Training courses are available to simulate real situations.

The following example outlines the general and specific considerations in the planning basis for emergency response at a truck-loading terminal handling volatile hydrocarbons.

Example
Emergency Response Planning Basis
ABC Truck-Loading Terminal

Description of Terminal Facilities

Multibay truck-loading terminal for gasoline-type hydrocarbons. A number of tanks are grouped in each dike area. Pumps are located outside dikes. Operations will be conducted both by terminal employees and truck drivers. Central observation office provided.

General Basis

Both water and foam fire-fighting systems will be provided on site. Backup is available from local municipality, but in-house response is to be adequate to control all but the worst case scenario (all tanks in a dike area on fire). The shift superintendent will be responsible for leading the fire-fighting effort and every shift operator will be trained as an emergency responder, operations, and will go to fire-fighting school. A diesel-driven fire water pump is provided, and water supply is very large.

Specific Risk Scenarios to be Considered

1. Truck overfilling, ignition, and fire at truck-loading station or electrostatic ignition during truck loading of nonene/toluene. Explosion and fire in truck vessel.
2. Pump fire on one product loading/transfer pump. Liquid spill on fire.
3. Leakage of product in diked area. Large spill fire.
4. Vapor space explosion in low-flash tank. Tank roof blown off and full surface on fire.

Combat Strategies (for the Above Risk Scenarios)

These strategies define both the facilities and human capabilities needed to deal with the scenarios outlined aboved.

1. Stop all product flows to the loading rack. Initiate foam attack on burning spill/truck. This should be done by a fixed foam sprinkler system.
2. Block flow of product to the pump(s) by closing all block valves on suction and discharge side(s). Apply foam on liquid spill.
3. Close all product block valves if possible. Start foam attack on spill fire. Cool adjacent tankage.
4. Start foam attack through foam lateral. Cool adjacent tankage, if water system still in place. If not, use foam truck and call for mutual aid.

Planning Considerations (for Implementation of Above Strategies)

Activation of the loading-rack protection should be based on how the rack is operated:

ALT. 1. If the rack is operated during off-hours or by the driver (without the assistance of terminal personnel), a fire detection system that activates the fire/water foam system automatically is recommended. The detection system should be designed to avoid nuisance trips. (Two out of three detectors would activate.) In addition, a series of manual activation alarm pushbuttons would need to be provided (as described under the next Item b below).

ALT. 2. If loading is done under terminal supervision and while the terminal is manned, activation may be done by push-button system only. There should be one push button at each loading bay, another push button at a safe distance from the rack, and another in the office (permanently manned room overlooking loading rack). These push buttons would activate a siren, start the firewater pump, and open the foam valve to the loading rack. If product lines to the loading rack can be blocked off by motor-operated emergency block valves (EBVs), they should also be closed by this push button (or the detection/activation system discussed above).

Foam System

The fixed foam mixing and distribution system cannot depend on electrical energy entirely.

The system capacity should be good for a tank fire or a fire in the diked area as discussed above. An appropriate basis would be that the tank on fire would get cooling water and foam and that sufficient water would be available to cool three adjacent tanks.

Training and Exercise Program

There should be an emergency handbook reflecting all information and procedures. The diesel-fire water pump should be operated once per week for the time needed to get the engine warmed up (about thirty minutes). Fire hoses should be pressure-tested once per year. There should be a fire-fighting exercise every quarter. It should include:

1. Starting the firewater pumps
2. Installing the portable monitor (connected to foam solution header) and throwing foam into a diked area for short time.
3. Activating the truck-loading sprinkler system.
4. Topping up the foam concentrate tank after finalizing the exercise.

In addition, it is recommended that a joint exercise each year with the public fire brigade and mutual aid from surrounding industry, first aid squads, and hospitals be carried out.

With the above bases in place, there remains the need to define the details of a response plan and coordinate these with the municipal fire department and its mutual aid plan. In addition, arrangements will be needed with local hospitals and first aid squads to transport and care for potential burn victims.

Since details vary considerably from place to place, I cannot get more specific here. However, the need for specifics, such as valve locations, telephone numbers, and individual responsibilities, should be obvious. It is important to realize that plans evolve with practice and that the requirements of community responders must be accommodated. Write down the best plan possible; then get on with an exercise, learn from it, and improve the plan.

CHAPTER

11

Managing Follow-up

Introduction

To this point, we have discussed some principles and systems aimed at controlling plant risks. In the "great risk management department in the sky," every employee will operate according to these principles and systems. However, there still remain the problems of interpretation of the principles and the implementation of the required actions. Those responsible for implementation may develop an action plan that tends to exclude other ideas in their fervor to make progress. Most people recognize that outside input is often extremely valuable in bringing new ideas and a more creative and productive approach to any course of action. In addition, in the real world, real people do not always do everything the system demands, particularly in this "lean and mean" age.

Many companies these days talk about the quality process in every aspect of their business. One of the main principles of the quality process is that progress can be measured. Specific actions are required and measurement of whether and how these actions are carried out is essential. The result of these measurements is then used to adjust the action plan and refocus it on the overall goal—thus the cliché "continuous improvement."

The essential point here is that, periodically, it is necessary to stop what one is doing and evaluate whether the planned actions are indeed advancing the program toward the desired goal. While some people are able to do this internally, without an outside viewpoint, most cannot. In addition, the transfer of what has been learned by others into a site-specific action plan cannot take place if only internal sources are used.

Thus, audits are required. While audit may be a dirty word to many people, I believe they are essential. In the days of energy shortages and rapidly rising prices, many companies initiated plant "energy audits." The reaction of many plant managers to this step was negative. They felt they had energy management under control. In response, energy auditors changed the name of their service to "energy efficiency study" or some other euphemism. My own experience was that this did not change the atmosphere very much. If the client viewed himself as the victim of some external meddling in his own responsibilities, his attitude was negative regardless of the label pinned on the task. If, however, the client felt that he would find ways to improve his plant efficiency and his own recognition, the climate was positive.

Aside from the philosophical aspects of the audit process, every plant undergoes various government audits authorized under safety and environmental regulations. In New Jersey we have not only the federally imposed inspections (OSHA, EPA, the coast guard, the Railroad Safety Agency) but also state-mandated inspections of environmental permits, risk management plans, hazard communication systems, and right-to-know labeling, to name just a few.

In addition to all of these, most plants now undergo scrutiny on other fronts. The SARA III emission data now required have prompted environmental organizations to ask tough questions about the health effects of these emissions and the responsibilities and liabilities of the emitters. The appointment of environmental prosecutors can result in adverse headlines over activities from the past that seemed perfectly legal at the time, and local legislators continue to see their path to reelection as enhanced by imposing even more stringent safety and environmental regulations on the industrial community.

My conclusion from all of this is that being audited is a way of life in the process industry. The challenge is to turn this potentially abhorrent process into a competitive edge. I believe those who can demonstrate a diligent effort to verify that they are in compliance with all regulations will reap a benefit both in the public eye and in the ability to forestall onerous regulation.

What Is an Audit?

Perhaps the easiest way to start the definition of an effective audit is to list some of the things it is not. An effective audit in the safety/environmental area is not a typical financial audit. There can be no grade for such an audit, nor can there be any "fault finding" mentality. The auditors cannot be inexperienced accountants who know nothing of their subject matter. Such people cannot begin to ask the type of probing questions necessary to identify fundamental problems. The audit cannot make a big presentation to management of any sort, much less a fuss over trivial objectives.

In addition, an effective audit cannot be a typical governmental inspection. There can be no violations, no fines, and no headlines.

An effective audit is a helpful consultation by knowledgeable people who bring with them the capacity for technology transfer and independent judgment. Its starting point needs to be a recognition of the good-faith attempts of the host organization to achieve compliance on all fronts. Since major helpful input from the auditors will be assistance with the interpretation of regulations and company standards, the members of the team must be knowledgeable experts capable of placing themselves in the situation of the host plant and suggesting creative solutions. An effective audit depends on open communication between the auditors and those audited. This can happen only after a period of time during which a level of trust is built up between the two groups. The surest way to accomplish this rapport is to avoid any negative observations and maximize helpful input. Another positive approach is to help transfer successful solutions from other sites to the one being audited. Care must be taken to ensure that other plants' solutions to apparently common problems do not become gospel, but rather a starting point from which to develop a solution specific to the plant in question.

The audit team needs to be very tolerant of different views of a problem and of debate about the definition of compliance. Occasionally, such a debate will result from a misunderstanding of requirements that are straightforward. In my experience, most debates originate in the "gray areas" where interpretation about what is required and, more importantly, what will be successful in the current plant culture are factors. The end result of such debate is often a consensus and buy-in for a path of action that upgrades compliance significantly. If consensus cannot be forged, then a clear presentation to management of the different views is essential. This presentation should have as its purpose providing a clear definition of the residual risks for every course of action so that management can make the necessary overall judgment. Note that my underlying assumption is that local management must decide on specific risk issues after having reviewed the best possible information about the requirements of corporate policy and government regulations.

To those with a jaundiced view of audits in general, this may sound idealistic or perhaps even naive. Having experienced both good and bad audits, I believe the above represents a practical expectation for an organization. In my view, the host plant cannot optimize its situation completely without outside input. It is in that plant's interest to obtain that input from helpful friends rather than from a punitive agency action.

Clearly, if senior management wants the assurance an audit provides, it must be prepared to provide the resources to carry out quality audits and those necessary to implement the follow-up activities identified. The plant staff must also realize that being audited involves a commitment of its manpower, both to organize itself for the audit process, but more importantly, to follow-up on its findings. There is nothing worse (or potentially more incriminating) than to have a long worklist of audit items against which no progress is being made.

This is not to say that every suggestion by the audit team will turn out to be efficacious. As implementation of ideas progresses and understanding grows, many audit recommendations may prove to be impracticable or in need of modification to provide a practical solution. It is important, however, to document the resolution of each item and the reason for each decision.

Internal Audits

In spite of my arguments for external participation, there is great value in carrying out frequent internal audits. Clearly, the best way to get audited by helpful, friendly consultants is for a plant to audit itself. The obvious objection to such a process is that the staff might be softer on itself than an outsider would be. Compensating for this is a much greater depth of knowledge about specific practices at the site and of the attitudes and philosophies of those charged with implementing compliance programs. All this presupposes that a good knowledge of company and regulatory requirements is available at the plant. What may be foregone in an internal audit is more detailed knowledge about how other plants have solved similar problems, unless specific efforts are made to acquire that information.

The NJ TCPA requires an annual safety review of each unit covered by the law. In recognition of the above, an internal audit is permitted. Since each plant's risk management plan has been approved by the state, the thrust of each safety review is to determine whether actual plant practice is what it was represented to be to the government.

Under these circumstances, many in-plant audit teams are more stringent than external auditors. Because of their knowledge base, they ask more probing questions, especially if one or more members of the team come from outside the specific unit being audited. The "dumb questions" of such people do a lot to add to the critical scrutiny of the audit. In addition, if it is clear that the plant wishes to do a good job at risk management, the in-plant audit team often takes a conservative approach—a much tougher one than usually mounted by a government agency team. They do not wish to overlook anything which might then be cited by an external audit. Experience shows that self-audits are very useful in establishing a correct assessment of status and in developing action plans to ensure compliance and continuous improvement.

This is not to say that one can simply pick a team, tell them to audit the *XYZ* unit, and expect success. The team needs to understand the basis and objectives of the audit and recognize its role as helpful consultant. In response to the latter, the team's objectives should be:

- To ensure that all that is acknowledged to be required is being done and is being done well
- To evaluate whether what is acknowledged to be required is accurate and complete
- To suggest ideas for improvement and solutions to problems identified
- To transfer information which might be pertinent
- To do all of the above in a noncritical and supportive way

Unit Safety Audits

There are some other problems with internal audits if they become too formal and the objectives become fuzzy. Exxon has done internal process safety audits of its refineries for years. The system has evolved into a checklist (several inches thick) of every possible refinery safety issue and a meticulous scoring system. Indeed, the score a given unit received on its audit often had an impact on the unit supervisor's salary. As a result, a lot of time was spent once per year to prepare for the audit *and* in lobbying the audit team leader after the fact for the best possible score. In many cases, temporary "fixes" were put in place just before the audit and abandoned soon after. The focus became totally on the score received rather than on the safety of the unit—a "cram-for-the-exam" philosophy. This tended to lend an air of cynicism to the entire process despite the best intentions of the audit team. The audit had an extensive basis (the checklist) and a dedicated and knowledgeable team, but because of the focus on grades, it was not as productive as it might have been.

This system was adapted in some plants to be more frequent, more specific to the unit in question, and much more friendly and useful. Grades were eliminated; remedial action was the focus. The annual frequency was modified to fit in better with the specific nature of the operation and work patterns. The scope of the audit was modified to reflect the specific risks at the unit by the people who worked there, and immediate remedial action became the objective. For example, if the team judged that a sign was needed at a certain tank or valve, a person will be assigned to get one, today!

The approach to doing this audit was also changed markedly. Over time, everyone on the unit participated in the audits, so it also served as an important training exercise. This approach created a great deal of ownership by individuals for the remedial action remaining when the audit was complete. Rather than cram for their supervisor's exam, the unit team undertook to eliminate and/or control as many of the risks in their work area as they could and then make a list of additional improvements to be worked on in the future as part of their day-to-day activities.

The sequence of events is something like this:

- A committee is appointed to formulate a basis for the audit. This takes the form of a checklist distilled from the general one and tuned to unit hazards. For example, if no light hydrocarbons are present on the unit, but chlorine is, much less emphasis is put on fire risks and much more on the impact of toxics, availability of self contained breathing apparatus and escape routes.
- The team then gathers in a general meeting to be split into groups to cover either specific sections of the unit or specific facets of the audit for the whole unit—for example, safety relief devices and the car-sealing of valves around them. Twelve to twenty-five people usually participate, split into two- or three-person teams.
- The teams then go into the field observing conditions, talking to workers, checking documents, until finished.
- The group reconvenes to discuss findings and develop a future worklist.

Things that can be corrected immediately are taken care of. Other items are assigned a responsible person and stewarded monthly.

This type of audit has been very valuable in developing ownership of remedial action and fast response. It has also been a valuable training aid as about half the group is different for each audit. The frequency of such audits has varied from three to four times per year at the most active unit to once per year at the least. Even in the latter case, some tasks were assigned for monthly and quarterly review to keep activity more consistent with normal work systems. To round out the value of the approach, after a few additions to the scope of the audit, NJDEP accepted this effort as an appropriate annual safety review for TCPA-covered units.

Overall, the workers at the unit were much more satisfied with the new approach. While not providing much in the way of cross-fertilization of ideas, the audits did provide a lot of ownership for remedial action and increased consciousness of the optimum conditions for incident prevention. The cross-fertilization issue still has to be addressed via external audits or by evolving the self-audits to include a deliberate effort in that direction.

Table 11.1 lists the general content of an audit for a feedstock preparation unit using C_3 and C_4 refinery streams as feedstock. This was a typical petrochemical purification unit with a very limited range of materials present. As such, it was easily adapted from the larger refinery-wide syllabus, but included a number of specifics for the unit such as the loading facilities.

Table 11.2 covers the report for a small section of the audit (Emergency Procedures, Section 2) as an example of the types of questions and document reviews carried out. Note that a number of items involved testing equipment and/or operator knowledge or performance in the field. Please also note that many of the items included for this particular audit required further action to bring things up to standard. There was no sign of leniency by those who worked on this unit. They had to live with it!

Clearly, the scope of each unit's audit will be different in some ways, but also will have many common items. The Emergency Preparedness outline should be common to many units and, therefore, was chosen as an example.

Mock OSHA Audits

Part of the cost of a major incident is a very stringent inspection by a team from OSHA after the fact. Almost invariably, these inspections result in significant fines and bad publicity. For example, after the Phillips polyethylene plant explosion in Texas, the OSHA inspection proposed penalties of $5.7M for alleged chemical and safety violations. Pepperidge Farms, Inc., was assessed over $600,000 in penalties after its inspection. In all, ten companies were given initial penalty assessments greater than about $1/2M in 1990.

The Occupational Safety and Health Reporter tabulated the most frequent types of violations in fiscal 1990 in its January 9, 1991, issue. A total of almost

Table 11.1. Feedstock Preparation Safe Operating Audit Table of Contents

Section	*Item*
1	Operating Procedures
	100 Operating manuals
	200 MSDS and NJ RTK law
	300 P&IDs and process equipment lists
	400 Critical safety system bypass
	500 Task book
2	Emergency Procedures
	100 Simulated emergencies
	200 Fire protection
	300 Safety equipment
	400 Miscellaneous
	500 Safety equipment supplies
3	Accident Investigations and Near-Miss Reports
4	Unit and Equipment Condition Checks
	100 Safety valves and rupture discs
	200 Other equipment inspection
	300 Platforms and catwalks
	400 Miscellaneous items
5	Permit System
	100 Permits
6	Housekeeping
	100 Propylene unit area
	200 Butene area
	300 Sphere area
	400 Flare area
7	Safe Working Practices
8	Miscellaneous and Plant-wide Items
9	Tank-Car and Tank-Wagon Facilities

200,000 violations were alleged in that year. The total list is reproduced in Table 11.3 to give the reader some idea of the most frequent types of violations.

Under such circumstances, it seems prudent to consider internal audits of operations vs. OSHA standards. Mock OSHA Inspections can bring to light many inadvertent violations in time to prevent accidents and forestall violations. A single well-trained auditor can highlight many potential field violations fairly quickly.

One approach to such a program contained the following elements:

- Area inspections are conducted on a nonscheduled basis, taking care to spread the schedule evenly among the units in the plant.
- Items inspected are those covered by the OSH Act and Code of Federal Regulations 1910 and 1926.
- At the end of each inspection, if deficiencies are found, a citation and notification of penalty is issued to the appropriate supervisor. These citations

Table 11.2. Typical Audit Results

Section 2—Emergency Preparedness

	Simulated Emergencies	Status	F/U Required
101	Unit simulated emergencies are held every two months by each shift with process and fire team participation. Check with operations coordinator for records	Simulated emergencies have not been held in several years.	Yes
102	Critique session is held after each simulated emergency	Not possible	
103	Feedback gets to all assigned unit personnel promptly	Not possible	
104	Recommendations from simulated emergencies are implemented within six months	Not possible	

Fire Protection

	Fire Training	Status	F/U Required
201	Training, including hands-on use, is provided annually for all process personnel and supervisors; check attendance	OK	No
	Fire Hydrants and Monitors		
211	Fire hydrants and monitors are in good operating condition		
	+ All valves operational	Yes	No
	+ Turret operational	Yes, except #435 had broken 2½-inch valve	Yes
	+ Hydrant and cap threads in good condition	Yes, except #407 had bad cap	Yes
	+ Caps are tight		
	+ Hydrant wrench present	Many wrenches were missing	Yes
212	Using task book, pick 5 hydrants and/or monitors and check condition	OK	
213	Personnel can demonstrate the proper use of a fire hydrant and/or monitor and how to avoid system freeze-ups; ask two operators to demonstrate proper equipment use	Yes, OK	No
214	Procedures are in use to control the use of fire water for other than fire fighting; permanent process to fire line connections is not used; any temporary connections have a current permit from the fire department; operating personnel are familiar with these procedures	Yes	No
215	Review task book to ensure that checks on hydrants and monitors are performed as scheduled	Checks are performed normally; included in task book	No

(Continued)

Table 11.2. *(Continued)*

		Status	F/U Required
	Hose Reel Boxes		
221	Hose reel boxes are in good condition		
	+ Ensure 250 feet of usable red fire hose inside box	Condition of boxes is very poor, boxes lacking proper materials (nozzles, adapters, reducers) Normally less than 250 feet in boxes	Yes
	+ Ensure that there is one coupling, one reducer, one wrench, and one nozzle in good working order, inside box		
	+ Check general condition of storage box and surrounding area		
	+ Nothing else should be stored inside box		
222	Review task book to ensure that checks on hose reel boxes are performed as scheduled	Have been written up since T/A	Yes
	Fire Extinguishers		
231	Fire extinguishers are in good operating condition		
	+ Check general condition of storage box or area	Good	No
	+ Check general condition of extinguisher	Good	
	+ Check charge	Good	
	+ Ensure inspection tag is up to date (monthly inspection and annual overhaul)	Good Fire extinguisher missing from compressor floor and 150# extinguisher south of compressor house had low pressure	Yes
232	Using task book, pick 8 fire extinguishers and check condition	OK	No
233	Review Task Book to ensure that checks on fire extinguishers are performed as scheduled	Included in task book	No

	Safety Equipment	Status	F/U Required
	Scott Air Paks		
301	Scott Air Paks and 5-Minute Escape Paks are available and in good condition		
	+ Check condition of storage box	OK	No
	+ Check cylinder pressure gauge	5710 ½ full	Yes
	+ Check condition of face mask and hose	OK	No
	+ Check condition of regulator	OK	No
	+ Ensure inspection tag is up-to-date (monthly operability check and annual performance check)	OK	

(Continued)

Table 11.2. *(Continued)*

	Safety Equipment	Status	F/U Required
302	Personnel know the location of Scott Air Paks and 5-Minute Escape Paks	Yes	No
303	Scott Air Pak training, including hands-on use, is provided annually for all process personnel and supervisors; check attendance	Yes	No
304	Check to see if spare Scott Pak cylinders are available in quantity and location is known to personnel.	Spares are available at Safety Department	No
305	Using task book, pick 5 Scott Air Paks and/or 5-Minute Escape Paks and check condition	OK	No
306	Review task book to ensure that readiness checks are performed weekly as scheduled	Yes	No
	Safety Showers and Eye-Wash Stations		
311	Safety showers and eye-wash stations are accessible and in good working order + Grounding: Both handles must be grounded to a ground rod + Identification: ID number, green light, green safety sign, electric tracing sticker + Shower: Good condition, adequate flow, pressure, temperature, handle operational, clear access to shower, area clean + Eyewash: Screens, flow control, handle operational, adequate flow + Electrical: Water temperature, if green light is out (after changing bulb), this is an indication the tracing is bad	All (4) showers checked; showers work but overall condition was poor; none of the safety showers' green lights were working, tracing in overall poor condition	Yes
312	Using task book, pick 4 safety showers and check conditions	OK	No
313	Review task book to ensure checks on safety showers are performed as scheduled	Included in task book	No

Table 11.3. OSHA Standards Most Frequently Cited in Fiscal 1990*

Standard Section 29CFR	Subject	No. of Alleged Violations
1926.59	Hazard communication/construction industry	22,578
1910.1200	Hazard communications/general industry	20,938
1904.2	Record keeping	6,109
1910.219	Mechanical power transmission equipment	5,899
1926.451	Scaffolding	5,399
1903.2	OSHA notice	4,993
1926.500	Guardrails	4,912
1926.404	Wiring design	4,879
1910.212	Machine guarding	4,706
1910.305	Wiring methods/general industry	4,555
1910.134	Respiratory protection	4,422
1910.20	Employee exposure/medical records access	4,252
1910.215	Abrasive wheel guarding	3,974
1910.147	General environmental controls/standards sources	3,602
1910.106	Flammable and combustible liquids	3,242
1910.95	Noise	3,106
1910.303	General electical rules	3,060
1926.405	Wiring methods/construction	2,889
1910.107	Spray finishing	2,823
1910.213	Woodworking machinery	2,750
Section 5(a)(1)	General duty clause	2,743

*Total number of alleged violations cited in fiscal 1990: 197,346.

and penalties were kept as close as possible to the likely citations and penalties issued during actual OSHA inspections.

- The citation contains the CFR standard, regulation, or section of the act violated; a description of the violation and location; the type of violation; and the penalty assessed. The following definitions of type of violation and penalty apply:

  Other-than-serious violations have a direct relationship to job safety or health but they would probably not cause death or serious injury. Penalties can range up to $1,000 for such violations.

  Serious violations require two conditions: a substantial probability that death or serious injury could result from the hazard and that it is known or should have been known that the hazard is present in the workplace. Penalties up to $1,000 are mandatory for serious violations.

  Imminent-danger violations result when an inspector finds a condition or practice that could reasonably be expected to cause serious injury or death before it can be eliminated through normal enforcement procedures.

  NOTE: Particular attention should be paid to previously cited OSHA violations and penalties as well as frequently cited OSHA violations. These may be declared "willful" by OSHA and incur much higher financial and publicity impact.

- The program also includes a closing conference between the supervisor and inspector to review the citations and penalties as well as to assist in whatever way possible to eliminate the violation.

Experience with such a program has been positive at one plant and others have expressed interest in following suit. A number of violations have been eliminated and others have been made more visible to the management by this approach. The net result is that the safety standards in the plant have been elevated while the potential for citations has been reduced.

In some cases, it was an educational experience for newly appointed supervisors. As indicated, the inspector cycles through each operating area about monthly to maintain continuous improvement momentum.

A General Approach

Defining the scope and organizing an internal site audit can be a real challenge if it is necessary to start fresh. In 1986, C. H. Vervalin, of *Hydrocarbon Processing Magazine,* wrote a special report dealing with environmental audits (October 1986, p. 41). In it, he reported on discussions with a number of environmental consultants about their approach to auditing. A fairly well-integrated approach was given by Thomas Kunes of RMT, Inc. (Madison, WI), which I have attempted to generalize to include all aspects of a risk management audit.

Three dimensions of an audit were identified in the article: compliance, risks, and management systems. There were also three areas to consider in judging the potential impact of any issue in establishing priorities for the remedial action program. These were: company organization, policies, and procedures; governmental laws, regulations, and permits; and the community aspects, environmental systems, and aesthetic values of the location. For each of these, the exposure of the company needs to be evaluated. A partial list of exposures is given below:

- Inefficiency, waste, unnecessary costs
- Lack of control, management surprises
- Loss of stockholder confidence
- Fines, penalties
- Forced plant shutdowns
- Criminal prosecution, civil action
- Difficult agency relations
- Responsibility for health impacts, accidents, environmental damage and repair
- Public complaints, nuisance
- Bad publicity, anticompany crusades

An attempt to capture the audit process is made in Figure 11.1, also adapted from the article.

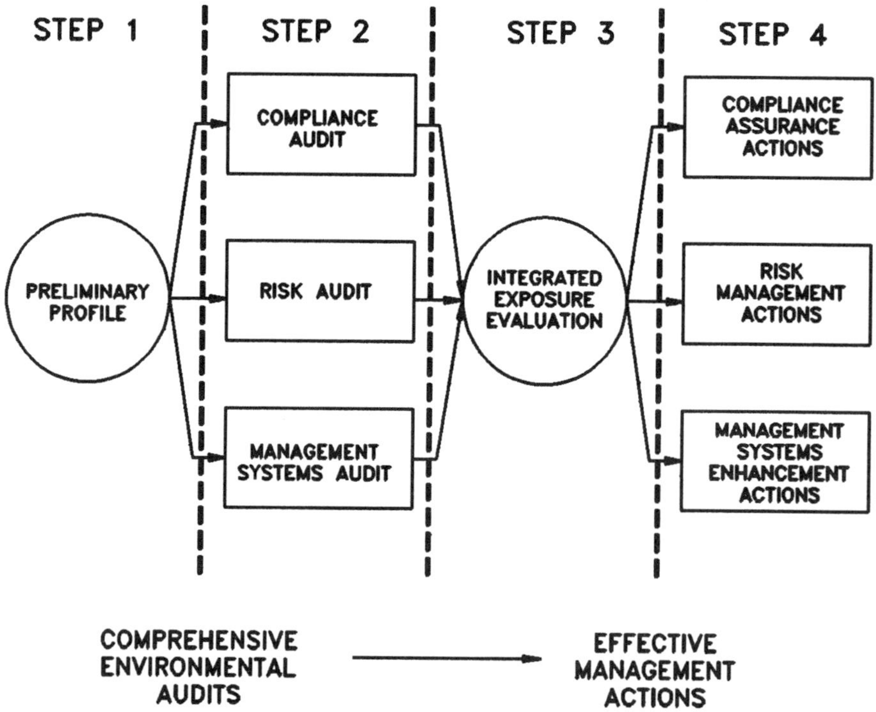

Figure 11.1. The audit process.

Four steps are envisioned: a preliminary risk profile, an assessment of each of the three audit dimensions, an exposure evaluation, and response actions. The latter is the final objective of the process. As indicated earlier, a commitment to remedial action is essential if the audit is to be more than mere posturing.

The information typically compiled for risk profile includes:

- Descriptions of raw materials, manufacturing processes, products, and residuals (emissions, discharges, wastes)
- Information on materials storage and handling, pollution control equipment, waste treatment facilities, equipment design practices, construction practices, and applicable codes
- Site plans, facility layouts, drainage and discharge systems (air, water, land)
- Natural and man-made features of the site surroundings, including community support systems
- Identification of the legal requirements that apply to the operations, and compliance systems
- Information on company policies, procedures, personnel responsibilities, and other management-related information applicable to the facility or operations

The profile is a key step in the management process. With it, audits and subsequent steps provide an efficient and thorough evaluation of exposures. Profile materials can also become an important reference document for managers, especially in companies having widely dispersed plants or facilities.

The second step, the on-site audit, should be "three-dimensional." It needs to address all three areas—compliance, risk, and the management system.

The compliance audit establishes the basis for evaluating administrative exposures subject to government intervention, including administrative enforcement and court action.

The risk audit assists a company in assessing possible risk events and resulting consequences. It first reviews the status of potential risk sources documented in the preliminary environmental profile. Other risk sources identified in the audit, which were not included in the profile, should be defined. The risk audit then investigates the pathways such as air, surface water, groundwater, and soil contamination.

Next, potential risk targets, such as plant personnel, neighboring residents, and natural ecosystems, should be identified and described. The risk audit may use environmental monitoring and modeling information. With this, it would estimate concentrations likely to be experienced by potential risk targets under various operating conditions, as well as nonroutine situations such as accidents or equipment malfunctions.

The third area, management systems, applies to the facility or operation. This includes an evaluation of the programs and procedures used to implement a company's internal objectives. Internal inspection and reporting-system documentation is examined to identify potential communication and control problems. Emergency preparedness is audited, as well as the status of emergency response systems.

The integrated exposure evaluation summarizes and integrates the findings in the three audit areas in terms of the exposures represented. It describes, in as much detail as possible, the potential consequences related to each exposure, such as government action, impacts on human health or the environment, economic effects either to the company or to the outside parties, legal problems, and so on. It also establishes a priority ranking for management response based upon the nature of the exposure, the potential consequences involved, the estimated likelihood of such consequences, and the time frame in which these consequences might occur.

The final important step is to outline the management actions necessary in response to the audit findings and corresponding exposure evaluation. Management response actions should be programmed and documented along with the rationale for all decisions in all three areas of the exposure environment as appropriate. As indicated earlier, the implementation of these principles will vary from site to site. In many cases, the overall exposure evaluation draws on many audits, some specialized, some general, some internal, and some external. In larger plants, a single comprehensive audit may not be manageable, but integrating a number of smaller internal audits and periodic external assessments can often be effective.

Legal Caveats

All of this internal scrutiny may make the legal staff nervous. They may fear that the plant's own records will be used to indict it (and its management) for some of the increasing number of criminal violations being defined. These kinds of advice prompt a "See no evil" approach to managing risks. I fail to see where that offers much protection to anyone, especially when regulations are including the language "knew or should have known."

This problem has recently been addressed by the U.S. Justice Department. *The Wall Street Journal* reported (9/25/91) that justice has issued a fourteen-page set of guidelines to all U.S. attorneys after a year of development. These describe a general policy and nine hypothetical examples of how a company's detection, prompt disclosure, and correction of violations can reduce or even eliminate criminal liability. Some lawyers felt the guidelines should have provided complete immunity for information collected from internal audits, but, of course, militant environmentalists do not agree.

I believe most people believe that internal company audits are the most effective way for those who care to ensure compliance with safety and environmental regulations. In the United States, some degree of protection can be achieved by carrying them out under attorney/client privilege or at least labeling them as "Critical Self Assessment". However, whether any violations found are reported to a government agency or simply corrected will be a matter for each individual management to decide. In any event careful documentation of followup activities is a must.

Short/Frequent Single-Topic Self-Audits

There are a number of subjects that need to be audited much more frequently than even every quarter and that can form a part of the internal audit system mentioned above. These include a number of items mentioned in Table 11.2 as "task book" items such as safety shower operation and fire extinguisher and hydrant checks. Work permit audits are also essential to ensuring that this highly critical control system is maintained constantly. In general, work permits and task book items should be audited monthly to achieve continuous compliance. The emphasis is on continuous here. There are other audits that are inherent in the operations stewardship process. These include monitoring changes that are covered by the change authorization system and those from the recommended separate system for monitoring critical safety-system by-passes.

The former should be monitored by the Change System manager, who will highlight any discrepancies between what was approved and what is installed and any overdue rectification of temporary changes. Any overdue safety system bypasses should be brought immediately to the attention of the operations manager if the system works properly. While these amount to one-person audits, they clearly are part of the system of verification procedures.

Investigation of incidents and near-misses is also a form of audit. These investigations should identify the fundamental cause of each incident regardless of its severity. In so doing, a great deal of information is developed on whether people are following the procedures and systems in place and on whether the facilities in place are adequate for the task for which they were designed.

Some corrective actions are very short term,—for example, counseling an employee, but others might require facilities changes or broader, longer-term training. Tracking the implementation of identified remedial actions for all of these matters until complete is critical to the fourth phase of the audit process.

For those plants using combustion processes, separate combustion safety and efficiency audits may be in order. This is particularly true when smaller gas-fired combustors are used, whether or not they are integral to the process. Worn, maladjusted, or bypassed combustion controls and devices can impact on both physical and environmental exposures inside and outside the plant.

G. J. Kauffman outlined Proctor and Gamble's combustion safety audit procedures in *Chemical Engineering Progress* (July 1987 pp. 23*ff*). The program was originally carried out by corporate engineering experts, but as the number of combustion devices grew, this became impractical. Checklists were prepared and plant people were trained to do frequent inspections, followed by an expert audit every two to four years. The audits concentrate on the proper design, maintenance, and operation of eight categories of combustion safeguards. These are:

- Proven, timed purge. A prelightoff sequence assuring sufficient air flow to purge the burners and downstream equipment of any potential for a combustible mixture. Four air changes are needed with at least one fuel shutoff valve proven closed.
- Proven combustion air flow for all firing conditions (at least twenty-five percent of maximum).
- Fuel monitoring. Appropriate flow, mixture, conditions and property monitoring, and shutoff capability.
- Proven ignitor. Verify pilot flame before main burner fuel is turned on.
- Enforced lightoff at minimum fuel rate consistent with past safe experience (ten to fifty percent of maximum).
- Timed ignition sequence to prevent fuel admission to a burner after ten to fifteen seconds if a flame is not proved.
- Flame monitoring.
- Critical process safeguards (such as low water level in a boiler drum) will cause shutdown.

In addition to verifying that these eight safeguards are appropriately designed and operating properly, an efficiency audit is required to ensure that excessive or dangerous emissions are not occurring. Flue gas analysis is usually required. Oxygen, carbon monoxide, oxides of sulfur and nitrogen, and flue gas dew point may all be of interest as well as those for potential fuel contaminants. For incinerators, special analyses may also be required to ensure complete destruction of the species being incinerated. All of these tests should be part of normal operating checks, but frequent verification is appropriate.

SOC Role

Many times, carrying out the investigations and tracking implementation falls to the SOC or other safety management committee. I do not believe tracking implementation is an appropriate role for such a committee. Their role should be that of helpful consultant and problem solver, standard setter, and accident prevention leader. Stewardship of audit follow-up items should fall to operations management as part of their line responsibilities. In this arrangement, the SOC may contribute to and approve the remedial plan, but line management owns the implementation. There may be times when the SOC, as an interested observer and counselor to top management, must step in to reiterate the risks being taken by failure to implement certain corrective actions, but a policeman role impedes the contribution the SOC can make to overall process safety. Self-policing by the appropriate operations group promotes ownership and self-esteem and minimizes defensiveness.

Unit Safety Surveys

Earlier, frequent accident prevention surveys were discussed as a means to demonstrate management participation in safety and promote communication. These surveys are also a very effective way to audit worker behavior and the overall condition and availability of equipment. The surveys generate a large amount of data that can be used to identify improvement needs and provide a way to reinforce safe behavior as well as correct that which is unsafe. Surveys focus on a limited area of the plant and are best when carried out daily.

Safety surveys are normally conducted using a planned schedule in small teams (2 or 3). Documentation should be used to get a better understanding of safety problems that may be occurring or to weigh safe vs. unsafe acts. It is important that the observers gear the survey toward the actions of people, correcting unsafe acts as they are observed while also reinforcing the safe actions of the employees. Length of survey should be determined by the number of observations made. Fifteen is a number often used as the amount of observations noted during a normal tour.

Documentation of observations should be done by using a safety survey form (Table 11.4) as provided by each department. A review of the week's observations will provide a clearer picture of a unit's safe work habits and of worker's understanding of the work procedures. Training needs and procedural changes can often be determined by these reviews. Copies of the safety surveys should be shared with designated individuals in the area and the plant safety coordinator.

All safety-survey observations should be forwarded to the area operations

Table 11.4.

SAFETY SURVEY WORKSHEET

AREA: ____________ RESPONSIBLE SUPERVISOR: ____________________ DATE: ________ TIME: ________

AUDIT BY: ________________

NO.	SAFE ACTS OBSERVED	UNSAFE ACTS OBSERVED	OTHER SAFETY COMMENTS FOR FOLLOW-UP

TOTAL NO. OF SAFE OBSERVATIONS: [ ]

TOTAL NO. OF UNSAFE OBSERVATIONS: [ ]

manager for review. Feedback to the observer should be given as reinforcement for the effort. Safety-survey information should be analyzed as part of the overall safety data base at the plant. Some typical examples of survey observations are:

- Workers working off a safe scaffold (reinforced)
- Goggles missing on a hard hat (corrective action)
- Employee barricading a slippery area (reinforced)
- Forklift operated safely (reinforced)
- Monitoring a tank wagon move to prevent an overfill (reinforced)
- Grinder being used with guard up (corrective action)
- Wheels chocked on unloading tank wagon (reinforced)
- Hose designed for nitrogen being used in steam service (corrective action)
- Sample being taken while wearing goggles and protective gloves (reinforced)
- Extension ladder tied off at the top (reinforced)
- Work permit improperly filled out (corrective action)
- Clean work area (reinforced)
- Employee working near an unguarded floor opening (corrective action)

The safety survey is simply the observation of what people do as they go about their daily job. It is known that over ninety percent of all on-the-job injuries are caused by an "unsafe act." Many so-called "unsafe conditions" are the result of an unsafe act (i.e., the unbarricaded floor opening). Numerous studies also show that for every on-the-job injury that occurs, there have been 1,000 or more unsafe acts. Put simply, eliminating injuries and incidents means eliminating unsafe acts.

The field observers are in a position to eliminate these unsafe acts by immediate reinforcement of safe, responsible behavior and correction of unsafe acts. Names of people are not important. The survey should be used as a tool to reduce unsafe acts, thus improving safety habits and reducing injuries and incidents. When the unsafe act or safe act is observed, it is important to talk to the worker (employee or contractor) and discuss why the job is being done that way and how to do it safely. Discussions should be constructive, supportive, and corrective. It is just as important to reinforce the safe act as to correct the unsafe act.

One thing that should be clear is that even though you already make tours of the plant and visits with the employees are carried out for other purposes, these cannot double as safety surveys. The time that is designated for a safety survey should be a tour specifically designed to evaluate conformance to safety policies and practices.

Some typical data from early safety surveys are listed now. In the early stages of this type of program, most of the surveys focus on personal protective equipment.

Also, the focus on contractors was more than twice that of employees. Some of the more-often-noted observations included:

Observation	Percent Compliance
• Use of personal protective equipment—employees	82%
• Use of personal protective equipment—contractors	73%
• Truck wheels chocked	82%
• Safe ladder use	40%
• Use of harness—employees	90%
• Use of harness—contractors	77%
• Safe scaffolding	61%

Observations that covered procedures, awareness, and housekeeping were most often noted negatively, making it more difficult to assess compliance. Infrequent use of reinforcement in these and other areas was typical of early surveys. The results concentrated on pointing out the areas where there is room for improvement. In addition, they have stopped many unsafe acts that could have resulted in a serious safety incident. As time goes on, more sophisticated interchange and a focus on more positive reinforcement becomes possible. In a recent tabulation of six months' data, 857 safe observations were recorded vs. 462 observations of unsafe acts in 237 separate surveys. The process evolves with the skill of the observers and the recognition of the workers that consistent standards are being applied consistently.

Training for Plant Safety

Developing capable auditors requires some training before introducing them to the process, even under the guidance of an experienced team leader. At least three areas of training will be required: (1) standards to be applied, (2) observation skills, and (3) interacting with the person/team being audited.

Of these, the first is the easiest, especially for someone who has a good background in the audit area. Where clear-cut laws, health effects, company policies, and rules exist, the applicable standards will be well defined. In other areas, acceptable standards may need to be developed during the audit process. For example, suppose a government public-health standard for chemical *X* is 100 ppm, but its odor threshold is ten parts per million and its TLV is twenty-five parts per million. This means that odors may well exist in the plant. The auditors may well wish to find out whether odors could exist outside the plant and raise the question to management of whether this is acceptable even if no health effects are likely.

Observation techniques are something that safety professionals have tried to inculcate into supervisors via films and field exercises. The following points and techniques are recommended in most training sessions, and exercises are provided to develop confidence in applying them. The techniques are:

- Be selective; know what to look for.
- Practice.

- Keep an open mind; guard against habit and familiarity.
- Do not be satisfied with general impressions.
- Record observations systematically.

Observation Techniques

In addition, to become a good observer, a person must:

- Stop for ten to thirty seconds before entering a new area to ascertain where employees are working.
- Be alert for unsafe practices that are corrected as soon as you enter an area.
- Observe activity—do not avoid the action.
- Remember: Look *A*bove, *B*elow, *B*ehind, *I*nside (*ABBI* principle).
- Develop a questioning attitude to determine what injuries might occur if the unexpected happened, and how the job might be accomplished more safely. Ask, "Why?" and, "What could happen if . . . ?"
- Observe using all senses: look, listen, smell, and feel.
- Maintain a balanced approach. Observe all phases of the job.
- Be inquisitive.
- Observe for ideas—not just to determine problems.
- Recognize good performance.

These techniques must still be applied in the survey itself if useful information is to be gathered. The key element is to observe what workers are doing and learn why they are doing it that way. Some hints about how to do this are:

- Compare the way employees are working with how the job is supposed to be done. Are employees:
  Showing good judgment
  Following rules and procedures
  Creating unsafe conditions
- Focus on workers' actions rather than on conditions.
- Note workers' attitudes toward their jobs.
- Find out:
  Why the job is being done in a particular way
  If the workers understand the job procedures
  If they have concerns about the job
  How they would do the job differently or more safely
- Commend those who are working safely.
- Show appreciation for workers' suggestions.
- Ask for a commitment to work safely.

When the observations are completed, document findings in two ways:

- Quantitative, a count of acts or conditions observed
- Descriptive, a record discussions and general observations

The objective here is to provide information that, when combined with that from past and future surveys, will allow soundly based recommendations for reducing the potential for accidents.

Communication Techniques

Communicating to employees about reducing risks is a critical skill. Some feel it is easy to reinforce safe behavior, but others do not feel comfortable with even a simple statement like, "We appreciate your wearing your safety glasses." On either side of the communication coin, sincerity is an essential element. Everyone truly is safer if each individual works as safely as possible. Supervisors and auditors should work hard at becoming comfortable with positive reinforcement. If they believe in it, their commendations will be received as sincere.

Correcting unsafe acts without raising animosity often requires some training. Employees can be enlisted in the correction process. If they can be made to feel part of a team working to make things accident-free, the process is eased. The following steps have been shown to be effective in striving for that objective:

1. Start a conversation about the job with an employee who is seen committing an unsafe act. Find out why he/she is acting unsafely. Ask him what problems he finds with the job.
2. Discuss the job procedure with an employee working on that job. Does the employee think the procedure needs to be changed?
3. When observing a job for which no procedure exists, discuss the need for a procedure with the employee doing the job. Ask for employee's suggestions. Listen to his/her reply.
4. Get an employee to draw up a procedure for his/her own job, go over the procedure with him/her, then have a safety meeting with others who do that job and establish a procedure together.
5. Use employees in audits. This practice:
   Shows employees your point of view
   Shows you their point of view
   Helps you and your employees to reach a consensus on safety requirements
6. When you go up to correct an employee, go alone. Then bring in the employees auditing with you for a discussion of remedies.
7. When correcting an employee:
   Do not confront.
   Avoid groups.
   Stay on the issue.
   Do not let employee sidetrack you, and do not drag in other grievances yourself.
8. Congratulate employees who are working well.

My observation is that first-line supervisors, who are more responsible for day-to-day counseling, may require significant formal training to reach peak skills. This training needs to include role-playing and other supervised practice to build confidence and skill in applying what has been taught.

External Audits

Basically, there are two kinds of external audits: a company audit and a government agency audit. While some variations of these basic types may be encountered, any audit carried out by a noncompany group will likely fall into the agency category. Both types of external audit involve a certain amount of salesmanship and defensiveness. As such, most people prepare for them in the same way. Aside from those audits that are legally entitled to be perpetrated upon you without warning, it is usually most important to define the scope of the audit in advance. This can include the geographical and/or organizational boundaries and the subject matter. In many cases, there will be a working checklist available to define the nature and degree of compliance being assessed. In addition, it is important to know who is coming by source, area of interest, and, if possible, name. As a follow-up to this, it is also important to know what approach the audit group will take and what facilities and/or support services it will require.

Most plants will do a self-assessment (or internal audit) as the final step in preparation for an external audit. In so doing, it is generally advisable to assemble whatever documentation is available that will help you put your best foot forward. Just as, in taking an examination, you will not get credit for the questions you did not answer because you ran out of time, it is appropriate to make sure to get credit for all compliance that you can demonstrate even if there are some areas in which you fall short. The way to do this is to have all of the documentation available ahead of time so that the only areas that may get forgotten are those in which you have some shortcomings.

The next step is to assign a skillful and knowledgeable coordinator at a high-enough level to be able to make commitments for the organization. This coordinator can identify appropriate interviewees and arrange for their availability when necessary. It is also important to train these interviewees to report only the facts and eliminate speculation. Volunteering information outside the range of questions being asked is generally frowned upon unless the coordinator fully understands the intent of the questions and is able to assess the risks. During an in-company audit, this rule might well be waived. One result that the audited entity may wish to pursue could be the support of the audit team for certain improvements that are apparently unable to be achieved without external intervention.

When the inspectors arrive, it is important to identify them personally and by their background/function. People from the company may or may not be well known to plant personnel. All agency people who are not personally known to the plant coordinator should be identified formally, and if necessary, verified with the parent agency.

Some agency people may seek an exception to plant safety rules for one personal idiosyncracy or another. The most common of these is to request a waiver of the "No Beard" rule in plants where the use of bottled breathing air may be required to escape in the event of a toxic gas leak. Refusing such people admittance to the plant invariably causes friction. However, it is my belief that staying with plant safety rules has the benefit of demonstrating commitment and integrity, providing they are universally applied. Obviously, the extent of embarrassment would be intolerable if the auditors were to find that a truck driver with a beard was allowed into the plant because he was bringing in an essential raw material while the auditors were excluded.

I believe all audits should open with a conference between the key people likely to be encountered in the audit and the audit team. At this conference, much more can be learned about the breadth and depth of the audit as well as the approach of the auditors. In some cases, you may find that the audit was prompted by a plant incident or a complaint from an employee. In such cases, the scope of the investigation may be much more narrow than anticipated. The key objective for this opening conference is to agree upon ground rules that will meet the inspector's needs and cause no unnecessary impact on operations.

A second objective is to set the best possible attitude on both sides. This can usually best be done by a high-level manager who demonstrates an interest in the results of the process and not anxiety or defensiveness. In addition, any obstructiveness will be immediately perceived by the team and be counterproductive.

The head of the inspection team will either bring or develop a plan for the audit and a schedule for its duration. If photographs or samples will be required, they should be taken in duplicate with a set remaining to allow plant corroboration of any agency observations. A complete list of any documents given to the inspectors should be kept with clear identification if any proprietary information is removed from the site. All documents studied by the inspection team should also be listed, and proprietary information should be clearly be identified on these as well. In addition, a list of all the people interviewed is also appropriate, whether or not plant management is allowed to sit in on the interviews.

In many insections, there is a heavy emphasis on records and documentation. It is generally considered appropriate to have these ready and organized for the inspector. In some cases, it will be necessary to dig into historical records that may not be immediately available. For example, OSHA requires that employee health records should be kept for thirty years after termination.

Unless the specifics of requirements for these historical records were presented in advance of the inspection, there is no reason to feel anxious about requiring a few days to retrieve them. However, it is important that you be able to retrieve them. In many cases, the investigator will provide you ample opportunity to demonstrate how your system works and that you are able to use it effectively.

Those who interact with the investigators should seek periodic feedback on their observations. This is important to provide the plant with maximum leverage in any upcoming negotiations. A clear understanding of the regulations and standards being applied and the basis for any remedial action suggestions should be gained at the earliest possible moment. Informal discussions of possible remedial actions may also provide useful background for the closing conference. If there are some simple solutions, you may be able to remedy the situation before the inspectors leave the site and thereby demonstrate good faith.

The closing conference is often a combined technical and legal affair. In unionized plants, both the management and union executives should probably be invited. The investigation team should cover the rights of the employer in detail as well as a summary of any violations found. They should also be able to provide information relevant to the abatement of the problem and be in a position to discuss the desired timing and potentially acceptable solutions.

At this point, the advantage of previously acquired information in these areas becomes clear. The discussions at the closing conference clearly involve negotiations of schedules and tolerable cost guidelines. In addition, there may be some legal questions about the basis for any violation, and the opportunity to contest any negative observations should be clarified.

If the audit is being done by the company itself, a more complete discussion of potential remedial actions and cost constraints should be possible. In addition, contacts at other plants within the company who have dealt with similar problems should be identified to assist in defining the solution. The latter should not be ignored in agency inspections either. In many cases, the remedial action taken at other plants is a matter of public record and therefore available.

Both the company and agency audits should be documented by a complete report that incorporates any decisions made at the closing conference. Invariably, this will include any violations and penalties assessed by the government agency. In many cases, the final report will contain some differences from the assessments presented at the closing conference. This will reflect the judgment of the area director of the agency. In many cases, this judgment will take into account the subjective perceptions of the investigators at the site. Obviously, it is appropriate to end the audit on as positive a note as possible for this purpose.

Once the final report from an agency audit is in hand, there are various decisions about appeals and/or contesting of findings that will have to be made. This may also be true in a company audit since functional experts often have little regard for profit/loss objectives of a business group. In such instances, the plant manager may find him/herself with two sets of conflicting requirements: one to reduce capital spending from the business group manager and the other to spend millions on additional compliance equipment from the functional management. In any of these cases, the next step will often be informal conferences with the appropriate level of the auditing body in an effort to negotiate solutions.

Having developed an agreement and a remedial action plan, it is important that it be documented comprehensively and progress be reported periodically. It is particularly important to make sure that both the abatement schedule and the consequences of any failure to meet the agreed upon (or imposed) dates are clear. Depending on the terms of your agreement, failure to meet the specified dates in a government agency agreement could be very expensive. The price in a company audit may have a direct impact on the management's career opportunity in some settings.

There is a caveat that flows from the above desire to define requirements in detail. In some cases, the quest for more details results is more stringent remedial action because the inspectors include some insurance to cover hypothetical problems. In these cases, it may be better to agree only on the principles to be covered in the solutions. Once the plant has studied the options, they will be in a much more structured position to negotiate with the agency or functional inspector on the specifics. In any event, internal communications about what is expected of whom and when must be models of clarity and precision. Regardless of the organization, those charged with implementing remedial action foisted upon them by "outsiders" will invariably choose the path of least resistance.

The audit coordinator is thus faced with two essentials. The first is to wrangle as much flexibility as possible from the inspection team. The second is to develop as detailed and precise an action plan as possible with those inside the plant who will implement the required remedial actions. In both situations, the coordinator is doing his or her best for the plant in the practical sense.

Summary

Audits are required to achieve an outside view of any compliance program and to assist in the transfer of solutions from other places to the plant in question. However, these audits need the right viewpoint to be successful. The most helpful approach is that of a friendly consultant with no grades or career implications attached. In any event, when a plant manager authorizes an audit, he or she must recognize that the plant is committed to following through with the remedial action program.

The best way to get a helpful consultant type of audit is to do it with in-plant personnel. The experience is that objectivity is not a problem in this process since any tendency to "go easy" on the home folks is tempered by the in-depth knowledge of the auditors and their inherent desire not to be found wanting when an external audit inevitably is carried out.

The objectives of the internal audit are several. The first question to be answered is whether the plant is doing what it thinks it should be doing for optimum compliance. The second is whether what it thinks it should be doing satisfies the regulations and/or the company policies that apply. The third objective should be to draw in as many people as possible to the process of generating ideas about how to do things better and cheaper.

In general, an audit should have three dimensions and four steps to completion. The dimensions include compliance with regulations, an assessment of residual risk, and the completeness of management systems for compliance.

The four steps include a preaudit risk profile, an assessment of compliance, an evaluation of whether the remaining risks merit remedial action, and the definition of a remedial action plan.

Invariably, legal worries surface when internal audits are carried out. Lawyers continue to fear that internally identified weaknesses in compliance and needs for remedial action will be held against the plant management should an incident occur. These days, more regulations are being written with the terminology "known or should have known" with reference to a need for remedial action. As such, a higher standard of risk assessment is being imposed on plants. The good will of a self-motivated action seems to be becoming of greater value in such legalistic debates, and is increasingly recognized in court decisions on legal privileges to withhold self-assessment conclusions from adversaries.

Various types of audits are possible. These have been discussed in some detail in the chapter. A common thread in all is the participation of various levels in the organization in an effort to develop true teamwork in the day-to-day process of preventing incidents.

Not everyone can be an effective auditor. A certain amount of training in the skills of observation, in the standards being sought, and in effective interviewing and communication is important. Needless to say, auditors must possess certain minimum personality traits to be effective.

External audits require a different approach. Whether these are voluntary (within the company) or involuntary (from a government agency), a certain amount of anxiety and defensiveness will be inevitable. Thorough preparation and the dedication of people skilled in negotiation will be critical to surviving them successfully and extracting some added value from the exercise. Careful follow-up on the agreed-upon remedial action plan and schedule is essential to getting the full benefit from the audit and avoiding any penalties.

My definition of a successful audit is one in which both sides ultimately recognize that the chance for an unhappy incident has been reduced in a way that is consistent with the profit-and-loss boundary conditions that exist in the plant.

CHAPTER

12

Creating a Working System

Introduction

"Creating" is a daunting word. It implies something special, something difficult. Have no fear, most of the creative work has been done. What is left is choosing. Choosing those elements of a system that is appropriate for the risks in your plant, and choosing an approach that will be effective in your plant culture.

There are four levels of the overall system that need to be addressed: elements, requirements, systems, and actions. The most general level is the *elements* of the program. A universally desired element is safety reviews. Each element has its *requirements* (or expectations). For the safety review element, the requirements would likely include items like the following:

1. All new facilities/designs will receive a safety review.
2. A pre–start-up safety audit will be made.
3. All existing facilities will undergo an annual safety review.

Notice that nothing has been said yet about *how* these reviews will be carried out. Thus, the next level of detail is to establish a system or approach to meet the requirement. As an example of a system to meet requirement 1 above, we might see the following:

> 1. The design of any new facilities will be approved by the Safe Operations Committee (SOC). The design subcommittee of the SOC will review the design against a safety checklist and identify any deviations from it or other design standards. The

subcommittee has the authority to approve the design details, but the layout, any deviations from design standards, and any unique process characteristics will be reviewed by the full SOC. The SOC will document its approval and/or any required modifications in its minutes.

Similar systems are needed for each requirement in each element.

The last level of detail is the *action plan* needed to implement each system. For the example above, such a plan might read as follows:

> The lead process designer will consult with the chairperson of the design subcommittee during the design to identify the appropriate standards and critical design features. It is expected that this will be an interactive process. The designer will deliver the process design specification and a completed "safety in design" checklist to the chairperson at least one week before the review date. This submission will highlight any known deviations from the standards required for the project (company or legal), as well as any unique process hazards and the methods proposed to control the inherent risks associated with them. The chairperson will decide on the approach for the review. All questions will be in writing, as will the responses. The subcommittee will document its approval status and the questions for full SOC review no later than the Friday before the scheduled SOC review meeting.

Note that this action plan spells out who is responsible for what and when what is to be provided to whom. Note also that the positions of the respective parties are identified, not the incumbents.

In November 1985, the U.S. EPA initiated its Chemical Emergency Preparedness Program (CEPP), many facets of which were legislated in SARA TITLE III in 1986. The study concluded that the prevention of accidental releases requires a holistic approach that integrates technologies, procedures, and management practices. It also concludes that "state of the art" cannot be defined generically; rather, it must be a function of the local plant situation. Thus each plant must understand its own situation comprehensively and synthesize an equally comprehensive system to manage its risks.

The CMA, CCPS, and a number of companies have listed their view of the necessary elements of a risk management system. The state of New Jersey and other government agencies have issued regulations specifying the requirements for each element of acceptable risk management programs. From all of this information, it is generally easy to choose the starting elements appropriate for a given plant and plant culture. Some are required by law. Some are required to deal with the risks at the site.

In addition, this book expands on the elements of a system, and the requirements to meet each element, and discusses some specific action items for achieving them. From Chapter 2 through Chapter 11 this is the focus of the text. Choices are still required, however. This is not a cookbook with a series of fixed recipes for success. Remember also that your system must evolve with both your understanding of the hazards and public expectations.

Characteristics of a Successful System

There are a number of characteristics of a program that are essential to success. These include:

- It must be a *system*.
- It must be comprehensive.
- It must fit the plant culture.
- It must have adequate documentation.
- It must provide for continuous improvement.

Figure 12.1 is an attempt to clarify the components and recycle loops of a process risk management system in a schematic way. The top box (1) represents the system to be operated. It consists of the equipment, the contents of the equipment, and the procedures by which it will be operated and maintained.

Following the system paths in numerical order, the system comes first to hazard identification (2). Remember that a hazard is a potential to cause harm, usually through a loss of containment. Next is step 3, an evaluation of what could go wrong to cause the loss of containment. Against these possibilities, we build a hazard control matrix (4), which should include (a) eliminating the hazard, (b) preventing loss of containment (accident) through equipment and procedures, and (c) steps to mitigate the consequences of a breach, should it still occur. This loop (1–4) recycles until the results of the risk analysis/assessment loop below (5–16) yield a satisfactory result (based on the judgment of local management and/or legal requirements). This result is maintained by ensuring that no changes take place that will increase the risks (15).

Implementation

Having an effective system may not be as easy as it first sounds. Systems do not depend on individuals, but rather on organizations. Whoever is currently filling the appropriate slot in the organization will get done everything that needs to be done.

A system defines objectives, standards, and requirements. At the same time the system needs to be flexible—tolerant of different ways to accomplish a given requirement. A good system will require that all the necessary activities be defined, however.

An effective system will provide for measurement of results and a mechanism to recycle these data so that the system can be improved. A system will also include a mechanism so that the management can verify that the plant staff is actually doing what the system requires.

In addition, a system must be appropriate for its function. An effective system will reflect the experience and capabilities of those who must use it day to day. People at all levels must be involved in system design and continuous evaluation. Providing the opportunity for broad participation in these functions not only

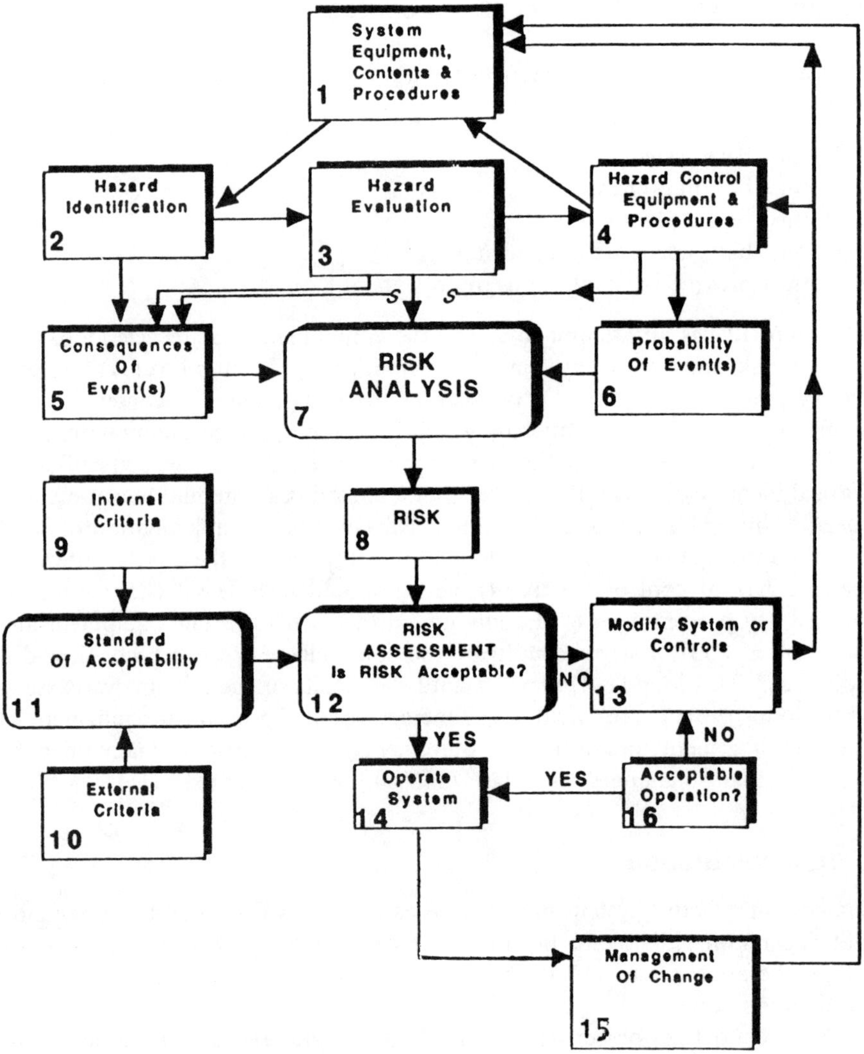

Figure 12.1. Schematic of a process risk management system.

results in more effective systems, it results in more ownership by the users. Ownership means that implementing the system will become part of the day-to-day work process, which is the only road to success.

Perhaps an example of system failure will illustrate the importance of a systematic approach. In a second TCPA inspection by the NJDEP a plant was fined $5,000 for what appeared to be an obvious typographical error in record keeping. The plant protested this as unfair: a typist had simply made a mistake. The agency replied that the fine was not for the error, but rather for the failure of the system to detect and correct it. In short, if the system were functioning, someone would have read the report, questioned the obvious error, and tracked down the truth. This did not happen, hence the fine.

A comprehensive system does not deal with all of the risks in the world. Rather it provides for understanding all of the hazards present at the plant. It then ensures that the risks resulting from those hazards are minimized through design, preventive maintenance, inspection, training, and testing. After everything feasible has been done to prevent an incident, a comprehensive system will provide for mitigation of credible incidents and appropriate emergency response. Education of the community and cooperative efforts with its emergency planning/response groups are part of the preparedness aspect.

A comprehensive system does not ignore (or pay lip service) to credible accidents because it would be too expensive to deal with them. In such circumstances, it is often the best approach to eliminate the hazard altogether. One of the things of which the NJDEP is most proud is that in response to its TCPA regulations many companies eliminated hazardous substances from their processing rather than develop the required risk management plan.

A comprehensive system does not need to deal with unrealistic scenarios (unless required to do so by law). If a plant is miles from an airport, a comprehensive plan does not have to cover a large plane crashing on its chlorine tank car. Similarly, if hurricanes or earthquakes are unknown in the region, they, too, can be eliminated from consideration. Obviously, the converse is also true.

Fitting the system to the plant culture is a continuous process because the culture must evolve to keep pace with the expectations and requirements of society. I believe incorporating current successful practices into the system as a foundation is the best place to start. Starting with a completely new approach is rarely the best thing to do.

Each plant has some things it does better than others in the area of risk management. Often there is significant pride in the work force around these functions. Perhaps the drafting group does a good job in keeping P&IDs up to date. Recognize this effort and expand it to include electrical one-line diagrams and the like. Perhaps the safety-valve testing system is excellent. Recognize it, and expand it to raise the level of other critical instrument maintenance to excellence. Perhaps your operations supervisors are proud of the simulated emergency scenarios they create to test their people. Recognize this and encourage them to document each exercise so that others can profit (and regulatory requirements can be met).

My belief is that using the people currently responsible for each system element is the best way to proceed in the long run. In short, let the inspection group leader be responsible for the inspection requirements. There is a temptation to set up a task force to deal with major requirements, be they a project or a risk management system. This involves a great deal of start-up learning and may lead to a "not invented here" attitude on the part of the line organization that is left to implement the new system year after year. Sometimes this can be diffused by including members of the task group in the line group. Ongoing ownership by the people who implement the system daily is clearly critical, whichever organization approach is chosen.

Often perceived major new responsibilities (even if only for documentation) cannot be added without providing help for the line leaders. In these days of the "lean and mean" organization no one has any spare time. Either certain duties must be eliminated or a way must be found to delegate them to someone else if time is to be provided for meeting risk management requirements. If a "catch-up" documentation effort is needed, provision of a temporary clerical/secretarial resource may be an answer. A comprehensive risk management system needs to become an ongoing part of a plant's modus operandi. It needs to be staffed accordingly, providing time for the various leaders of the system's elements to do the job correctly day after day.

Documentation

Documentation is often a problem area for many program elements. It is commonplace to blame the government agencies for much of this, but the general industry move toward "quality" also mandates more documentation. The precepts on which "quality" is built are measurement of performance and continuous improvement—that is, a report card. In many cases a report card is all that is called for by government regulations.

Another precept of "quality" is to eliminate waste—chores that do not add value. In the area of documentation, this often includes duplicate records, multiple formats, and the use of paper when computer records would be more efficient. I know of one plant where all of the operating manuals are maintained in computer files. With proper authorization new procedures are available in all parts of the plant immediately. In many plants, MSDS sheets are available electronically, eliminating the need to update numerous paper copies every time a revised sheet is issued.

The needs for documentation are many. Some stem from the need to have up-to-date plant data and procedures. While a plant manager may correctly observe that the operators know the plant and its procedures from years of experience in *normal* operations, he/she is often at a loss to describe how the necessary information is obtained in *abnormal* situations. These may not be emergencies, merely infrequently performed operations such as producing a special product or loading a special container. There is no way to face an inquiry with impunity if the information necessary to operate the plant was not up to date at the time of an incident.

Another documentation need is to understand the maintenance history of the equipment. One of the fundamental requirements of risk management (Chapter 1) is containment. Equipment and containment failures need to be prevented. To do this, one must track failure and corrosion rates and have an inspection program to verify the suitability of equipment for service. Further, a plant must show that its maintenance information is *used* to set inspection and replacement intervals. As a result, *the analysis* of inspection and failure data must also be included in the documentation so that the basis for maintenance activities is clear.

Mechanical procedures are often lacking in plants. These are sometimes covered by general training given the mechanical forces before they are permitted to work on real plant equipment. Documentation of this training (content and recipients) cannot be overlooked. People do change crafts in a plant. Operators become mechanics and vice versa. Appropriate training records are important to ensure that all tasks are assigned to qualified people.

Two other aspects of mechanical procedure documentation also bear mention. The first is special procedures—those that happen infrequently, are not done by a specialist, and/or may involve unusual hazards. For example, a once-a-year clean-out of a sewer sump may require a special write-up and retraining session.

The second special consideration applies when contractor personnel will do a job, hazardous or not. After the Phillips polyethylene plant incident in Texas, the training of contractors received much attention. The Gray Institute report by Wells et al. "Managing Workplace Safety and Health: The Case of Contract Labor in the U.S. Petro Chemical Industry" Beaumont, TX, July 1991) concluded (amid much controversy) that contractor personnel did not receive as much safety-related training as direct-hire people, that they were often asked to perform the more hazardous tasks, and that they were perhaps less able to assimilate what training they did get. Various legal complexities are involved in the United States, but clearly a renewed emphasis on training of contract personnel, and on the documentation of what training has been given and on any qualification tests is in order.

Operator training and qualifications testing are another essential, as mentioned earlier. Added to this are records of hazard communications and other safety training, records of which operators participated in which simulated emergencies, OSHA emergency-preparedness training, HAZMAT and fire training, cardiopulmonary resuscitation (CPR) and other first-aid training, and so on. Required operating licenses, exams, and the proper display of earned (and current) certificates is another documentation issue. In short, if you did it, write it down and save it. I know of no other way to face inquisitors.

An area often forgotten or excessively decentralized is the documentation of engineering and equipment standards. It is easy to say that all vessels are designed to ASME VIII, but can the calculations, or even a copy of the code, be produced quickly for an inspector? Piping design standards, industry practices, and so on can also be a problem—even when they are used all the time. Providing some sort of central file system for standards and compliance documents is generally necessary *and useful.* Be aware, however, that these documents will soon disappear again unless the file room is supervised.

An issue that is currently receiving much attention is internal audits and management's response to them. The question is whether the recommendations of such an audit represent a legal requirement to implement them. At the very least such recommendations should be written with advice from local legal experts. In addition, management must document its response to each recommendation and the reasons for it. Under these circumstances the plant manager will, at least, be able to substantiate why something was done or not done should this be required.

A Specific System Example

All of the published systems intend to cover the same general (and well-known) principles of responsible risk management. Different boundary conditions are imposed by the enabling legislation (such as the list of covered substances) and/or the viewpoint of the institution attempting to define the requirements. Fundamentally, each plant (or plant unit) must define, and carry out, the actions necessary to prevent incidents within its borders.

At the simplest end of the scale are the eight elements in the NJDEP regulations for a risk management plan under NJ TCPA. They are:

1. Safety reviews for new and existing facilities
2. Documented operating procedures
3. Preventive maintenance program
4. Documented operator training
5. Required Accident investigations
6. Risk assessments
7. Emergency response planning
8. Internal or external audits of compliance

Of course, there is a book of regulations thicker than this one specifying the acceptable approaches to implementing these principles. The NJDEP requires each plant subject to the law to write a one-page summary of how it plans to comply with each element. We can characterize this as the *system* level. The action plans have yet to be defined, In an attempt to elucidate how one organizes ongoing plant programs into a compliance system, a summary of one plant's system to satisfy the requirements of these eight elements, which was found acceptable in the first inspection, follows. *Note the emphasis on qualifying existing in-house programs and approaches as covering the requirements of the regulations.*

1. Safety Reviews of Design for New and Existing EHS Facilities
   - General

     The Plant has specific written procedures and design criteria for new and existing extraordinarily hazardous substances (EHS) facilities. These procedures/criteria are given in a set of manuals for Process Design and in the Engineering Standards for detailed or mechanical design. In addition to these written standards, technical and operations personnel consult with experts within the corporation and refer to other well-known industry information sources as required. These information sources include, but are not limited to, API, ASME, ANSI, the Chlorine Institute and NFPA. These reviews include comparisons with state-of-the-art and other criteria for design and operation, but are not necessarily in the same format as specified in the regulation.

- New Facilities

  Safety reviews of proposed new facilities are carried out by the Safe Operations Committee (SOC) and its Design Sub-Committee. A "safety in design checklist" is used. The Design Sub-Committee reviews the process design and P&IDs in detail against applicable codes and design practices. The whole SOC evaluates potential hazards, how the design approaches the state-of-the-art and any exceptions to the appropriate standards which are requested. Approval by the SOC is required before any project can be progressed. Small improvements and temporary changes are documented by a Field Change Authorization and reviewed by the SOC.

- New Products & Procedures

  Before any new product is produced or any new material is introduced into existing facilities, procedures and conditions are reviewed by the plant Safe Operations Committee (SOC). This is typically a "what if" analysis covering the details of the new operation.

- Existing Equipment

  Safety reviews of existing equipment are initiated by the Operations Department at frequencies consistent with the requirements of TCPA. These reviews follow the requirements set forth by the Company safety management practices, safety review and audit plan which include TCPA requirements. For facilities handling TCPA substances, safe operations inspections are conducted at least once per year. These safety reviews involve both mechanical and operations people and serve as an inspection, training exercise and an awareness promoter. HAZOP reviews for manufacturing and storage facilities are conducted at 4-year intervals.

- SOC Role and Organization

  The SOC reports directly to the plant manager and is the final review on "safety" matters relating to facilities and operations. However, "safety," in the current environment, encompasses a broad range of issues. Not only is it necessary to provide for the well-being of the plant workers and equipment, but also to protect the community from any harm resulting from an in-plant incident. The role and objectives of the SOC recognize this broader involvement. The chairman of the SOC is the Manager of the Safety, Environment and Risk Management Department, which currently also encompasses the inspection and materials function. Membership consists of representation from senior process design and technical personnel, industrial hygiene, maintenance, plant operations, inspection and the plant safety coordinator.

  The general objectives of the SOC are as follows:

  - *Help anticipate* hazards and work with operations to avoid them.
  - *Promote* improvements in "safety" matters throughout the plant.
  - *Help develop and maintain* a realistic and credible plant-wide "safety" policy which reflects this broader definition.
  - *Monitor and provide continuity* for "safety" standards in facilities and procedures, and monitor progress in plant programs aimed at a more secure workplace.

2. Standard Operating Procedures

   Standard Operating Procedures are of two types: plant-wide and unit specific. Both are available in each control house and are in plain English; all operations staff are fluent in English. The plant-wide manuals consist of general procedures regarding safe work practices, emergency response, fire protection, etc. A typical example is the Accident Prevention Manual which gives procedures for work permits, traffic, protective equipment requirements, electrical lock-out procedures, tools, railroad, reporting emergencies, siren signals, housekeeping, etc. These procedures apply in all parts of the plant.

Unit specific procedures apply to a single processing or storage area. They include the Unit Operating Manual, a compilation of MSDS sheets for the unit, manning schedules and computer control procedures where appropriate. These procedures are updated as changes occur. An overall update is scheduled every three years, assuming major changes to the unit have not prompted an earlier review.

The contents of these manuals include at least the following information:

(i) Simplified process flow sheet
(ii) Process description
(iii) Procedures and conditions for normal operations
(iv) Procedures and conditions for abnormal conditions
(v) Multiple emergency conditions and procedures
(vi) Pre-start procedures, including leak tightness
(vii) Start-up procedures
(viii) Shutdown procedures, normal and emergency *and* details of condition of equipment to be maintained after S/D
(ix) Type, location, purpose of safety relief devices, alarms, interlocks
(x) Sampling procedures
(xi) Safety procedures
(xii) Procedures for maintenance preparation *and* inspection after maintenance
(xiii) Plot plans covering safety and emergency response facilities, electrical classification and utility supplies

No EHS facility is operated without the established operating posts being filled by a qualified person.

Many operations have various degrees of computer control, data logging and information retrieval. The extent of computerization affects the format in which some of this above information is available in the various control houses.

3. Preventive Maintenance and Inspection Program

The Preventive Maintenance (PM) and Inspection Program is centered around a computer database and commercial software. This system was installed several years ago and the transition from a system of paper inspection and maintenance records to the computer system has recently been completed. The computer contains technical data on piping, fixed equipment, machinery, pressure relief devices, and instrumentation, and specifies the preventive maintenance frequencies required for each individual piece of equipment. The historical records for some equipment remain in the paper-based file, as the new computer database is being built.

Fixed equipment inspection intervals for all equipment are developed on a systematic basis which includes historical data and current non-destructive inspection results. A quarterly inspection alert report is generated showing the equipment that is scheduled for inspection during the current year and has not been completed. This report requires management sign-off.

Safety valves, rupture disks, and pressure/vacuum vents are handled similarly in the system. The intervals between inspection of these devices are set on the basis of criticality of service and analysis of the historical data. Monthly reports are generated to track progress. Quarterly PM alert reports indicate overdue equipment and require management sign-off.

Preventive maintenance on instrumentation systems designated as critical for either safety or environmental purposes are also given separate attention. The list of safety and environmentally critical instrumentation is periodically reviewed to ensure effective coverage.

The fixed equipment inspection techniques in use are state-of-the-art. New applications of technology, such as acoustic emission testing, are developed in concert with experts. A network of metallurgical and inspection experts exists which encompasses all of the corporate sites in the United States and abroad. This network exchanges information on effective technology applications. Inspectors attend outside training courses to facilitate the implementation of new technology, and the plant holds an ASME Certificate of Authorization for use of the Code Symbol Stamps "A," "PP," and "U" from the National Board of Boiler and Pressure Vessel Inspection.

The inspection program ensures that all of the requirements necessary for approvals by the various code agencies and the National Board are maintained.

4. Operator Training

The Operator Training Program at the plant is in the process of being updated. The format chosen does not exactly follow the format outlined in the TCPA questionnaire, but the training manuals currently being developed will cover all of the essential elements required.

In the meantime, training of new EHS operators is being carried out with a combination of existing and new materials and in supervised on-the-job assignments. Informal updates are carried out weekly at safety meetings at which discussions of incidents and simulated emergencies are used to maintain awareness and prepare for the handling of emergencies. Qualified operators receive refresher training periodically and are requalified via documented oral and written reviews. Training documentation is maintained by a central coordinator in the Operations Department to ensure that the necessary qualification/requalification activities are completed and documented.

Requalification procedures also include training in general safety techniques (e.g., Scott Air Pak use, firefighting, etc.), DOT requirements, emergency response procedures and other areas not uniquely applicable to EHS handling, but generally useful in preventing and/or mitigating any emergency.

5. Accident/Incident Investigations

Because an accident is a consequence of a breakdown in the control systems, investigation of accidents will help prevent recurrence. It is the policy at the plant to investigate all Safety and Environmental incidents. The magnitude of the investigation will depend on the seriousness, or potential seriousness, of the incident.

The Operations Department is notified of all incidents and initiates the appropriate investigation. By reporting those accidents which do not result in injury or loss, we have an opportunity to learn the causes and communicate them to others, thus preventing recurrence of this and similar incidents and a possibly more serious outcome. A trained pool of accident investigators is available to assist in the investigation.

All accidents are investigated immediately. This is important in order to conduct interviews with people who were involved while the incident is still fresh in their memory. The longer the delay in examining the scene of the incident, interviewing the injured and witnesses, the greater the possibility of obtaining erroneous or incomplete information. Prompt accident investigation also expresses concern for the safety and well-being of employees.

Accident investigations focus on the possibility of more than one cause. Often, the most obvious cause contains other underlying causes which contributed to the incident.

The general format of the investigation report includes:

- *SUMMARY*—A short description of the job activity and the resulting incident.
- *NARRATIVE*—A description of the detailed factual sequence of events.
- *OBSERVATION*—Key facts found in the investigation.
- *CONCLUSION*—Based on the facts found during the investigation.
- *RECOMMENDATIONS*—Made to address each cause or deficiency found in the investigation. Major recommendations are highlighted for future follow-up.

Communication of the accident investigation is given to all affected personnel with additional distribution to other plants which have similar concerns. Follow-up of the recommendations continues until completion with periodic feedback on their status given to the organization. Safety meetings are frequently the forum in which incidents and their causes are discussed.

All incident reports, minor or major, are reviewed by the Safe Operations Committee (SOC).

6. Risk Assessments for Specific Pieces of EHS Equipment or Operation Alternatives

A hazard analysis, utilizing the failure mode and effect analysis (FMEA) technique, has been completed for each of the EHS registered chemicals at the plant. These hazard analyses were performed in conformance with the requirements of the TCPA program. Hazard reviews (HAZOPS) are conducted on a four-year cycle for EHS units.

Subsequent to these hazard analyses, the identified potential release scenarios were reviewed against TCPA criteria and none were found to require quantitative risk assessment. A company safety specialist based external to the plant was employed to make this analysis.

As part of the plant's ongoing process of managing change, any new facilities which have the potential to release an EHS to the atmosphere are required to provide air dispersion calculations as a part of the process design specification for review by the SOC. In this way, the TCPA rule that any modifications or new facilities with the potential to release five times the registration quantity for any EHS should be identified. To facilitate this, a checklist and documentation form entitled "Analysis of Proposed Change as Required by the TCPA" is utilized to evaluate proposed changes to the equipment, the process or the operating procedures of equipment handling EHS.

7. Emergency Response Planning

Overall emergency response at the plant is a combined Refinery/Chemical Plant process. The site maintains extensive firefighting equipment, a command post vehicle, two ambulances, numerous self-contained breathing systems and back-up air supplies, acid suits, rescue equipment, etc. A well-equipped chemical plant team is trained to respond to transportation emergencies in the community as part of the CMA Chemnet/Chemtrec Programs. The site Emergency Manual is very comprehensive. It includes very specific external notification instructions.

Members of both staffs are trained in various phases of emergency response procedures and take part in competitions to test their skills against the best in the country. A separate staff of trained emergency medical technicians are available 24 hours/day and the Medical Department is staffed about 50 hours/week. Arrangements are in force with Alexian Brothers Hospital to provide medical services at other hours.

Periodic meetings are held with the Municipal Fire Department to coordinate procedures and educate officers about the site. The Commander who arrives at the site for any emergency will be escorted to the scene to communicate with the Plant Commander, give advice and render an independent judgment of the situation.

The Chemical Plant Shift Foreman is in charge of any plant emergency until Operations management arrives. He has been trained to identify incidents which require agency notifications and has the authority to call out the Chem Plant Emergency Response Organization (ERO). Assistance is available on-site from the Refinery Shift Superintendent and Refinery emergency resources, if needed.

The ERO consists of the plant management and senior staff people in Risk Management, Environmental, Technical, Medical, Industrial Hygiene, and Personnel Safety. A manual is available to guide this group to the proper agency contacts. It supplements the material in the site-wide Emergency Manual with additional information specific to chemical accidents. The plant is a member of a mutual aid group through which it can obtain backup material in a major incident.

An on-site meteorological station is available to provide data to help evaluate the consequences of any release of an EHS.

Special communication systems and procedures are in place and are periodically updated to ensure that accurate information is available to both decision-makers and the public.

Emergency response teams are exercised often. About twice-per-year drills, some involving outside agencies, are carried out to test coordination with public response groups. In addition, the plant participates in local emergency response planning activities.

8. Internal or External Risk Management Program Audit Procedures

In-plant compliance reviews are conducted on an annual basis and on a three- to four-year cycle by a senior team formed from personnel external to the plant. Typical membership of the team would include the manager(s) of another site(s), a senior safety engineer and a technology representative.

In addition, separate environmental compliance audits are carried out on the same frequency. These are done by corporate environmental staff and environmental coordinators from other plants. Again, the results are reported to the Plant Manager.

Periodic audits of fire protection and other major safety systems are done by external safety specialist consultants approximately every three to four years.

Internal safety audits are carried out on each unit on a very frequent basis: at least weekly. These are done by a cross-section of site management, unit supervision and the safety coordinator. Primary emphasis is on procedures and personnel. Results are documented and discussed at weekly safety meetings.

The SOC also reviews all safety and environmental incident reports (see accident investigation section). The Near-Miss reporting system summarizes information about incidents and near-incidents. The SOC reviews these incidents, with all atmospheric releases and spills reported given particular attention. Trends are analyzed and the SOC's recommendations for improvements are brought directly to the site management.

Mock regulatory compliance reviews are conducted on a scheduled basis for each unit approximately once per quarter. Findings are reviewed with each unit team leader for information and follow-up. Focus is on both safety and environmental compliance.

Taken altogether, these specific audits, along with other programs of stewardship, fulfill the basic requirements of the TCPA but not in the same format. Work is underway to incorporate the specific TCPA requirements into the overall audit procedures for the Risk Management Program.

Some Feedback

Both informal and formal feedback on TCPA compliance issues is becoming available. From the industrial side, even though common teams participate, the approach of the NJDEP audit team leader is perceived as a significant variable. Some leaders are more flexible in judging compliance; others quote chapter and verse from the regulations and are less tolerant of alternate approaches to achieve the same requirement. All team leaders fulfilled an advisory role in the first audit, but are assessing fines in the second year.

To begin the second year NJDEP began a series of more formal feedback sessions to help the regulated community understand its practical requirements. The second annual inspection of sites covered under NJ TCPA began in the late-1991, 1992 time frame. At the end of 1991, eight industrial sites and three sewage/water treatment plants had been reinspected and found deficient in one way or another. Of twenty-five fines levied, 17 were for inadequacies in training and procedures. An additional five fines were assessed for weaknesses in emergency drills and critique documentation. One plant was fined for twelve separate deficiencies, while most had only one violation. If the one weak program is ignored, inadequate write-up and critique of emergency response drills becomes the largest single category of deficiency. Philosophically, documentation of any requirement is likely to be the most sensitive area. From these very preliminary data one might conclude that compliance with NJ TCPA is not particularly troublesome for those of goodwill. As will be demonstrated in the following pages, the similar regulations passed by other jurisdictions should be a little easier to meet, even though the list of chemicals covered may be longer.

Other U.S. Regulatory Actions

Other laws and the OSHA process safety regulations take relatively similar approaches to achieving the same results. So far, in addition to New Jersey, the states of California, Texas, and Delaware have passed some form of regulations related to risk management in plants handling certain hazardous substances. The number and type of materials covered by each state vary and are listed below.

State	Toxics	Explosions
California	366	—
New Jersey	109	—
Texas	46	—
Delaware	89	46

These lists are dynamic—subject to update periodically. At the federal level, the OSHA process safety regulation covers 137 substances above various threshold amounts plus light hydrocarbons above a 10,000-lb threshold (if not used as a fuel on the site).

A. E. Burk of DuPont (Wilmington, Delaware) presented a comparison of the direction of actual and proposed legislation in June of 1990. Aside from the number of chemicals covered, he listed four issues which varied slightly from law to law. These were:

- Performance based vs. specification based
- The role of quantitative risk assessment
- How much information must be submitted to the state agency
- Approval process

Three states (New Jersey, Delaware, and California) require the performance-based approach of defining a risk management program and auditing for compliance with it. The requirements of the three states are comparable, except that New Jersey requires that state-of-the-art (SOA) technology be used. The definition of SOA caused some discomfort until it became clear that what was meant was commercially available technology that provided an effective cost/benefit ratio.

Texas regulations require owners to carry out consequence analysis on worst case scenarios. If the worst case has the potential to cause a fatality outside the plant, the owner must develop risk reduction actions and have these approved by the state with a permit.

The use of quantitative risk assessment (QRA) is related to each state's requirements for hazard analysis. The group of three states mentioned above require hazard analysis periodically for existing facilities (three to five years) and for all new facilities. Texas requires risk reduction reviews for new and modified facilities with potential for a fatality outside the plant. All states require some form of consequence analysis. Texas and Delaware do not require quantitative estimates of the probability of an incident. California mentions it in agency guidance documents but not in the law. New Jersey currently (1992) requires more quantitative risk assessment only when there is the potential for a release of five times the registration quantity of a given substance in one hour. If this situation can be avoided, QRA is not required. At this point, we may conclude that quantitative risk assessment is an option. The OSHA process safety regulation preserves this optional status.

The issue of how much information must *be submitted* to the agency (as opposed to being available for inspection at the site) is a matter of concern regarding confidential data. The requirements vary from state to state with California requiring most and Delaware the least.

The paperwork approach for approval varies from state to state as well. All states, except Texas, audit for compliance. Texas uses a permit approach for the approval of new or modified facilities. New Jersey memorializes the results of its audit of both the plan and the plant in a consent agreement, which requires some additional legal review.

At the federal level, the OSHA principles closely parallel those of the state laws. Specific rules are more flexible than those of New Jersey and many thresholds are higher. Explosion hazards are included in the list of covered substances. Some coordination with EPA's assessment of the impact of plant incidents on the community and environment under the Clean Air Act Amendments is to be hoped for, but some laws introduced in Congress may cause conflicts (HR 3030 and 2585 and SR 1630). As the regulations implementing the Clean Air Act amendments are developed, the potential for conflict needs to be carefully followed because the EPA is required to issue regulations governing risk management plans.

At present, the fundamental requirements of the OSHA rules, which are aimed primarily at protecting workers, are as follows:

- Employer must develop and maintain in formation pertaining to technology of the process (block flow diagram, chemistry, inventory).
- Employer must compile information concerning equipment used in process.
- Employer must document that existing equipment operated under old codes is safe.
- Employer must perform process hazards analysis (PHA)—OSHA's *cornerstone* of hazards management, using one or more of *6* listed methodologies. The PHA must include the potential for human error, and participation by all employee groups
- Employer must establish a system to address PHA team's findings.
- Employer must assess potential impact of catastrophic release on surroundings.
- Operating procedures must be clear, consistent, and communicated—employer must develop and implement written operations procedures.
- Employer must certify that employees have received and successfully completed required training.
- Employer must ensure that contract employees are trained.
- Employer must perform pre–start-up review (for new and modified facilities for which modification required change in process safety information).
- Employer must establish written maintenance procedures.
- Employer must issue permits for hot-work operations performed in, or near, processes.

- Employer must establish written procedures to manage any changes—process chemicals, technology.
- Employer must investigate every incident resulting, or that *which reasonably could have* resulted, in a major accident. (Employer must prepare report.)
- Employer must implement emergency action plan in accordance with existing regulations.
- Employer must evaluate compliance with the regulation via safety audits.

The Chemical Process Safety Report, Vol. 2, No. 4, April 1992, (Thompson Publishing Group, Washington, D.C.) discusses the requirements of the final OSHA rule and critical changes from earlier versions concisely. The rule becomes effective May 26, 1992. At this point it is not clear what impact it will have on NJ TCPA and similar regulations in states not certified by OSHA.

There are some who feel that the tone of the OSHA regulations is more respectful and indicates more faith in management's capacity and willingness to work the problem. In a paper given at the summer meeting of the AIChE in 1988, Alan Edwards, one of the architects of the NJ TCPA program, acknowledges that "the philosophy embedded in the TCPA reflects a belief that regulatory oversight is *necessary* to encourage facility owners to make safety-related expenditures when necessary." In short, New Jersey has taken an aggressive posture deliberately in an effort to set the standard for process safety legislation. Yet, living with TCPA has not been intolerable to date. As a result, I spent a fair amount of space in this book discussing TCPA requirements on the basis of its worst case status.

A Technical Approach

The Center for Chemical Process Safety (CCPS) of the AIChE has developed some guidelines for the management of chemical process safety. These were published in a 1989 monograph entitled *Guidelines for Technical Management for Chemical Process Safety,* one of a series of guidelines publications from CCPS. The twelve elements discussed in this approach are as follows:

- Accountability: objectives and goals
- Process knowledge and documentation
- Capital project review and design procedures (for new or existing plants, expansions and acquisitions)
- Process risk management
- Management of change
- Process and equipment integrity
- Incident investigation
- Training and performance
- Human factors
- Standards, codes and laws
- Audits and corrective actions
- Enhancement of process safety knowledge

Many of these are the same as listed in NJ TCPA, and, as we shall see, in the CMA program of Responsible Care and specific company programs. Highlighting "human factors" in the list is unique. However, the coverage is scant. The topics covered briefly include:

- The human/machine interface
- Administrative controls vs. hardware
- Human error assessment

Further guidelines are planned for the human factors area and many consultants and research groups are active in trying to apply learning from the nuclear and aviation industries and various biological studies to the process industries. Since many of the incidents now being analyzed show that lack of procedures or failure to follow procedures is a major cause of incidents, we can expect considerable interest in the human aspects of incident prevention in the future. A number of approaches to improving the management systems that impact human performance are discussed in Chapter 9.

A second highlight of the CCPS list that is not explicitly listed elsewhere is the last: enhancement of process safety knowledge. This is a listing of information sources and a description of some ongoing research, in which CCPS has a leading role. As indicated in Chapter 4, much more sophisticated understanding of chemical reactivities is possible these days. When coupled with more standard engineering analysis, a better definition of possible risks and preventive measures is not only possible, but likely, to be required.

The CCPS series is a good starting point for developing a technical basis for meeting program requirements when in-house experience needs strengthening.

CMA's Responsible Care System—an Example of a Voluntary Approach

The CMA system for upgrading the incident prevention performance of the industry involves three elements: guiding principles, codes of management practice (requirements), and self-evaluation. The first two components were developed with the input of a public advisory panel. In general, the CMA approach has a heavier emphasis on public involvement and communication than other systems. In addition, it has an explicit code of practice dealing with the shipping of products.

As of March 1992 five codes of management practice have been approved. These include:

- Community awareness and emergency response
- Process safety
- Pollution prevention
- Distribution
- Employee health and Safety

The product stewardship code will soon be issued. This code is aimed at defining the remaining hazards of the product and at informing the company's customers of these hazards and of safety methods of using the product. This code has been the subject of much legal debate, but is expected to be issued by mid-1992.

The employee health and safety code parallels much of the process safety code but focuses on the human rather than the equipment side. The code requires hazard identification and evaluation; appropriate training for workers; and worker participation in setting health and safety standards, in emergency preparedness, as well as in the more conventional requirements for personal protective equipment, fire training, proper maintenance, and so on. The clear intent is for all of these CMA codes to become a voluntary standard for the industry, perhaps in hopes of forestalling onerous legislation.

To follow the breakdown given earlier, the elements of the Responsible Care codes, as stated in the purpose section, include:

- Community and employee awareness
- Preparedness for plant emergencies
- Incident prevention in plant
- Incident prevention and emergency preparedness in product distribution
- Waste and release reduction
- Personnel injury prevention

The codes spell out specific requirements for compliance, but not the methods by which they must be accomplished. The requirements[1] of the first four approved codes are listed in Tables 12.1 through 12.4 primarily to show the level of detail given in the requirements. Clearly these requirements must still be converted to systems and actions at each plant to achieve the stated purpose.

Table 12.1 Responsible Care® Community Awareness and Emergency Response Code of Management Practices

Purpose

The goal of the Community Awareness and Emergency Response (CAER) Code of Management Practices is to assure emergency preparedness and to foster community right-to-know. It demands a commitment to openness and community dialogue. The code has two major elements: first, to assure that member facilities that manufacture, process, use, distribute, or store hazardous

[1] This material is used with permission of the Chemical Manufacturers Association (CMA). Responsible Care® is a health, safety, and environmental performance improvement initiative of the chemical industry and is a registered service mark of CMA.

materials initiate and maintain a community outreach program to openly communicate relevant, useful information responsive to the public's questions and concerns about safety, health, and the environment; and second, to help protect employees and communities by assuring that each facility has an emergency response program to respond rapidly and effectively to emergencies.

The community outreach component will communicate program activities and performance under all codes of management practices and will promote an open, ongoing dialogue with employees and the community. Information should be provided about such activities as waste minimization, emission reduction, health effects of chemicals, and efforts to ensure the safe transport of chemicals.

The CAER Code of Management Practices is supported by, and will build on, CMA's CAER process. CAER supports the community's right to know about chemical industry operations and their effect on safety, health, and the environment. CAER was originally a voluntary initiative focused on emergency response issues. The new CAER Code of Management Practices broadens the facility-community dialogue to cover the full range of safety, health, and environmental issues.

Management Practices

Member companies that manufacture, process, use, distribute or store hazardous materials shall have a communicated outreach program that includes:

A. Community Awareness and Outreach

For Employees:

1. An ongoing assessment of employee questions and concerns about the facility.
2. Communications training for key facility and company personnel who communicate with employees and the public concerning safety, health, and environmental issues.
3. Education of employees about the facility's emergency response plan and safety, health, and environmental programs.
4. An ongoing dialogue with employees to respond to their questions and concerns and involve them in community outreach efforts.
5. A regular evaluation of the effectiveness of the ongoing employee communications efforts.

For Community:

6. An ongoing assessment of community questions and concerns about the facility.
7. An outreach program to educate responders, government officials, the media, other businesses, and the community about the facility's emergency response program and risks to the community associated with the facility.
8. A continuing dialogue with local citizens to respond to questions and concerns about safety, health, and the environment and to address other issues of interest to the community.
9. A policy of openness that provides convenient ways for interested persons to become familiar with the facility, its operations, and products and its efforts to protect safety, health, and the environment.
10. A regular evaluation of the effectiveness of the ongoing community communications efforts.

B. Emergency Response and Preparedness

1. An ongoing assessment of potential risks to employees and local communities resulting from accidents or other emergencies.
2. A current, written facility emergency response plan which addresses, among other things, communications, and the recovery needs of the community after an emergency.
3. An ongoing training program for those employees who have response or communications responsibilities in the event of an emergency.
4. Emergency exercises, at least annually, to test operability of the written emergency response plan.
5. Communication of relevant and useful emergency response planning information to the Local Emergency Planning Committee.
6. Facility tours for emergency responders to promote emergency preparedness and to provide current knowledge of facility operations.
7. Coordination of the written facility emergency response plan with the comprehensive community emergency response plan and other facilities. If no plan exists, the facility should initiate community efforts to create a plan.
8. Participation in the community emergency response planning process to develop and periodically test the comprehensive community emergency response plan developed by the Local Emergency Planning Committee.
9. Sharing of information and experience relating to emergency response planning, exercises, and the handling of incidents with other facilities in the community.

Table 12.2 Responsible Care® Process Safety Code of Management Practices

Purpose

The Process Safety Code is designed to prevent fires, explosions, and accidental chemical releases. The Code is comprised of a series of management practices that reflect this goal, with the expectation of continuous performance improvement for each management practice. The practices are based on the principle that facilities will be safe if they are designed according to sound engineering practices, built, operated, and maintained properly, and periodically reviewed for conformance.

Process safety is an interdisciplinary effort. Consequently, the Code is divided into the following four elements: management leadership, technology, facilities, and personnel. Each element is composed of Management Practices. Individually, each Practice describes an activity or approach important to preventing fires, explosions, and accidental chemical releases. Collectively, the Practices encompass process safety from the design stage through operation, maintenance, and training. The scope of this Code includes manufacturing, processing, handling, and on-site storage of chemicals. This Code must be implemented with full recognition of the community's interest, expectations, and participation in achieving safe operations.

The process safety management program in each facility is complemented by workplace health and safety programs, as well as waste and release reduction programs which address and minimize releases and waste generation. These three programs, and others, will help assure that CMA member facilities are operated in a manner that protects the environment and the health and safety of personnel and the public.

Management Practices

Management Leadership

Each member company shall have an ongoing process safety program that includes:

1. Leadership by senior management through policy, participation, communications, and resource commitments in achieving continuous improvement of performance.
2. Clear accountability for performance against specific goals for continuous improvement.
3. Measurement of performance, audits for compliance, and implementation of corrective actions.
4. Investigation, reporting, appropriate corrective action, and follow-up of

each incident that results or could have resulted in a fire, explosion, or accidental chemical release.

5. Sharing of relevant safety knowledge and lessons learned from such incidents with industry, government, and the community.
6. Use of the Community Awareness and Emergency Response (CAER) process to assure public comments and concerns are considered in design and implementation of the facility's process safety systems.

Technology

7. Current, complete documentation of process design and operating parameters and procedures.
8. Current, complete documentation of information relating to the hazards of materials and process technology.
9. Periodic assessment and documentation of process hazards, and implementation of actions to minimize risks associated with chemical operations, including the possibility of human error.
10. Management of changes to chemical operations to maintain or enhance the safety originally designed into the facility.

Facilities

11. Consideration and mitigation of the potential safety effects of expansions, modifications, and new sites on the community, environment, and employees.
12. Facility design, construction, and maintenance using sound engineering practices consistent with recognized codes and standards.
13. Safety reviews on all new and modified facilities during design and prior to startup.
14. Documented maintenance and inspection programs that ensure facility integrity.
15. Sufficient layers of protection through technology, facilities, and employees to prevent escalation from a single failure to a catastrophic event.
16. Provision for control of processes and equipment during emergencies resulting from natural events, utility disruptions, and other external conditions.

Personnel

17. Identification of the skills and knowledge necessary to perform each job.
18. Establishment of procedures and work practices for safe operating and maintenance activities.
19. Training for all employees to reach and maintain proficiency in safe work practices and the skills and knowledge necessary to perform their job.

20. Demonstrations and documentation of skill proficiency prior to assignment to independent work and periodically thereafter.
21. Programs designed to assure that employees in safety critical jobs are fit for duty and are not compromised by external influences, including alcohol and drug abuse.
22. Provisions that contractors either have programs for their own employees consistent with applicable sections of this Code or be included in the member company's program or some combination of the two.

Table 12.3 Responsible Care®Distribution Code of Management Practices

Purpose

The purpose of the Distribution Code of Management Practices is to reduce the risk of harm posed by the distribution of chemicals to the general public; to carrier, distributor, contractor, and chemical industry employees; and to the environment. Adherence to the Code will lead to continually safer chemical distribution and help member companies to:

- Evaluate the risks associated with chemical distribution and methods to reduce those risks
- Meet or exceed all regulations and industry standards governing chemical distribution
- Provide emergency advice and/or assistance to people on the scene in the event of a chemical distribution emergency
- Develop new technologies and methods to improve chemical distribution safety

The Code will also promote improvements in:

- Employee preparedness and awareness in preventing distribution emergencies
- The safety performances of carriers and other providers of distribution services
- The public's preparedness in responding to chemical distribution emergencies
- The public's understanding of, and confidence in, industry efforts to improve chemical distribution safety

The Distribution Code of Management Practices applies to all modes of transportation (highway, rail, marine, air, and pipeline) and to the shipment of all chemicals, including chemical waste. The Code also applies to distribution activities (storage, handling, transfer, and repackaging) while chemicals are in transit between member companies and their suppliers and customers. The implementation of a number of practices of the Code will vary according to the characteristics of the chemical being distributed, the mode of transportation, and the type of distribution activity involved.

Management Practices

Each member company shall have an ongoing chemical distribution safety program that includes senior management commitment through policy, communications, and resources to improvements in chemical distribution safety. The program should include the following elements:

1. Risk Management

1.1 Regular evaluations of chemical distribution risks which consider the hazards of the material, the likelihood of accidents/incidents, and the potential for human and environmental exposure from release of the material over the route of transport.

1.2 Implementation of chemical distribution risk reduction measures that are appropriate to the risk level.

1.3 Internal reporting and investigation of chemical distribution accidents/incidents and implementation of preventive measures.

2. Compliance Review and Training

2.1 A process for monitoring changes and interpretations of new and existing regulations and industry standards for their applicability to the company's chemical distribution activities and for implementing those regulations and standards.

2.2 Training for all affected company employees in the proper implementation of applicable regulations and company requirements.

2.3 A program for providing guidance and information to carriers, distributors, and contractors who perform distribution activities for the company on the company's training, and compliance requirements for the activities.

2.4 Regular reviews of company employee, carrier, distributor, and contractor compliance with applicable regulations and company requirements.

3. Carrier Safety

3.1 A process for qualifying carriers of all modes and types (common, contract, private, and customer controlled) that transport chemicals to and from company facilities, that emphasizes carrier safety, fitness, and regulatory compliance, and includes regular reviews of their performance and compliance.

3.2 Feedback to carriers on their safety performance and suggestions for improvement.

4. Handling and Storage

4.1 Documented procedures for the selection and use of containers that are appropriate for the chemical being shipped, in compliance with testing and certification requirements, and free of leaks and visible defects.
4.2 Documented procedures for loading chemicals at company facilities that will reduce emissions to the environment, protect personnel, and provide securement of the lading during transit.
4.3 Documented procedures for unloading chemicals at the company's facilities that will reduce emissions to the environment, protect personnel, and provide for safe unloading into proper storage facilities.
4.4 Defined criteria for the cleaning and return of tank cars, tank trucks, marine vessels, and returnable/refillable bulk and semi-bulk containers, and for the proper disposal of cleaning residues.
4.5 A program for providing guidance and information to customers, distributors, and other receivers on proper procedures for unloading and storing the company's chemicals.
4.6 A process for selecting distributors and other facilities that store or handle the company's chemicals in transit that emphasizes safety fitness and regulatory compliance and includes regular reviews of their performance and compliance.
4.7 Feedback to distributors and operators of other facilities that store or handle chemicals in transit on their safety performance and suggestions for improvement.

5. Emergency Preparedness

5.1 A process for responding to chemical distribution accidents/incidents involving the company's chemicals.
5.2 Documented procedures for making information about the company's chemicals in distribution available to response agencies.
5.3 A program for making facilities and/or training materials available to emergency response agencies.
5.4 Dialogue with state and local emergency planning organizations on the distribution and hazards of the company's chemicals to improve community preparedness to respond to chemical distribution emergencies.
5.5 Dialogue with the public on their concerns about chemical distribution safety, actions taken by the industry and the company to improve the safety of chemical distribution and effectiveness of emergency preparedness and emergency response assistance.

Table 12.4 Responsible Care® Waste and Release Reduction Code of Management Practices

Purpose

This Code is designed to achieve ongoing reductions in the amount of all contaminants and pollutants released to the air, water, and land from member company facilities. These reductions are intended to respond to public concerns with the existence of such releases and to further increase the margin of safety for public health and the environment.

The Code is also designed to achieve ongoing reductions in the amount of wastes generated at facilities. These reductions are intended to help relieve the burden on industry and society of managing such wastes in future years.

In implementing the Code, each company should strive for annual reductions, recognizing that production rates, new operations, and other factors may result in increases. Despite these fluctuations, however, the goal is to establish a long-term, substantial downward trend in the amount of wastes generated and contaminants and pollutants released. Quantitative reduction goals will be established for giving priority to those pollutants, contaminants, and wastes of highest health and environmental concern.

This Code complements, and should be implemented in conjunction with, current and future Codes of Management Practices. Key terms are defined in the Glossary, which should be consulted for assistance in interpreting the provisions of this Code.

Management Practices

Each member company shall have a waste and release reduction program which shall include:

1. A clear commitment by senior management through policy, communications, and resources, to ongoing reductions at each of the company's facilities, in releases to the air, water, and land, and in the generation of wastes.
2. A quantitative inventory at each facility of wastes generated and releases to the air, water, and land, measured or estimated at the point of generation or release.
3. Evaluation, sufficient to assist in establishing reduction priorities, of the potential impact of releases on the environment and the health and safety of employees and the public.
4. Education of, and dialogue with, employees and members of the public about the inventory, impact evaluation, and risks to the community.
5. Establishment of priorities, goals, and plans for waste and release reduction, taking into account both community concerns and the potential health,

safety, and environmental impacts as determined under Practices 3 and 4.

6. Ongoing reduction of wastes and releases, giving preference first to source reduction, second to recycle/reuse, and third to treatment. These techniques may be used separately or in combination with one another.
7. Measurement of progress at each facility in reducing the generation of wastes and in reducing releases to the air, water, and land by updating the quantitative inventory at least annually.
8. Ongoing dialogue with employees and members of the public regarding waste and release information, progress in achieving reductions, and future plans. This dialogue should be at a personal, face-to-face level, where possible, and should emphasize listening to others and discussing their concerns and ideas.
9. Inclusion of waste and release prevention objectives in research and in design of new or modified facilities, processes, and products.
10. An ongoing program for promotion and support of waste and release reduction by others, which may, for example, include:
    a. Sharing of technical information and experience with customers and suppliers;
    b. Support of efforts to develop improved waste and release reduction techniques;
    c. Assisting in establishment of regional air monitoring networks;
    d. Participation in efforts to develop consensus approaches to the evaluation of environmental, health, and safety impacts of releases;
    e. Providing educational workshops and training materials;
    f. Assisting local governments and others in establishment of waste reduction programs benefitting the general public.

Company Programs

A number of companies have developed, or rather continue to develop, risk management systems for their own use. At least one company (DuPont) sells this service to others, as will numerous consultants. These systems all have the same characteristics.

- They state an intent.
- They define requirements/standards.
- They allow the local plant to determine the specific action items necessary for compliance.
- They evolve as experience grows.
- They provide for evaluation and feedback.

Many CMA member companies have formalized their programs in response to the Responsible Care guidelines. There are differences in the number of elements in each program, and in the labels and terminology, but the fundamental objectives are common. As an example, in March of 1991, Exxon

Chemical put out a brochure giving a status report on its safety management system. It describes the Safety Management Practices (SMPs) in place at the time and acknowledges that more are being developed. The following excerpt from the brochure describes the twelve elements of the program as constituted at that time:

1. Safe Operations Program—This practice outlines the general requirements in each of the five key categories of operational safety.
2. Incident Reporting—Requires timely investigation of incidents to find the underlying causes and to determine the appropriate follow-up to avoid future incidents.
3. Safety Review Plan—Requires that all Exxon Chemical facilities conduct safety reviews to assure that the SMPs are in place and effective.
4. Training—Requires that all employees who might affect the safety of operations be assured of having the knowledge and skill to do their jobs safely.
5. Process and Instrument (P&I) Flow Diagrams—Requires that accurate P&I diagrams defining process facilities be readily available.
6. Operating and Maintenance Procedures—Requires that procedures be readily available, accurate and up-to-date.
7. Safety-Critical Controls Reliability—Requires that reliability be ensured for critical devices, whose failure could result in a hazardous situation.
8. Inspection—Requires an inspection program with follow-up documentation to ensure that plant equipment is in safe condition.
9. Work Permit System—Requires a work permit system to document and communicate that the work site is safe. Hazards are to be identified, evaluated and eliminated whenever possible.
10. Control of Defeat of Safety-Critical Systems—Requires that safety-critical systems, which control hazards, are not "defeated" or made inoperable without proper approval.
11. Emergency Preparedness—Requires that systems be in place to handle emergencies. Responsibility for all phases of emergency response must be clearly defined and assigned.
12. Personnel Safety—Requires that procedures exist at each site to prevent incidents between people and their work environment.

Since the brochure was printed, more SMPs have been developed. In addition, Exxon Corporation formalized a similar program for the entire organization. Such are the dynamics of corporate programs. Needs are recognized, starter programs are initiated, and as people begin working each system, it evolves. The programs will continue to evolve as the expectations of society and concerned management continue to become better defined.

Developing "Comprehensive" Requirements

As with all other sections of this book, there is no complete answer to the challenge posed in the heading—hence the quotation marks. Each company and each plant in a given company, must decide what list of requirements/expectations is needed to manage its particular risks. Then it must translate the list of requirements into a system of specific actions to meet them.

Many plants in the United States will soon begin complying with the OSHA rule. To develop program requirements a good place to start is the lists of CMA management practices given in Tables 12.1, 12.2, and 12.3. *Additional* requirements can then be drawn from those listed in Table 12.5. Regardless of the list of elements needed at a specific plant, I believe these lists represent a good starting point from which management can extract a list of requirements/expectations for its own operations. Please recognize that no list can be totally complete. Plant management is not absolved from thinking about the particulars of its own hazards by presentation of any list generated by others, no matter how much expertise it represents.

Table 12.5. Supplemental Requirements/Expectations for a Comprehensive Risk Management Program

Hazard/Risk Elimination/Management

- Eliminate all nonessential hazards starting at the business analysis stage and progressing through process research, development, and design
- A system is in place to minimize risk in site selection and anticipate encroachment of the community into what currently may be an isolation zone around the site
- Minimize risk in plant location/layout within the site
- Residual risks are considered in decision economics
- A system is in place to deal with risks as they are identified and avoid procrastination
- A system is in place to analyze existing and potential operations with a goal of maximizing the inherent (intrinsic) safety of each operation
- A system is in place to ensure the consideration of natural phenomena in plant design/layout
- A system is in place to ensure that reaction chemistry hazards are understood early in process development for new chemistry and for all existing operations
- Periodic risk assessments are performed and documented by qualified personnel including expertise from outside the unit
- Procedures are established to update risk assessments at specified intervals or as changes occur
- Assessed risks are addressed by designated levels of management and resulting decisions clearly documented
- Similar types of incidents are analyzed for repetitive problems
- A site risk assessment plan is updated annually
- A follow-up system is in place to ensure that decisions have been implemented
- Roles, responsibilities, authorities, and accountabilities within the system are known and exercised for employees, contractors, and third-party services
- A system exists for each objective, and expectations/requirements are translated into procedures and practices
- Employees are involved and participating across the organization, individually and through committee structures, in application of the system. Efforts are collaborative and learnings are shared
- Sufficient resources are committed to ensure that expectations are met
- Management systems must specifically address all requirements. A broad and detailed awareness of all facility operations that impact compliance exists. Compliance responsibilities are clearly identified

(*Continued*)

Table 12.5. *(Continued)*

- Calculations, chemical analyses, and monitoring are aggressively employed to ensure compliance with operating procedures. Proper records and files are maintained
- Compliance verification of all systems is carried out by the facility on a routine, periodic basis. Areas of noncompliance are promptly corrected
- A system is in place to share compliance information and problems with appropriate neighboring facilities to share experiences and to ensure that plant practices do not lag industry standards

Community Awareness and Emergency Preparedness

- Equipment, facilities, and trained personnel needed for emergency response are defined and readily available
- Facilities exist from which to facilitate effective management of an emergency

Inspection

- Critical containment systems are prioritized and administered based on the degree of risk
- Training plans and periodic reassessment are in place to assure an appropriate knowledge level among inspectors
- Welding procedures and welder qualifications are tested according to job requirements
- Inspection is an integral part of the management-of-change system
- Recommendations are documented, communicated, and stewarded

Critical Devices Reliability and Defeat

- Critical alarm, control, and shutdown systems are identified and tested and undergo preventative maintenance. A system is in place that controls their temporary defeat
- Systems are temporarily defeated only to prevent a greater safety problem or for repair
- Personnel and equipment safeguards are in place during defeat
- The system stipulates the level of authority required to approve defeat
- Defeats are communicated and made highly visible
- Scheduling and records are ever-green
- Devices are tested to the final trip element connected to the process. Relays, wiring, and control device are exercised

Incident Investigation and Analysis

- Incident/near-miss investigation and analysis procedures exist which:
- Provide for timely investigation
- Identify root causes and contributing factors
- Determine actions needed to reduce the risk of this and related incidents
- Ensure that appropriate action is taken and documented
- Reflect legal input
- Measure the effectiveness of actions taken
- Findings are retained and periodically analyzed:
- For common cause to determine where improvements to practices, standards, procedures, or management systems are warranted
- As a basis for improvement
- Systems are in place to share lessons learned from incidents/near-misses and to interact with others as appropriate to facilitate improvements in performance
- The system addresses responsibility for review, approval, and distribution
- The review level required reflects the severity of the risk or hazard involved

Operations and Maintenance Procedures

- Safety Critical and nontypical tasks have written procedures with checklists. The level of written detail for procedures for other tasks is based on safety, operability, and reliability
- Any safety critical portions of procedures are highlighted
- Procedures identify risks and hazards relating to the task and describe how they are managed
- Procedures are included in management-of-change and updated accordingly
- The work permit system documented:
- Describes work type, specific location, and involved equipment

(Continued)

Table 12.5. *(Continued)*

- Confirms a safe work area, inspected, and, if appropriate, gas tested
- Sets a time limit
- Specifies special precautions involving personnel or equipment
- Provides signature space for responsible parties
- The systems deal separately and directly with cold work, hot work, confined space entry, electrical lockout, and gas testing
- Gas testing is done only by employees formally trained and certified
- Plant management periodically reviews the system to assure that needs are being met

Facilities Design and Construction

- Project management systems are procedures that are documented and well understood
- Quality control and inspection systems are in place that ensure that facilities meet design specifications and that construction is in accordance with the applicable standards.
- Deviation from approved design practices and standards or the approved design is permitted only after review and approval by the designated authority. The rationale for the decision is documented

Training

- Holders of key positions and their stand-ins are trained in safety and emergency aspects of the position when newly appointed
- Supervisors are trained in safety supervision, legal, and regulatory requirements
- A system is in place for periodic refresher training, which includes updated assessments of employee's knowledge and skills

Management of Change

- Systems are in place for the management of both temporary and permanent changes. Included changes are:
- Hardware that is not repair in kind
- Operating parameter set points outside of design basis
- Alarm and interlock set points
- Process control software including computer control applications
- Materials being processed or utilized in the process
- Operations and maintenance procedures
- Procedures for managing change address:
- Authority for approval of changes
- Analysis of safety and environmental implications
- Acquisition of needed permits
- Specification, including reason for change
- Communication of potential consequences and required compensating measures
- Time limitations
- Training prior to implementation
- Update of plant documentation
- Criteria are in place to ensure that, when personnel changes are made, necessary levels of individual and collective experience and knowledge are maintained
- A system is in place to ensure that temporary changes do not exceed initial authorization for scope or time without review and approval
- Personnel who can impact hazards or compromise protective systems are trained in the management-of-change

Process and Facilities Information

- A system is in place based on risk analysis to identify those drawings and other pertinent documentation that are necessary for safe, environmentally sound operation and maintenance of facilities. These include:
- Process and equipment design
- Process and instrumentation flow diagrams (P&IDs)
- Electrical one-line drawings

(Continued)

Table 12.5. *(Continued)*

- Infrastructure
- Process flow diagrams
- Plot plan with electrical classifications
- Process monitoring and control system hardware, software, settings, parameter values, loop diagrams, and so on
- Procedures
- Applicable regulations, permit requirements, codes, workplace standards, and practices are identified, documented, and communicated in a form understood by those affected.
- Pertinent records covering operations, maintenance inspections, and facility changes are maintained for existing and abandoned facilities
- Major risk assessment decisions and their basis are documented

Personnel Safety and Health

- Safety- and health-related organizational roles and responsibilities are assigned, documented, and communicated widely. The basis for this shall be a broad participation in the definition and adoption of programs by affected employees
- Adequate qualified resourcing is available to provide the site/plant with support for personnel safety and health programs
- A system is in place to recognize individual and team participation in and contribution to improving the safety and health process
- A drug, alcohol, and substance abuse policy is in place at each site/plant
- These must be identification of those activities that require special measures to ensure the protection of personnel. Reducing risks to acceptable levels must first be pursued. If still necessary, appropriate personnel protective equipment must be provided along with necessary instructions on training and usage
- Off-the-job safety- and health-related training and information are included as part of the site/plant program
- Any identified health problems are communicated to responsible management
- A monitoring program should exist, which as a minimum ensures that all regulatory requirements are met, that any overexposure of personnel is detected, and that monitoring records are permanently retained
- A system is in place to analyze jobs and tasks for potential health risks

Turning Requirements Into Actions—an Example

Obviously, this book cannot define action programs or systems for all its readers. This has to be done individually. However, a more extensive example of such a translation might be helpful. The example covers part of one element of a system: personnel and training. It defines the fundamental principle, three of the specific requirements, the actions required, and the segment of the organization responsible for implementing them. Note that means of measuring the success of the system are included where appropriate, and that specific documentation actions are spelled out.

The actions listed remain a little general so as to apply to a variety of positions in the plant. Clearly, the specific requirements for each position must be documented. Much of this may be common for all positions, but each position or unit will have its particular requirements that must be specified. The value of computer management for both the requirements and the records should be obvious.

Personnel and Training Element—A Partial Example

Principle/Intent

Control of operations depends on people. Maintaining operations that are safe, environmentally sound, and conform to laws and regulations requires the careful selection, placement, ongoing assessment, and proper training of employees.

Requirements	*Actions*
1. A system is in place for selection and placement of qualified employees to to meet specified job requirements.	1.1 Written criteria for the performance of all jobs shall be in place. These criteria include a job description and responsibilities, and the skills, knowledge, and experience required to perform the job. Where applicable, required licenses and/or other certification shall be specified operations management.
	1.2 Responsibilities are defined, by function, for the following areas of job placement: – regulations/policies – candidate searching, both internally and externally. – initial interviewing, or other screening criteria, such as job performance. – credential verification and general testing, if applicable. These areas are the responsibility of the human resources function.
	1.3 The responsibility, by function, is defined for the determination that candidates are physically capable of performing a job, including screening for substance abuse. Medical typically has this responsibility.
	1.4 The responsibility for selection for placement is a management decision. The specific management level responsible for selection shall be included in each job criteria. Records of the selection process, including interviews, medical testing, credentials (e.g., diplomas), and so on, shall be retained for each employee.

1.5 The placement program shall be evaluated by measuring employee turnover, absenteeism, and performance, including incidents. Feedback shall be solicited from all levels of employees. The criteria for the selection and placement should be reviewed periodically. Responsibilities shall be designated for the reviews and evaluations.

2. Criteria are in place to ensure that necessary levels of individual and collective experience and knowledge are maintained and are carefully considered when personnel changes are made.

2.1 In addition to job descriptions and responsibilities noted above, the site shall have procedures to examine site manning on a collective basis to ensure that adequate skills and experience are in place to minimize risks. Criteria include having only one new person in position at the same time in a vertical organizational reporting line, and a limit of one new unit operations supervisor on a unit.

2.2 As part of ensuring adequate site staffing, replacements should be identified such that suitable training can be given as required. In addition, replacement planning is necessary to allow evaluating personnel changes at a given point in time relative to future expected changes. Replacement planning on an organizational basis should be integrated with individual career planning.

2.3 Where personnel are moved among lateral positions—for example, field operator positions—criteria should be established for the length of time an individual is considered qualified for a position that one is not currently working. These criteria should include requalification requirements as appropriate.

3. A system is in place to provide initial and ongoing training to meet job and legal requirements.
   This includes:
   —mechanisms for assessing program effectiveness, training documentation

3.1 A site shall have a documented, standardized introduction program for all employees. This program should include, as a minimum, orientation to the site and to the organization general site policies, site safety documentation, and working practices. Specific introductory material tailored to a job position should include applicable work procedures, legal requirements, reward and recognition system, conditions of employment including disciplinary offenses and discipline procedures, and organizational position and interfaces. The responsibilities for preparing and updating introductory material for new employees are with the training group.

3.2 A site shall also have standardized introductory materials, for employees moving to new positions, including the initial movement of an employee to a supervisory position. These training requirements can include leadership and coaching skills; team building; applicable safety, industrial health, and environmental requirements; legal requirements; hazard and risk analysis; employee motivation and performance assessment; incident reporting; emergency response. The necessary training shall be specified in the job description along with a specified maximum amount of time allowed in a position while completing the minimum specified training requirements. Responsibilities for the preparation and updating of this training material are with the training group.

3.3 A program for skills acquisition for initial hires shall be specified by job position. This program shall include the necessary training required before placement in an initial position and in moving to a new position when applicable. The program will specify minimum times required in various phases of initial training, and the type of tests for determining satisfactory completion of a phase shall be stated. The program shall specifically include the use of appropriate personnel protective equipment. The responsibilities for preparing and updating this material shall be in the training group.

3.4 Ongoing job training is that training that enhances the performance of an employee in a position. The site shall specify ongoing training requirements. A site has a mutual means of determining, in conjunction with an employee, desirable training needs.

3.5 Training records shall be maintained for all employees. The responsibility for maintaining these records shall in the training group. These records shall include initial, ongoing, and refresher training. Tests or other means of determining skills acquisition shall also be retained.

3.6 A periodic review of the effectiveness of training material and personnel shall be conducted.

Verifying Compliance

There are several approaches to verifying that a system will meet regulatory and company requirements. The first step, obviously, is to compare your actions to the requirements as shown in the example. This done, an outside opinion is often helpful. This might come from a consultant experienced in the area, or by networking. This means, quite simply, comparing notes with another plant(s) in your company or with neighbors subject to the same regulation. These networking sessions must be geared to two-way communications and should be nurtured as long-term relationships.

Some agencies hold seminars and issue guidance documents to clarify the intent of their regulations. These represent good opportunities not only to hear and debate the word "from the oracle," but also to get to know the regulators and how they think. Occasionally, direct approaches to regulators on company specifics will be an important option to have available. The decision to talk directly to an agency about a company problem is never an easy one. When some personal relationships are established with key agency people, more confidence in this option can be developed.

Sooner or later, each plant will be ready for an internal audit prior to the real thing. This has been talked about earlier, but I wish to emphasize the criticality of the attitude in place for the internal audit. The auditee must be *open,* and the visiting experts must be determined *to help.* They need to dig deep, and plant people need to probe the nuances to make the most of this opportunity to test the system.

Once the real compliance audit comes, do not panic. While some auditors like to find problems, most are interested in improving a plant's performance and, hence, the public safety.

Summary

Creating an overall risk management system for a plant begins by choosing the system elements and requirements/expectations pertinent to the plant's risk situation. Obviously, the latter needs to be defined and assessed prior to the choosing. (See Fig. 12.1.) These then need to be converted to systems and action plans that are appropriate for the plant's personnel and culture.

The choosing can be done from a number of existing lists to start. It is important to recognize that the list of requirements and the components of the systems and action plans will evolve as the understanding and experience of the plant does, and also with the expectations of the local society. As with other aspects of risk management, there is no fixed, correct position (except where required by law). The needs are for feedback on efficacy and continuous improvement to keep abreast of requirements and, in addition, to fashion a competitive edge from the activity.

Several lists and examples are presented in this chapter. Provided other requirements are not mandated by law, the recommendation is to start with the CMA codes of management practice and to select the additional requirements needed to meet the plant situation from those in Table 12.5. Such a list should provide compliance with many of the existing and proposed regulations, but that remains the responsibility of the local management to ensure. The NJ TCPA regulations are often the most demanding, but variations do exist and the lists of chemicals covered are very different from state to state.

Once the list of objectives and requirements/expectations is established, the plant needs to convert it into implementation systems. The characteristics of a system include comprehensiveness, independence from personalities, measurement of performance, documentation, and feedback for improvement.

Since most plants will have many of the necessary actions already going on, it is recommended that the leaders of these efforts be charged with the task of fleshing them out to meet all of the requirements/expectations in their areas. A steering committee can adjudicate any interface problems. Task forces are an alternative approach if the long-term ownership of the result can be ensured.

At some point verifications of compliance will be required. A thorough, open, friendly, expert self-assessment should be done before exposing the plant to an external audit. The critical element will be a system for evaluating the findings and improving performance on a continuing basis. Ultimately, the vision is for enough community interaction to ensure that their expectations are satisfied by the plant's record of performance and recognition of their input.

Index